T0262759

Recent Trends and Best Practices in Industry 4.0

RIVER PUBLISHERS SERIES IN MATHEMATICAL, STATISTICAL AND COMPUTATIONAL MODELLING FOR ENGINEERING

Series Editors:

MANGEY RAM
Lecturer (Asst. Prof.) in Automotive Engineering,

TADASHI DOHI
Hiroshima University, Japan

ALIAKBAR MONTAZER HAGHIGHI
Prairie View Texas A& M University, USA

Applied mathematical techniques along with statistical and computational data analysis has become vital skills across the physical sciences. The purpose of this book series is to present novel applications of numerical and computational modelling and data analysis across the applied sciences. We encourage applied mathematicians, statisticians, data scientists and computing engineers working in a comprehensive range of research fields to showcase different techniques and skills, such as differential equations, finite element method, algorithms, discrete mathematics, numerical simulation, machine learning, probability and statistics, fuzzy theory, etc.

Books published in the series include professional research monographs, edited volumes, conference proceedings, handbooks and textbooks, which provide new insights for researchers, specialists in industry, and graduate students. Topics included in this series are as follows:-

- Discrete mathematics and computation
- Fault diagnosis and fault tolerance
- Finite element method (FEM) modeling/simulation
- Fuzzy and possibility theory
- Fuzzy logic and neuro-fuzzy systems for relevant engineering applications
- Game Theory
- Mathematical concepts and applications
- Modelling in engineering applications
- Numerical simulations
- Optimization and algorithms
- Queueing systems
- Resilience
- Stochastic modelling and statistical inference
- Stochastic Processes
- Structural Mechanics
- Theoretical and applied mechanics

For a list of other books in this series, visit www.riverpublishers.com

Recent Trends and Best Practices in Industry 4.0

Editors

Abhinav Sharma

University of Petroleum and Energy Studies, India

Arpit Jain

Qpiai India Private Limited, India

Paawan Sharma

Pandit Deendayal Energy University, India

Mohendra Roy

Pandit Deendayal Energy University, India

River Publishers

Routledge
Taylor & Francis Group

NEW YORK AND LONDON

Published 2023 by River Publishers
River Publishers
Alsbjergvej 10, 9260 Gistrup, Denmark
www.riverpublishers.com

Distributed exclusively by Routledge
605 Third Avenue, New York, NY 10017, USA
4 Park Square, Milton Park, Abingdon, Oxon OX14 4RN

Recent Trends and Best Practices in Industry 4.0 / by Abhinav Sharma, Arpit Jain, Paawan Sharma, Mohendra Roy.

Routledge is an imprint of the Taylor & Francis Group, an informa business

ISBN 978-87-7022-805-3 (print)
ISBN 978-87-7022-997-5 (paperback)
ISBN 978-10-0096-438-7 (online)
ISBN 978-10-0344-171-7 (master ebook)

While every effort is made to provide dependable information, the publisher, authors, and editors cannot be held responsible for any errors or omissions.

Contents

9 Bio-inspired Multilevel ICHB-HEED Clustering Protocol for Heterogeneous WSNs **225**
Prateek Gupta, Amrita, Himansu Sekhar Pattanayak, Gunjan, Lalit Kumar Awasthi, and Vachik S. Dave
9.1 Introduction 226
9.2 Review of Literature 227
9.3 Proposed Work 229
9.3.1 Network model for multilevel energy heterogeneity 229
9.3.2 MLICHBHEED protocol 231
9.3.2.1 Clustering and CH election process 232
9.3.2.2 ICHB algorithm 232
9.3.2.3 Process of data transmission and collection 233
9.3.2.4 Energy depletion model for MLICHB-HEED protocol 233
9.4 Results and Discussions 234
9.4.1 Network lifetime 236
9.4.2 Total energy consumption 238
9.4.3 Number of packets sent to the base station 242
9.5 Conclusion 242

10 IoT Enabled by Edge Computing for Telecomm and Industry **247**
Manoj Kumar Sharma, Ruchika Mehta, and Rajveer Singh Shekhawat
10.1 Introduction to IoT and Edge Computing 248
10.1.1 IoT and edge-computing scenarios 250
10.1.2 IoT versus edge computing 253
10.1.3 Industry 4.0 254
10.1.4 Cloud computing and edge computing 256
10.1.5 Cloud computing and edge computing: use cases . . 257
10.2 IoT and Edge Computing Framework 261
10.2.1 IoT and edge computing integrated framework . . . 263
10.2.2 Pros and cons 264
10.2.3 Opportunities and challenges 264
10.3 Edge Computing Devices 265
10.4 Telcos and Edge Computing 266
10.4.1 Telco network modernizations challenges 268
10.4.2 Multi-access edge computing 268

Preface

In the 21st century, world experienced rapid change of technology across the industries which leads to the evolution of Fourth Industrial Revolution interchangeably used with Industry 4.0. It revolutionized the way industries design, manufacture, and optimize their products and represents a new stage in the organization and control of the industrial value chain.

Industrialists are integrating digital technologies such as internet of things (IoT), artificial intelligence (AI) and machine learning, blockchain technology in the production and throughout their operations. These modern technologies lead to improvement and automation of the industry. Since Industry 4.0 is multidisciplinary in nature, this book combines multitude of technologies from different domains. Keeping this in mind the book has been organized into four sections.

CPS: Advancements in Connected Industrial Systems and Robotics

CPS are not limited to industrial floors anymore. They now comprise of complex systems such as state-of-the-art robotics systems, drones, self-driving cars, and many more. The applications of these systems are changing how the industry functions. Chapter 1 provides application of AI techniques in digitizing modern chemical industries and how it aids in sustainability goals. In the chemicals field, digital transformation success necessitates a complete, holistic approach that considers the entire asset lifespan, from design to operations and maintenance. The chapter not only discusses the applications but also the practical challenge that arise during digital transformation. Chapter 2 provides data collected from semi-structured interviews with employees of an industrial manufacturing enterprise who are involved toward migrating old-generation industrial systems to a more modern platform. The chapter lays down a systematic plan as adopted/suggested by industry practitioners. Chapter 3 provides a guide of how to deploy container as a service (CaaS) in cloud environments. Authors provide the architecture and implementation of CaaS, along with the advantage and disadvantages of the technology. Chapter 4 provides an automated framework for detecting unknown activity in vehicular ad hoc networks. This improves traffic security and efficiency by providing low-latency traffic security applications. The authors suggest an improved hybrid cooperative malicious node detection approach combined with automated predication offer security against external attackers in vehicular ad hoc network (VANETs). Chapter 5 presents the control of mobile manipulator with object detection for explosive ordnance disposal (EOD) application for a remotely operated mobile manipulator. The robot prototype is an unmanned ground vehicle (UGV) with multi-terrain tracked wheel chassis and a 6-degree of freedom (DOF) robotic arm mounted on the top along with FPV camera for visual feedback.

AI: Machine Learning, Deep Learning Algorithms, and Applications

Smart agriculture and precision farming techniques are some of the key applications of current industrial revolution directly effecting sustainable development goals (SDG). Smart agriculture is a perfect example of interplay of IoT/sensor networks, AI, cyberphysical systems (CPS), and blockchain for various stages of farming due to the highly complex nature of the subject. Chapter 6 reviews the existing and emerging sensors along with wireless network and other communication technologies, IoT and AI focused on agricultural industry. It also addresses the challenges of the latest technologies and extends toward the discussion on future direction of smart agriculture. Chapter 7 discusses the application of machine vision in the field of plant disease identification. The study focuses on a thorough examination of machine learning-based extraction methods, as well as their benefits and drawbacks. It covers a wide range of features, based on shape, texture, and color for different diseases in diverse cultivations.

IoT and WSN bringing Enhanced Connectivity to Edge Devices

IoT brought a disruption in industrial communication. The edge devices suddenly saw a huge multi-fold rise in resourceful and the industry had tons of datapoints based on which the various operations can be optimized and synchronized to a higher degree of accuracy. Chapter 8 provides a guide to system development for scientists/engineers working in the field of IoT. The chapter provides detailed insights from the secure-software development methodology for different IoT applications. Various methodologies are analyzed for two different parameters, that is, physical infrastructure, emerging technologies. The chapter explores the various aspects such as lifecycle, technologies, architecture, and frameworks. Chapter 9 proposes a novel multi-level intelligent cluster head selection based on bacterial foraging optimization (ICHB-HEED) (MLICHBHEED) protocol consisting of varying energy levels in heterogeneous model based on ICHB-HEED protocol.

The hybrid energy-efficient distributed (HEED) protocol is a clustering technology that is frequently used in WSNs. ICHB-HEED is a smart cluster head election protocol using BFOA (bacterial foraging optimization algorithm). The authors report an improvement in energy efficiency while delivering improved network performance. Chapter 10 presents a comprehensive literature review of IoT applications with edge computing and their respective architectures. Chapter 11 summarizes recent developments in low-energy network protocols. The chapter also proposes a novel self-powered network based on varying node energy levels. Hardware design and simulation results have been provided to verify the optimized performance of the proposed network.

Cybersecurity: Blockchain Technology and Applications
With more complexity of communication systems in Industry 4.0 and dependency on internet and cloud technology security was never more critical than in current generation of industrial systems. Blockchain is state–of-the-art cybersecurity technique that not only disrupted the financial markets but is significantly altering the supply chain, logistics, energy management, and various other facets of today's connected industries. Chapter 12 is focused on detailed application analysis of consensus algorithm in blockchain to provide an efficient and secure network by using emergence of cryptocurrencies as an example. Chapter 13 provides a systematic study of various blockchain protocols and algorithms. Here, the authors investigate a wide range of blockchain algorithms and protocols' applications in literature to summarize their properties in a comprehensive classification under different criteria in the form of a tree structure which acts as a quick guide to diverse set of researchers looking to employ these. Chapter 14 provides various security threats to self-driving cars discussing via case studies and possible techniques for security management and prevention of the attacks/threats.

Editors:

Abhinav Sharma
University of Petroleum and Energy Studies, India

Arpit Jain
QpiAI India Private Limited, India

Paawan Sharma
Pandit Deendayal Energy University, India

Mohendra Roy
Pandit Deendayal Energy University, India

Acknowledgement

The Editor acknowledges River Publishers for this opportunity and professional support. My special thanks to Mr. Rajeev Prasad, River Publishers for the excellent support, he provided us to complete this book. Thanks to the chapter authors and reviewers for their availability for this work.

List of Figures

List of Tables

List of Contributors

Akiti, Narendra, *Design & Engineering, Jubilant Pharmova Limited, India*

Alzahrani, Yazeed, *School of Computing and Information Technology, University of Wollongong, Australia*

Amrita, *Banasthali Vidyapith, India*

Antony, Cecil, *School of Biotechnology, National Institute of Technology Calicut, India*

Athavale, Vijay Anant, *Walchand Institute of Technology, India*

Awasthi, Lalit Kumar, *National Institute of Technology, India*

Bansal, Ankit, *Chitkara University Institute of Engineering and Technology, Chitkara University, India*

Bavarva, Arjav, *Department of Information and Communication Technology, Marwadi University, India*

Boudour, R., *Embedded Systems Laboratory, Annaba University, Algeria*

C. S., Meera, *Advanced Remanufacturing and Technology Centre, A* STAR, Singapore*

Chaudhari, Dinesh N., *Department of Computer Sci & Engineering JDIET, India*

Chowdary, Vinay, *School of Engineering, University of Petroleum and Energy Studies, India*

Dave, Vachik S., *Walmart Global Tech, USA*

Dey, Amar Kumar, *Department of Electronics & Telecommunication Engineering, Bhilai Institute of Technology, India*

Echchaoui, H., *Embedded Systems Laboratory, Annaba University, Algeria*

Ghodke, Praveen Kumar, *Department of Chemical Engineering, National Institute of Technology Calicut, India*

Gunjan, *SRM University Delhi-NCR, India*

Gupta, Medini, *Amity Institute of Information Technology, Amity University, Uttar Pradesh, India*

Gupta, Mukul Kumar, *School of Engineering, University of Petroleum and Energy Studies, India*

Gupta, Prateek, *University of Petroleum & Energy Studies, India*

Jadhav, Avinash P., *Department of Computer Sci & Engineering JDIET, India*

Kapil, Anil Kumar, *College of Technology, Surajmal University, India*

Kathiriya, Hiren, *Department of Electronics and Communication, R K University, India*

Kathole, Atul B., *Department of Computer Engineering, PCCOEIndia*

Kilari, Hemalatha, *Davidson School of Chemical Engineering, Purdue University, USA*

Lal, Niranjan, *Computer Science and Engineering, SRM Institute of Science and Technology, India*

Mehta, Ruchika, *Manipal University Jaipur, India*

Miloud-Aouidate, A., *Embedded Systems Laboratory, Annaba University, Algeria*

Mishra, Akhilesh Kumar, *Panipat Institute of Engineering and Technology, India*

Patra, Manoj Kumar, *Department of Computer Science and Engineering, National Institute of Technology, India*

Pattanayak, Himansu Sekhar, *Bennett University Greater Noida, India*

Pawar, Ritvik, *Mahatma Gandhi University, Meghalaya, India*

Reddy, P. Swapna, *Department of Chemical Engineering, National Institute of Technology, India*

S., Sathiya, *Department of Instrumentation & Control Engineering, Dr. B. R. Ambedkar National Institute of Technology Jalandhar, India*

Sahoo, Bibhudatta, *Department of Computer Science and Engineering, National Institute of Technology, India*

Sharma, Abhishek, *Department of Electrical & Electronics, Ariel University, Israel*

Sharma, Manoj Kumar, *Manipal University Jaipur, India*

Shekhawat, Rajveer Singh, *Manipal University Jaipur, India*

Sorathiya, Vishal, *Department of Information and Communication Technology, Marwadi University, India*

Tanwar, Sarvesh, *Amity Institute of Information Technology, Amity University, Uttar Pradesh, India*

Turuk, Ashok Kumar, *Department of Computer Science and Engineering, National Institute of Technology, India*

List of Abbreviations

AMQP	Advanced message queuing protocol
ANN	Artificial neural networks
ANPBFT	Advanced PBFT-based consensus
API	Application program interface
APTEEN	Adaptive periodic TEEN
BFOA	Bacterial foraging optimization algorithm
BFT	Byzantine fault tolerance
CA	Certification authority
CaaS	Containers as a service
CAE	Convolutional auto encoder
CNN	Convolutional neural network
CoAP	Constrained application protocol
CPI	Chemical process industry
CPS	Cyber-physical system
CR	Cognitive radio
CSP	Cloud service provider
CWSI	Crop water stress index
Dapps	Decentralized applications
DcBFT	Democratic Byzantine fault tolerance
DDPoS	Delegated proof of stake with downgrade
DHCRA	Distributed hierarchical clustering routing algorithm
DOF	Degrees of freedom
DOS	Denial of service
DPoS	Delegated proof of stake
DRBFT	Delegated randomization Byzantine fault tolerance
EESSC	Energy-efficiency semi-static cluster
HER	Electronic health record

EH-WSN	Energy harvesting wireless sensor network
ENO	Energy-neutral operation
EOD	Explosive ordnance disposal
FAO	Food and Agriculture Organization
FBA	Federated Byzantine Agreement
FLS	Fuzzy logic system
FPGA	Field- programmable gate array
GAN	Generative adversarial network
GDP	Gross domestic product
GHG	Greenhouse gas
GLCM	Gray-level co-occurrence matrix
GNSS	Global navigation satellite system
GPDCNN	Global pooling dilated CNN
GPS	Global positioning system
HEBM	Hierarchical energy balancing multipath
HEED	Hybrid energy-efficient distributed
HTE	High-throughput experimentation
HTTP	Hypertext transfer protocol
IaaS	Infrastructure as a service
ICPS	Industrial cyber-physical systems
ICT	Information and communication technology
IHCMNDA	Improve hybrid cooperative malicious node detection approach
IIoT	Industrial Internet of Things
IoS	Internet of services
IoT	Internet of Things
IRT	Infrared thermography
ISCP	Improved SCP protocol
IWN	industrial wireless network
JSON	JavaScript object notation
KPI	Key performance indicator
LCA	Life cycle assessment
LEACH	Low-energy adaptive clustering hierarch
LEICP	Low-energy intelligent clustering protocol
LIDAR	Light detection and ranging
LMP	Locational marginal pricing

LoRaWAN	Long-range wide area network
LoWPAN	Low-power wireless personal area network
LPWAN	Low-power wide area network
M2M	Machine to machine
MAEC	Multi-access edge computing
MANET	Mobile ad hoc network
MEC	Mobile edge computing
MEMS	Micro-electromechanical systems
MLC	Manufacturing Leadership Council
MODIS	Moderate resolution imaging spectroradiometer
MpoC	Metaheuristic proof of criteria
MQTT	Message queuing and telemetry transport
MSigBFT	Multisignature Byzantine fault tolerance
MTEEN	Modified TEEN
MVP	Minimum viable product
NDVI	Normalized difference vegetation index
NEM	New Economy Movement
NFV	Net function virtualization
ONPOB	Online benefit generating
OS	Operating system
PaaS	Platform as a service
PEGASIS	Power-efficient gathering for information systems
PFL	Power fluctuation level
PKI	Critical public infrastructure
PM	Physical machine
PoB	Proof of burn
PoC	Proof of capacity
PoCO	Proof of contribution
PoCt	Proof of concept
PoET	Proof of elapsed time
PoEv	Proof of evolution
PoEWAL	Proof of elapsed work and luck
PoI	Proof of importance
PoPL	Proof of play
PoR	Proof of reputation

PoRX	Proof of reputation X
PoS	Proof of stake
PoSe	Proof of search
PoSn	Proof of sincerity
PoSP	Proof of service power
PoTS	Proof of TEE-stake
PoW	proof of work
PSO	Particle swarm optimization
QoS	Quality of service
RDV	Register, deposit, vote
RFID	Radio-frequency identification
RL	Reinforcement learning
ROI	Region of interest
RSSI	Received signal indicator
RTO	Real-time optimization
RTT	Round-trip time
SaaS	Software as a service
SDN	Software-defined net
SGX	Software guard extensions
SMOS	Soil moisture and ocean salinity
SN	Sensor node
SoC	System-on-chip
SSCM	Site-specific crop management
TCP/IP	Transmission control protocol/internet protocol
TEA	Techno-economic analysis
TEE	Trusted execution environment
TFOD	Tensorflow object detection
UGV	Unmanned ground vehicle
UID	Unique identification
UNL	Unique node list
URDF	Universal robot description format
VANET	Vehicular ad hoc networks
VDA	Variable data acquisition
VM	Virtual machine
VRT	Variable rate technology

WAN	Wide area network
WSN	Wireless sensor network
XMPP	Extensible messaging and presence protocol
YAC	Yet another consensus

1

Artificial Intelligence in the Digital Chemical Industry, its Application and Sustainability

Praveen Kumar Ghodke[1], P. Swapna Reddy[2], Narendra Akiti[3], and Hemalatha Kilari[4]

[1,2]Department of Chemical Engineering, National Institute of Technology, India
[3]Design & Engineering, Jubilant Pharmova Limited, India
[4]Davidson School of Chemical Engineering, Purdue University, USA

Abstract

The chemical manufacturing industries are at the forefront of innovation, exploring novel technologies through "smart manufacturing" and "industrial modernization" using computational approaches to cater to increasing market demands and stringent regulations. Digital design enables the manufacturers to understand and streamline the process development and production using process systems engineering tools. The present drive toward industrial automation advancements such as Industry 4.0, Pharma 4.0, and smart manufacturing also enables technological developments in research to percolate into industries to meet product quality, safety, and profitability challenges. It is envisaged that the promises of continuous manufacturing will be realized with the evolution of digitalization technologies in various stages of design, development, and implementation. Further, artificial intelligence (AI) and machine learning-based intelligent systems are gaining popularity in all engineering and science disciplines due to their ability to solve real-world challenges. In the chemicals field, digital transformation success necessitates a complete, holistic approach that considers the entire asset lifespan, from design to operations and maintenance. However, it has also led to far-reaching impacts on technology and the workforce worldwide. In the present work, we concentrate on practical applications of AI in chemical manufacturing industries in the digital transformation of the chemical industry, discussing the challenges and technological impacts.

Keywords: Artificial intelligence, machine learning, chemical industry 4.0, digital transformation, smart manufacturing.

1.1 Introduction

The chemical sector is extremely important to the world economy. The chemical sector is expected to contribute \$10.7 trillion to the world's gross domestic product (GDP) in 2025. The chemical process industry is the most energy-intensive and largest emitter of greenhouse gases (GHGs) industry [1]. The petrochemical and chemical industries emitted 764 million tons of greenhouse emissions (CO_2-equivalent tons) in 2018, an increase of 8% from 2016. (IEA, 2020b) [2]. Chemical manufacturing also frequently involves potentially dangerous ingredients and high-pressure/high-temperature settings, resulting in fires, explosions, and other chemical accidents. Chemical mishaps could result in deaths and financial and social losses. Many attempts have been made in the chemical industry to develop and deploy technologies to decrease environmental burdens, improve energy efficiency, and improve operational safety [3].

AI is a new technology that has gotten much attention over several decades. AI is a collection of cutting-edge technologies that can execute tasks similar to human intellect. AI divides into deep learning, IoT, heuristics, and machine learning [4]. The process of human cognition reveals via AI cognition and logic deduction. Another sort of AI is machine learning, which can learn and enhance the performance of specific tasks based on previous experiences. Artificial neural networks (ANN), support vector machines, and random forests are just a few machine learning approaches developed. Heuristics are frequently employed to solve high-dimensional issues by simulating natural biological evolution or animal societies' collective behavior (e.g., particle swarm optimization, ant colony optimization). Hybrid approaches and agent-based modeling are two other AI strategies [5].

The chemical industry is increasingly interested in adopting AI to solve problems, including process modeling, optimization, control, and fault identification and diagnostics. The chemical industry's environmental, economic, and social sustainability are intertwined. According to an Accenture poll, 94% of executives in the chemical and advanced materials industries predict an industry-wide digitalization, with AI playing a pivotal role in facilitating the digital revolution [6]. Many research studies looked at AI applications, digital transformation, impacts, and obstacles in the chemical sector. On the other hand, these assessments focused on AI applications and looked at the long-term ramifications of deploying AI in the chemical sector.

Some studies have shown that AI could help achieve sustainable development goals, but little industry-specific research, particularly in the chemical industry, has been done [7]. A few studies looked at the effects of AI on higher education, government administration, and policymaking. All three studies are concerned with social and policy ramifications, but none have built frameworks for assessing sustainability. The present study analyzes the research on AI in the chemical industry, digitalization, and the quantitative/qualitative assessment of AI technologies' sustainability-related consequences. Identifying the new chemical molecule/drug or catalyst process is described based on the ML models, generative network methods, and learning models such as RL and ANN. Additionally, the sustainability of AI-based learning was emphasized in the future prospectus.

1.2 AI in the Chemical Industry and Applications

AI has had a significant impact on all sections of the chemical industry, with tremendous potential that has revolutionized value chain management, enhanced efficiency, and opened up new ways to market [8]. There has been much excitement about transforming the business by combining cutting-edge technology for creating, collecting, and storing data at reduced prices with advances in computational power to tackle previously impossible difficulties. Companies have demonstrated a readiness to use technology to increase quality, service level, and operational efficiency.

The chemical study explores the structure and properties of matter and the chemical reactions that transform things into other substances and so play a "central" role in other disciplines. As a result, chemistry is a data-rich discipline of science that contains complicated information derived from hundreds of years of experimentation and, more recently, decades of computational analysis. Chemical research is still a labor-intensive process due to the infinite complexity of the variety of material compounds [9]. Chemicals offer a unique chance to accomplish substantial advances through AI due to their high complexity and large amounts of data. To begin with, the types of molecules that are structured from atoms are nearly limitless, resulting in an infinite chemical universe. The interconnectedness between these molecules and all potential combinations of variables, such as temperature, substrates, and solvents, is enormous, resulting in an infinite reaction space [10].

Exploring the infinite chemical and reaction space and navigating to the best ones with the desired attributes is thus almost impossible with only human efforts. Second, in the chemical study, the vast number of molecules and their interactions with their surroundings introduce a new degree of complexity that cannot be predicted merely using physical laws [11]. Figure 1.1 shows the possible application of AI methods in chemical industries. In the petroleum industry, AI is applied in surface geological frameworks, reservoir engineering, and product quality detection. AI has many applications in wastewater treatment plants, such as assessing water and wastewater quality, wastewater monitoring system, and quality detection. In manufacturing, fault detection and diagnosis are performed using an AI-based model and also used in digital designs. Similarly, AI is applied to automated system design process operation, modeling and optimization, and real-time control in process system engineering.

Many notions, principles, and theories have been generalized during centuries of research on trivial (i.e., single-component) systems. When the scale changes, symmetry breaks down in larger, more complex systems, and the rules shift from quantitative to qualitative, nontrivial complications emerge. Chemical research is thus incorrectly guided by heuristics and fragmentary

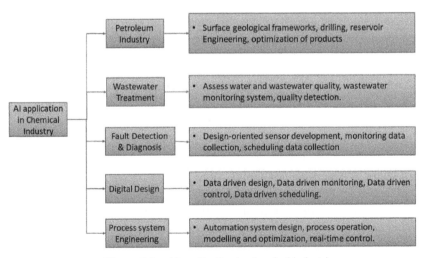

Figure 1.1   AI application in chemical industries.

rules accumulated over the previous centuries, yielding progress that only proceeds through trial and error due to the lack of systematic and analytical theory toward the structures, properties, and transformations of macroscopic substances [12]. ML will be able to recognize patterns in vast amounts of data, providing an unparalleled means of coping with complexity and altering chemical research by revolutionizing the way data is handled. Currently, AI is being used in every sub-field of chemistry, including research and data creation tools like analytical chemistry and computational chemistry and applications in organic chemistry, catalysis, and medical research. While AI has a wide range of applications in technology, it also has a wide range of applications in chemical science. Let's look at some of the ways that AI can be used in the chemical sciences [13].

1.2.1 AI in chemical science

The detection of molecular characteristics is the first and most important use of AI in chemical science. Scientists have been manually detecting the chemical characteristics of molecules since determining a molecule's attributes is a time-consuming process. However, AI has aided this process and allowed scientists to detect chemical features. It has simplified the manual detection process, making chemical treatments more efficient [14]. Furthermore, scientists have been able to assess the potential of a hypothetical molecule by identifying molecular characteristics. As past data empowers computers to interpret current data, AI algorithms have aided the study of chemistry. Machine learning technologies can now sift through massive datasets of existing molecules and their attributes, generating new possibilities based on the data. It has the potential to speed up and reduce the cost of finding new medication candidates. While molecular property detection is helpful in chemical science, AI in molecular design has led to groundbreaking discoveries in the discipline. Scientists have collected historical data and manufactured chemical bonds by creating molecules. Scholars have advanced in their discovery of molecules by incorporating AI algorithms, which will undoubtedly aid them in making groundbreaking discoveries in AI chemical synthesis. Furthermore, creating molecules leads to various practical applications that have significantly advanced the area of chemistry [15].

The process of drug discovery is one of the essential uses of AI in chemical research. Identifying medications has shown to be highly beneficial in healthcare and science. MIT researchers used a machine-learning system to discover a novel antibiotic molecule. The drug destroyed several of the world's most troublesome bacteria in laboratory tests, including some kinds resistant to all known antibiotics. Scientists are working hard to identify pharmaceuticals to build new molecules and formulate excellent medicines for curing fatal ailments when new diseases emerge on the surface [16]. When molecules are broken down to determine their building blocks, this is known as a retrosynthesis reaction. As the phrase implies, the synthesis process (to generate something) will carry backward to uncover a molecule's building blocks. Unlike now, when scientists employ AI to perform retrosynthesis reactions, scientists performed this process manually in the past. The process was lengthy that required a significant amount of time and resources. However, with the advent of artificial intelligence, this procedure can now be performed with the assistance of computers, making it more accurate and efficient [17].

Finally, the use of AI in chemical science has enabled scientists to perform predictive analyses. Computers produce advanced AI algorithms and patterns that indicate predictive analysis for the future using past data and understanding current data [18]. Computers are fed data that improves their interpretation skills and empowers them to work along the lines of human intellect using machine learning or deep learning algorithms. This application is critical because it emphasizes the potential repercussions or influence of specific chemical bonds, compounds, or even medications, which can help steer future steps in the right path.

1.2.2 AI in research and development

R&D is critical to industry innovation, especially in organizations that deal with sustainability. AI has been used to improve chemical synthesis design by predicting and optimizing chemical reactions. Catalyst screening and design have both been investigated using machine learning. Several studies have emphasized AI's potential in assisting the production of environmentally friendly chemicals and materials [19]. A study looked into the usage of artificial neural networks to evaluate and increase job happiness in research labs and those technical aspects. Every business has a plan to implement procedures, tools, and techniques. Let's look at how AI can be applied to the chemical sector in a larger sense [20].

Most gamers concentrate on AI-assisted research that can deliver quick and precise results. Advanced research uses machine learning technologies and computerized permutations and combinations to recognize molecules, develop formulas, and determine the quantity of a chemical [21]. AI aids in predicting whether the combinations utilized will result in a breakthrough in innovation. Chemical industry innovations can help bolster the efforts of numerous ancillary sectors that rely on the chemical sector. AI tools provide several insights into preventative measures and predictive forecasts, allowing industries to be proactive in responding to a crisis. Advanced analytics can assist predict raw material demand, optimizing the supply chain to reduce delays, and avoiding price rises at the last minute. Through AI forecasting methodologies, several modifications at every level of molecule development in the chemical industry are achievable. Laboratory experiments, clinical trials, and analytical expertise are frequently used to understand complex products better [22]. AI tools churn and feed massive amounts of data that have been methodically organized and analyzed. AI and machine learning in the chemical sector also help speed up efforts to mitigate climate change by predicting the effects of dangerous pollutants. The AI process motivates businesses to improve their machinery and methods and decrease pollution in water bodies and the environment.

In general, both economic and human investments in AI are enormous. AI is evolving at a rapid pace. Adoption in the chemical industry, on the other hand, can be difficult due to a lack of practical support for innovation and research. Many industry-specific AI solutions will significantly increase the demand for digitally trained operators to monitor, develop, and maintain AI operations [23]. For companies that use AI, investing in skills and training is always a top concern. Chemical companies must develop technologies that generate profits, reduce energy consumption, fulfill circular economy obligations, and have a low environmental impact. Saving money without sacrificing quality or safety is another factor to consider.

1.2.3 AI in catalyst design spaces

Catalytic chemistry arose from catalyst technologies used in the chemical industry to produce chemicals and fuels more efficiently and sustainably. Because a catalyst's performance is dependent on many properties, such as composition, porous structure, surface termination, support, particle size, atomic coordination environment, particle morphology, and reactor during the reaction, developing novel heterogeneous catalysts with good performance (i.e., stable, active, and selective) has remained a difficult task (Figure 1.2). The intrinsic complexity of catalysts makes the finding and producing catalysts with desired qualities more reliant on intuition and experimentation, which is both expensive and time-consuming [24]. From

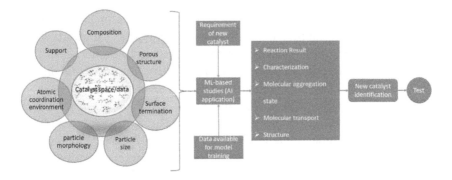

Figure 1.2   Catalyst identification process using ML-based technique.

the extensive combination of libraries with experimental and chemical composition, AI technologies such as ML can promote catalyst discovery by assisting in the search through huge design spaces. AI will expose the intricate relationship between catalytic structure and catalytic performance using a well-defined structure and standardized data, such as reaction results and in situ characterization results. It was also possible to anticipate an accurate description of the action of molecules, molecular aggregation states, and molecular transport on catalysts. With these methods, researchers can create virtual laboratories to test new catalysts and catalytic processes [25]. Figure 1.2 shows the process of finding a new catalyst using the data available in the literature and applying the ML-based technique.

"Insilico Medicine" is an AI startup that aims to speed up to three aspects of drug discovery and development: illness target identification, generative chemistry, and synthetic biological data (generative biology) [26]—in addition, working on clinical trial outcome prediction. In 2015, the company was the first to use generative adversarial networks (GANs) and reinforcement learning (RL) to create novel molecular structures with the parameters supplied. Insilico Medicine works on internal drug development projects in many disease areas and collaborates with large pharmaceutical corporations. Deep learning and other AI techniques' capacity to discover novel compounds with desirable features will revolutionize drug research. It has the potential to generate novel medicines significantly more quickly and effectively, and it is an essential new tool in the search for better therapeutics. Insilico Medicine has been a pioneer in using some of the most cutting-edge AI approaches to drug discovery, such as GANs [27]. Figure 1.3 shows the pathway of finding new drug using AI-applied techniques.

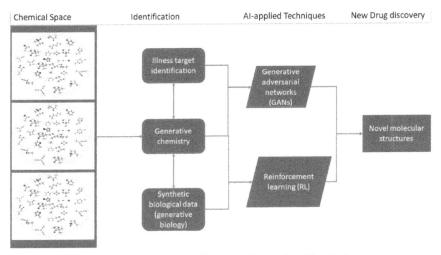

Figure 1.3 Pathway to finding new drugs using AI-techniques.

1.3 Sustainability of AI Applications

The present section contains studies that look at the effects of AI on the environment. Environmental, energy, economical, time, and safety were recognized as five implications. Although reducing energy usage and computing time may improve economic and ecological performance, such gains must be proven by further life cycle assessment (LCA) or techno-economic analysis (TEA) [28]. As a result, research focused solely on energy and time without considering economic and environmental outcomes was categorized.

In the chemical industry, 63 experiments on AI are being investigated. None have comprehensively looked at AI's environmental, economic, and societal effects. Most research offered a quantitative assessment of AI's advantages. The present chapter solely considers peer-reviewed papers which are sustainable that analyze the potential benefits of AI [29]. Studies on discoveries need quantitative and holistic assessments to understand the long-term consequences of AI applications in the chemical industry.

1.3.1 Environmental aspect

A few researchers have studied the environmental effects of using AI in the chemical industry. Since the chemicals are connected in upstream processing and production, they affect energy consumption and the environment.

AI studies show the possible reduction in the number of experiments by predicting the chemical ingredient or composition, which reduces the number of experimentation and chemical disposal by 20%. Thereby, environmental impact can be reduced.

The complete process of chemical discovery needs life cycle assessment (LCA) since the chemical used in R&D efforts has much impact on the environment. To understand the sustainability of AI, a comparison of life cycle implications of chemical used in the tests before and after AI adoption need to be performed. Thus, tools like LCA are required for conducting organizational LCA before and after AI implementation. To perform LCA, emissions from different sources are to be considered. Examples are CO_2, SO_X, and NO_x emissions from the steam boiler plant, emissions for production of naphtha from the ethylene cracking, and the ethylene production plant. Considering the supply chain of chemical GHG emissions, few studies are measured using LCA and AI implementation to show the reduction in GHG emissions by optimizing and automation controls. In AI, sorting genetic algorithm models were used to optimize the chemical supply chain, and LCA studies were implemented to study the reduction in emissions [30].

The LCA needs to understand AI's possible environmental consequences and determine the most sustainable AI adoption paths. AI can be hard to use with LCA because of the unavailability of data for new technologies [31]. As a result, it may be more challenging to figure out how physical processes and specific technical characteristics might change because of AI. The problem can only be solved with the help of a conceptual framework.

1.3.2 Energy aspect

Most AI experiments examined a quantitative examination of energy efficiency similar to the environmental impact advantages. The existing energy/exergy analysis methodologies and energy data are used in these quantitative assessments. Energy and exergy analyses were performed to improve the energy consumption in chemical industries. Most studies combined AI with analytical techniques to improve energy usage. There was a reduction in specific or total energy usage from 10% to 45%, and their study mainly depends on previous plant energy data. In a few studies, agent-based modeling was used to merge several models and blocks in polymer-based industry and observed the possibility of reduction in energy usage up to 10% in heat flows through unit operations. Machine learning was applied to boost energy efficiency in terephthalic acid manufacturing. It was concluded to improve the plant's overall fuel, water, steam, and power usage. Most of the AI studies in chemical industries are applied to unit operations or equipment used for processing. Finally, the energy consumption data was recorded to minimize the energy utilization and grow the economics [32].

None of the AI experiments are reported in the literature to examine the energy efficiency in R&D efforts. Few studies have found that R&D operations reduce the number of experiments trials. AI has the potential to reduce energy consumption in R&D operations by reducing the number of trials. To assess the advantages of AI, some generic energy metrics, such as energy-saving potential and energy consumption, could be employed in the supply chain. It is worth considering the energy consumption in information technology and infrastructure for day-to-day programming [33]. Researchers tallied the number of arithmetic operations and communications during the training phase to assess the energy consumption of a feedforward neural network. Few calculated the energy consumption of a convolutional neural network by IT components such as the CPU, RAM, and other peripherals. Depending on the type of AI and supporting technologies, estimating the energy consumption of AI and machine learning might be challenging.

1.3.3 Economic aspect

The economic impact of chemical manufacturing is the economic component such as profit and loss. Most research focuses on economic benefits in the chemical industry [34]. The studies are focused on using AI to lower chemical production costs. All the studies use different measures to estimate profit and loss costs. Researchers use AI to improve chemical processes and plants, with concerns about lowering the cost of the imbalance between projected and utilized resources (fuel, water, and power). Few studies have found that using AI reduces energy, capital, and raw material costs. A genetic algorithm-based approach for AI supply chain applications was provided in one study to optimize the various indirect cost components for the chemical supply chain [35].

Several AI projects targeted increasing chemical manufacturing profits using quantitative methodologies and process data generated from the plants. Few researchers investigated profit rate and annual operating profit using AI. Many used net present value, projected profit, average profit metrics, and self-defined profit functions to understand the chemical supply chain management using AI. AI studies on bio-based chemical plants are not found due to their noncommercial scale.

In addition to costs and profits metrics, few studies have used plant production capacity, product yields, and plant throughput to measure the profitability of the chemical process using AI. Some AI research focuses on boosting reaction yields and boosting revenues if chemical production is scaled up. However, none of these studies have offered validations of possible economic benefits at the unit or process level. Some studies employed AI to boost yields by creating model predictions for unit operations and using heuristics to optimize results. Reinforcement learning is being used to improve process control and optimize high-quality goods [36]. In conclusion, numerous studies have measured the economic benefits of AI in the chemical industry. Depending on the AI methodologies, the cost reduction, profit enhancement, and capacity/yield increase can be performed. The findings can be a foundation for future economic assessments using AI.

1.3.4 Time aspect

The time aspect in AI refers to reducing the time for experiment design, modeling, and optimization. However, by reducing resource consumption and fostering the development of more sustainable chemical processes and products, time reduction can help lessen environmental, economic, and social impacts. Previous studies have shown that computing time is strongly related to AI and machine learning's energy and carbon emission consequently [37].

AI was explored to reduce the time in chemical process and product design, experimental time, and computation time. Most researchers have emphasized the benefits of AI in terms of time savings and quantified them in their studies. Heuristic algorithms were used to find near-optimal supply chain solutions concerning time and established the optimization. AI technologies have been extensively applied in other industrial sectors to minimize computational time. Green supply chain design and management is a field incorporating environmental impact and time aspects into the supply chain. AI has been used to assess risk management in the supply chain; moreover, big data analytics to improve the green supply chain. Overall, AI could help overcome computational hurdles by giving near-optimal solutions to supply chain management [38].

1.3.5 Safety and human factor aspect

The human and plant safety factors are considered to improve process safety and staff productivity. Few studies developed AI-based fuzzy logic controls integrated into a chemical plant for product design to increase chemical plant and human safety. Based on the reports, AI improved the accuracy and reliability of detecting abnormal situations, process monitoring, and fault diagnostics to improve chemical product safety. A distillation column-based rectification process has been designed with AI process monitoring. AI was also utilized to monitor reciprocating compressors' polymerization operations, refineries, and defect detection [39]. In all the studies, indicators like defect detection rate and dependability on safety consequences parameters are considered in AI-based models.

Labor safety and productivity were considered as two impacts of AI. AI-based models for hazardous chemical facilities were developed to estimate and improve human labor safety and productivity. Based on the report, resilience engineering measured the productivity of laborers working in hazardous chemical facilities. In the supply chain, AI can help with supplier selection and analyzing the efficacy of various supply chain management parameters. Some researchers used backpropagation neural network and case-based reasoning approaches to solve supply chain issues. During the last decade, an enormous amount of data accumulated by the petroleum industry has been used in AI to create a strategy to improve labor safety and productivity. Safety and human aspects are inextricably linked in chemical manufacturing. However, just a few researchers have studied the connection between these factors. The indicators were added in most AI-based models that offered quantitative analysis. Studies show that AI could improve labor safety, supply chain decision-making, and productivity from a practical standpoint and boost formal risk assessment [40].

1.4 Digital Transformation of the Chemical Industry

Chemical industries are far from the bottom in digitalization by applying AI interventions. Recently, COVID-19 has accelerated the digitalization of chemical plants. Many factories were forced to implement the digital transformation in work habits, supply chain, and responding to shifting customer demands. According to an industry survey, digital transformation in the chemical industry has accelerated by 48%. According to Manufacturing Leadership Council (MLC) study, 90% of respondents felt that COVID-19 had "generated a new sense of urgency" to drive new technology and digitalization investment [41]. Different manufacturers implemented varying degrees of digital maturity. This section will report the analysis of digital transformation in various process industries.

1.4.1 State of digital transformation

Innovations and digital transformation have occurred during the last decade. Labor interactions, maintenance routines, supply chain operations, and customer demand were all disturbed by the pandemic, while expectations for sustainability, personalization, and efficiency grew. The importance of digital transformation is well acknowledged in the current situation. According to the studies, about 65% of CEOs consider digitalization for the next two years. Keep priorities for innovation, cost reduction, product modification, and overall cash flow protection. Another study discovered that 66% of chemical industry executives foresee revolutionary transformation during the next five years and intend to invest an average of 7% of their annual turnover. In the current situation, the manufacturing business has a digital maturity profile of 40%, while the chemicals industry has 45%. Although chemical facilities appear to be having difficulty scaling up to implement digital transformation, overall sectors average, 45% of companies are rolling out new products and implementing digital transformation.

Compared to more digitally mature plants and implementing digital transformation strategies, those are still in the early stages of widening. A substantial gap exists between digital leaders of chemical industries and those that lag in digitalization. Industries in Asia-Pacific see strategic digitalization prospects, while others concentrate on operational improvements, market and consumer access, and cost reduction.

1.4.2 Key trends in digitalization

The chemical industries are considered complicated for digitalization due to their multi-disciplinary nature. Few sectors saw a drop in demand due to the pandemic, while others grew and shifted toward digitalization. Extensive facilities boost computation speed, while specialized chemicals are concerned with improving quality [42].

All the plants are looking for maximum output, improved safety, reduced waste, and be sustainable and ready to respond to rapid changes in demand and supply. AI, industrial internet of things (IIoT), and cloud computing will be the most popular digital transformation applications in the next two years.

1.4.3 Optimizing production

Digitalization in chemical plants focuses on enhancing equipment efficiency, automation, predictive and remote monitoring, and streamlining maintenance. Digitalization's top three concerns are data analysis, integration and optimization of the process, and integration of data management. Advanced data analysis, such as predictive maintenance, is manufacturers' top priority [43].

Machine learning (ML) and AI-based predictive analytics can detect abnormalities that suggest impending part failures, fouling, product quantity, and product quality. Control systems generate warnings till a prompt response. Unlike old systems that go unreported and are necessary to replace the part, repair, or possibly shut down production. Maintenance crews can also schedule repairs at the most convenient times, thanks to early alerts.

Chemical facilities, which typically supply base chemicals for usage in other sectors, must maintain high consistency in product quality. However, it is more complicated when raw material supply and quality can be inconsistent. Process engineers need adjustments to maintain the quality feedstock and reaction conditions to make acceptable products. Thus, end-to-end digital transformation will enable industries to foresee low demand, reduce fixed costs, or reallocate demand across the plant using data and analytics. Digital transformation can quickly scale up production when demand increases [44].

1.4.4 Supporting remote operations

After the epidemic, plants or industries have to rethink their ability to handle remote and hybrid operations. Remote teams can operate together with better communication and collaboration systems [45]. Digital twins are also created by chemical plants, which digitally replicate specific systems, operations, or the complete plant. Remote visibility into machinery, processes, remote diagnostics, and remote maintenances are possible with digital twins. Maintenance workers can analyze and troubleshoot the systems, reducing plant downtime. Advanced sensors enable factories to reduce the number of people required onsite without compromising production output.

1.4.5 Reducing waste

Plants that have undergone a digital transition have reaped the benefits of lower costs. Raw material pricing can be variable in the chemical production business, yet customers want continuous low prices. On the other hand, plants face significant energy expenses, making it critical to limit waste as much as possible. Maintenance teams can use early warnings about inefficiencies and potential part failures to perform a low-cost repair instead of replacing a part. The more efficient operations consume less energy and fewer raw materials, thereby producing less waste. Analytics tools that track changing raw material prices can assist factories in negotiating the best deal with suppliers and planning ahead of time when prices fluctuate dramatically. Plants can prepare the proper amounts of various products with more accurate demand forecasting, reducing the danger of oversupply.

1.4.6 Unlocking new growth opportunities

As early as the 1990s, digital transformation was primarily used to make operations run more smoothly. However, digitally mature plants are now using digital technologies to open up new growth opportunities, drive innovation, and improve their competitive advantage. Now, AI is used in chemical R&D to find new synthesis pathways and build new materials or chemical structures that are more sustainable. Scientists can use multivariate analysis to determine the impact of specific constituents in the mix, improving product quality. Automation reduces the time to develop new products from three to four years to six to nine months, allowing companies to satisfy demand much more quickly.

1.4.7 Increasing supply chain visibility

The vulnerability of chemical factories was revealed when the pandemic destroyed supply lines. As a result, businesses have concentrated on increasing supply chain visibility and integration. Over 80% of manufacturing executives stated they would focus more on supply chain robustness. Digital twins enable plants to acquire a unified view of the entire supply chain, from basic raw materials to finished products to market forecasting. They will be able to respond to changes and bottlenecks more quickly.

Additionally, plants are reshoring and nearshoring supply chains, resulting in a digitally connected ecosystem with their new suppliers. Manufacturers can produce goods in line with consumer resources available for raw materials. Thanks to a streamlined digital supply chain that enables complete data sharing on an upstream and downstream level [46].

1.4.8 Safety, compliance, and sustainability

The chemical industries are heavily regulated as they produce many hazardous raw materials. Many big industries rely on raw materials produced. Companies use digital transformation to improve their safety and reduce emissions, fire incidents, and accidents. AI, such as predictive maintenance solutions, digital twins, and remote monitoring, can reduce the risk of human labor entering potentially dangerous conditions to perform repairs and maintenance and run operations smoothly. The lesser the emissions, the more smoothly and effectively the plant runs. Finally, digital information is more precise and reliable than conventional paperwork, which is more valuable during safety or compliance checks [47].

1.5 Digital Chemical Industry 4.0

"Digitalization" now has a much more broad implication. It is a term that refers to the implementation of any computer hardware or software solution. Although digitalization is a critical enabler of smart manufacturing, it is frequently used interchangeably with the term. It is most commonly used to define a technology that connects the entire system (e.g., equipment, instruments, processes, models, analyzers, and people) by allowing them to interact and utilize digital data streams. On the other hand, digitalization has expanded to incorporate cutting-edge technologies such as connectivity, predictive analytics, mobility, robots, and artificial intelligence. Further, chemical process industries (CPI) have practiced plant digitalization for safety, automation, and optimization. Digitalization in the CPI is enabled by:

► Cloud computing: Plant operational data exchanged over the cloud can be used for various purposes, including remote technology professionals troubleshooting real-time.
► Online optimization and model-based control: Model-based control and real-time optimization (RTO) play a crucial role in operating the plant within permissible limits and optimizing the process parameters. The RTO utilizes steady-state models to optimize the overall plant, whereas advanced process control uses dynamic models of different units (e.g., reactors, distillation columns).
► Online monitoring: The operational data can be investigated to predict the process's performance and equipment failure by predictive analytics.
► Process automation: The application of smart sensors, radio-frequency identification (RFID), robotics, augmented and virtual reality are being practiced in chemical process industries to ensure safer operation of the plants and enhance profitability. Smart sensors are proven to give high accuracy of measurements for the desired operation. Furthermore, the preprocessing of measured signals by smart sensors improves

measurement consistency. Robotics can be used in many R&D or plant operations in the plant critical and emergency (Hazards).

Digitalization can help a chemical process industry increase quality, perceptibility, and effective optimization. Further, digitalization in the chemical process industry has accounted for many benefits, such as:

In-plant operation includes:

▶ Minimum operational costs
▶ Early fault identification based on predictive analytics ensures asset availability
▶ Predictive maintenance replaces preventative maintenance, resulting in decreased maintenance expenditures
▶ more efficient supply chains
▶ Using production planning systems that can reduce inventory
▶ early detection of flaws and disturbances improves reliability
▶ Advanced process control reduces waste while maintaining quality standards

In R&D include:

▶ Experiments in the lab that are run at a faster pace
▶ Enhanced collaboration in research and open innovation networks
▶ Technology advancements and market intelligence
▶ The time it takes to get a product to market has been cut in half.

Nontechnical areas include:

▶ Customer interactions have improved
▶ a better experience for suppliers and distributors
▶ a better way to find and hire talent
▶ a boost in business development

On the other hand, faster experimentation could benefit research and development, but CPI firms must conduct lab tests to develop and assess various catalysts and establish kinetic reaction rate characteristics when creating new technologies. These tests are usually carried out on a few single-reactor configurations. Several experiments must be conducted to verify suitable temperature, pressure, and species composition ranges and create credible kinetic models. This can take anywhere from a few months to over a year. Robotics is used in high-throughput experimentation (HTE) systems to execute up to 50 concurrent tests. HTE makes it possible to complete months of work in a few weeks.

Overall, the advantages to plant operations, R&D, and nontechnical sectors are apparent, and the CPI will be forced to accept this new generation of digital technologies as a must. Digitalization is seen as desirable but not necessary by some CPI decision-makers. Digitalization can be costly and time-consuming, but both can be reduced with the right expertise, a realistic approach, and thorough preparation. The first challenge is to begin utilizing the data, and the data can then be turned into measurable key performance indicators (KPIs) that can be used to guide and support decisions. Companies will be more knowledgeable, nimble, and competitive as a result of digital transformation.

1.6 Challenges of AI in the Chemical Industry

The chemical industry operates in a globalized environment undergoing significant transformation, with process performance influenced by plant operations and consumer expectations and environmental, social, political, and economic factors. The process industry should adapt to the changes due to market fluctuations, demand, and economics in the current scenario. In order to overcome these problems, the chemical industry should be dynamic and flexible through optimization and control.

As aforementioned, those demands are starting to exceed human capacity. It is anticipated that by integrating new Industry 4.0 technologies into the physical system, a more efficient, safer, environmentally friendly, and competitive system will be developed. The disruptive nature of these emerging technologies was highlighted at the 2016 Davos Forum. As a result, deploying cyber-physical system (CPS) in the chemical industry is a once-in-a-lifetime chance to meet the industry's current and future difficulties. AI techniques are frequently employed in the chemical industry to solve a variety of challenges on a regular basis, such as soft sensing [48], modeling [49], and control. On the other hand, this application is limited to standard AI technologies like feedforward neural networks and fuzzy logic. Intelligent systems are currently driving the development of AI technologies. However, as previously said, these drives are primarily in computer science and mechanical engineering. As a result, there is a growing demand in chemical engineering to translate breakthrough AI discoveries into chemical systems tools [11].

However, few authors have looked into the possibilities of CPS in chemical processes (17 hits on Web of Science in the last 20 years). Ji *et al.* [50] proposed an intelligent rectification column based on CPS. The automation of a batch distillation column has been developed by Budiawan *et al.*

[51] using the CPS technique. Furthermore, the current research exclusively considers CPS for automation. To progress from automation to autonomy, a technological gap must be bridged.

Cognitive competence is required to fill this void. In this regard, AI has the potential to be a critical enabler [52]. As a result, it is a problem that must be resolved for the chemical industry to use the promise of emerging technology fully. This could lead to a safer and more efficient chemical industry, with systems that can run without human intervention. Firms in developed countries find it difficult to compete with businesses in Eastern Europe, owing to the significantly differentiated labor in those countries, both in terms of number and cost, compared to the workforce in developed countries [53]. Industries' efficiency levels must be pushed to their limits to address this issue while keeping safety and environmental restraints in mind. As a result, there is an apparent demand for enhanced technology. On the one hand, the new digital industrial revolution promises more age-friendly work environments in this scenario [54]. On the other hand, it has the potential to increase the efficiency of mechanical and repetitive tasks, which could be performed by intelligent systems under human supervision rather than by humans.

On the other hand, the search for high-performance production systems as a more efficient and cost-effective means of production is critical in addressing these issues. The demand for products increases along with the human population. As a result, production systems must be able to provide the necessary scale to meet demand. The 2030 Agenda for Sustainable Development expresses the urgent need for significant changes in consumption and production. According to some experts, the most pressing concern for humanity's future is the demand for more efficient and sustainable production methods [55]. As a result, urgent studies that address this issue from the standpoint of chemical process systems and chemical plant efficient management are required in the chemical sector. In this scenario, the future potential of CPS systems enabled by AI cognition is evident, not just for the chemical industry but also for society.

1.7 Industry 4.0 Impact Technologies

Industrialization is a critical step in the manufacturing process. Many businesses struggle to stay afloat because they have not embraced information technology (IT). Regardless of their sectors or operations, businesses today have the same outlook on the future and do not want to be caught off guard. As a result of the possibilities of the quality gained in product customization, Industry 4.0 is being adopted by every sector. Despite the multiple benefits of Industry 4.0, there is still a high consumption of resources, raw materials, information, and environmentally unsustainable energy. As a result, society and the public sector have become more aware of the risks and environmental issues [56].

With the usage of Industry 4.0, this industrialization sector has concentrated chiefly on production and maximizing profits, which has resulted in a slew of difficulties in other areas. Depletion of natural resources, negative environmental impacts, unequal wealth distribution, and unsuitable working conditions, for example, are all factors that may eventually lead to an unsustainable consumption pattern in terms of the environment, economy, and even society [57].

Industry 4.0 has lately been a popular term, and it was included in a German government initiative in 2011 as a plan for industrial output in that country [58]. Industry 4.0, in this picture, is part of a more integrated and linked world that has emerged due to the information and communication technology (ICT) revolution. Industry 4.0 is characterized by technical advancements such as the internet of things (IoT) and the internet of services (IoS), which electronically connect industry through a supply chain network. This paved a new trend in the industry as smart manufacturing and cyber-physical systems (CPSs) help to support it.

Through sensors and other devices, CPSs connect the physical parts of the industry, such as machines and other components, to an electronic layer of information systems. This technology allows information to be exchanged and procedures carried out and act as a link between humans and machines. The internet of things (IoT) is a new concept that refers to the Internet's connection to the physical world via sensors and actuators. All signals from the physical production system can be monitored and documented as "big data" for subsequent use in other processes, such as creating new value, thanks to this connectivity, particularly in the framework of varying demand for innovative services and new types of employment. "Cloud computing," [59] which is described as computing resources and visualization services, is a mix of innovation and implementation of new technologies like Industry 4.0 that have a meaningful and profitable impact on enterprises' operations to save costs and increase efficiency. Finally, Industry 4.0 benefits from integrating all existing technologies with physical elements in manufacturing and all linked stockholders (suppliers, customers, consumers, and employees). This aids decision-making because all relevant information is readily available, reducing the official hierarchy.

The environment's sustainability in Industry 4.0 is an important topic that has received much attention in the past literature. Although there are differing perspectives on this relationship in the prior research, there is no apparent consensus that Industry 4.0 has a long-term influence on environmental sustainability. According to other studies, startups and new businesses should integrate their strategies and visions for integrating Industry 4.0 and manufacturing with environmental sustainability in their business plans. Studies by Ford and Despeisse [60] and Jelonek and Urbaniec [61] demonstrated the uses of implementing new technology in manufacturing (such as 3D printing) for environmental sustainability but also indicated that the technology is still in its early stages, posing various hurdles. On the other hand, Stock and Seliger [62] contend that industrial value should be oriented toward sustainability and that Industry 4.0 offers a significant opportunity. According to Burritt and Christ [63], Industry 4.0 impacts environmental sustainability by providing more precise, superior organization, and real-time supervision.

Overall, these studies exploit the problems and opportunities connected with Industry 4.0 and the technologies linked with this industry in terms of environmental sustainability, which have not been effectively deployed because of innovative technologies. As a result, there is still a gap in how to connect the efficient use of finite resources, feedstock, data, energy consumption, and usage with long-term solutions for sustainable development. The 4Rs—reduce, reuse, recycle, and replace—can be utilized to reduce environmental pollution and promote sustainability. As a result, in Industry 4.0 and the environment, efficiency, and eco-innovation will be attained.

1.8 Conclusion

- ▶ A number of AI applications in the chemical industry were discussed in the chapter, highlighting five main sustainability-related implications of AI adoption.
- ▶ Although AI is not a new concept in the chemical industry, quantitative examination of AI's effects on the environment, the economy, and society are still in their early stages.
- ▶ For a comprehensive assessment of artificial intelligence's impacts, a conceptual framework based on a literature review and incorporating methods from industrial ecology, economics, and engineering will

be developed to guide the selection of performance indicators and evaluation methods for a holistic assessment of AI's impacts.

▶ Real-world case studies with appropriate process information are required for practical implementation of the framework, which may be difficult for researchers such as the authors to obtain due to the confidentiality of most AI projects declared by chemical firms, as the authors have discovered.

▶ Researchers in AI and chemical engineering could use the framework to quantify the big-picture impacts of AI and identify the most beneficial pathways of AI applications.

References

[1] O. Hollins et al., "Towards a circular economy - waste management in the EU Study," 2017. [Online]. Available: https://www.europarl.europa.eu/RegData/etudes/STUD/2017/581913/EPRS_STU(2017)581913_EN.pdf.

[2] IEA, "Assessing the effects of economic recoveries on global energy demand and CO2 emissions in 2021," 2020. DOI: 10.1787/90c8c125-en.

[3] K. Volkart, C. Bauer, and C. Boulet, "Life cycle assessment of carbon capture and storage in power generation and industry in Europe," Int. J. Green. Gas Control, vol. 16, pp. 91–106, 2013, DOI: 10.1016/j.ijggc.2013.03.003.

[4] J. Nayak, K. Vakula, P. Dinesh, B. Naik, and D. Pelusi, "Intelligent food processing: Journey from artificial neural network to deep learning," Comput. Sci. Rev., vol. 38, p. 100297, 2020, DOI: https://doi.org/10.1016/j.cosrev.2020.100297.

[5] K. K. Sahu et al., "Artificial Intelligence and Machine Learning: New Age Tools for Augmenting Plastic Materials Designing, Processing, and Manufacturing," Elsevier, 2021.

[6] Accenture, "Are You Ready For What's Nextă? The Post-Digital Ready For What's Nextă?," 2019. [Online]. Available: https://www.accenture.com/_acnmedia/PDF-94/Accenture-TechVision-2019-Tech-Trends-Report.pdf.

[7] G. Del Río Castro, M. C. González Fernández, and Á. Uruburu Colsa, "Unleashing the convergence amid digitalization and sustainability towards pursuing the Sustainable Development Goals (SDGs): A holistic review," J. Clean. Prod., vol. 280, p. 122204, 2021, DOI: https://doi.org/10.1016/j.jclepro.2020.122204.

[8] M. Pournader, H. Ghaderi, A. Hassanzadegan, and B. Fahimnia, "Artificial intelligence applications in supply chain management," Int. J. Prod. Econ., vol. 241, p. 108250, 2021, DOI: https://doi.org/10.1016/j.ijpe.2021.108250.

[9] K. Nagapudi et al., "Microstructure, Quality, and Release Performance Characterization of Long-Acting Polymer Implant Formulations with X-Ray Microscopy and Quantitative AI Analytics," J. Pharm. Sci., vol. 110, no. 10, pp. 3418–3430, 2021, DOI:https://doi.org/10.1016/j.xphs.2021.05.016.

[10] Y.-W. Lee, "A stochastic model of particulate matters with AI-enabled technique-based IoT gas detectors for air quality assessment," Microelectron. Eng., vol. 229, p. 111346, 2020, DOI:https://doi.org/10.1016/j.mee.2020.111346.

[11] V. Venkatasubramanian, "The promise of artificial intelligence in chemical engineering: Is it here, finally?," AIChE J., vol. 65, no. 2, pp. 466–478, 2019, DOI: 10.1002/aic.16489.

[12] A. Chakraborty, A. Sivaram, and V. Venkatasubramanian, "AI-DARWIN: A first principles-based model discovery engine using machine learning," Comput. Chem. Eng., vol. 154, p. 107470, 2021, DOI: https://doi.org/10.1016/j.compchemeng.2021.107470.

[13] L. Li, S. Rong, R. Wang, and S. Yu, "Recent advances in artificial intelligence and machine learning for nonlinear relationship analysis and process control in drinking water treatment: A review," Chem. Eng. J., vol. 405, p. 126673, 2021, DOI: https://doi.org/10.1016/j.cej.2020.126673.

[14] V. Venkatasubramanian and V. Mann, "Artificial intelligence in reaction prediction and chemical synthesis," Curr. Opin. Chem. Eng., vol. 36, p. 100749, 2022, DOI:https://doi.org/10.1016/j.coche.2021.100749.

[15] N. Choudhary, R. Bharti, and R. Sharma, "Role of artificial intelligence in chemistry," Mater. Today Proc., vol. 48, pp. 1527–1533, 2022, DOI: https://doi.org/10.1016/j.matpr.2021.09.428.

[16] A. Haleem, M. Javaid, R. P. Singh, and R. Suman, "Applications of Artificial Intelligence (AI) for cardiology during COVID-19 pandemic," Sustain. Oper. Comput., vol. 2, pp. 71–78, 2021, DOI: https://doi.org/10.1016/j.susoc.2021.04.003.

[17] S. He, L. G. Leanse, and Y. Feng, "Artificial intelligence and machine learning assisted drug delivery for effective treatment of infectious diseases," Adv. Drug Deliv. Rev., vol. 178, p. 113922, 2021, DOI: https://doi.org/10.1016/j.addr.2021.113922.

[18] J. Hu, C. Kim, P. Halasz, J. F. J. Kim, J. F. J. Kim, and G. Szekely, "Artificial intelligence for performance prediction of organic solvent nanofiltration membranes," J. Memb. Sci., vol. 619, p. 118513, 2021, DOI: https://doi.org/10.1016/j.memsci.2020.118513.

[19] A. Sircar, K. Yadav, K. Rayavarapu, N. Best, and H. Oza, "Application of machine learning and artificial intelligence in oil and gas industry," Pet. Res., no. xxxx, 2021, DOI: https://doi.org/10.1016/j.ptlrs.2021.05.009.

[20] Z. Said, P. Sharma, L. Syam Sundar, A. Afzal, and C. Li, "Synthesis, stability, thermophysical properties and AI approach for predictive modelling of Fe3O4 coated MWCNT hybrid nanofluids," J. Mol. Liq., vol. 340, p. 117291, 2021, DOI: https://doi.org/10.1016/j.molliq.2021.117291.

[21] L. J. Catania, "3 - The science and technologies of artificial intelligence (AI)," L. J. B. T.-F. of A. I. in H. and B. Catania, Ed. Academic Press, 2021, pp. 29–72.

[22] M. Wever, M. Shah, and N. O'Leary, "Designing early warning systems for detecting systemic risk: A case study and discussion," Futures, vol. 136, p. 102882, 2022, DOI: https://doi.org/10.1016/j.futures.2021.102882.

[23] P. L. González Ramírez, J. Lloret, J. Tomás, and M. Hurtado, "IoT-networks group-based model that uses AI for workgroup allocation," Comput. Networks, vol. 186, p. 107745, 2021, doi: https://doi.org/10.1016/j.comnet.2020.107745.

[24] O. A. Moses et al., "Integration of data-intensive, machine learning and robotic experimental approaches for accelerated discovery of catalysts in renewable energy-related reactions," Mater. Reports Energy, vol. 1, no. 3, p. 100049, 2021, DOI: https://doi.org/10.1016/j.matre.2021.100049.

[25] N. Lee, J. Jeong, and D. Shin, "A System for Supporting Generation of Optimal Synthetic Pathways Based on Chemical Reaction Big Data," in 31 European Symposium on Computer-Aided Process Engineering, vol. 50, M. Türkay and R. B. T.-C. A. C. E. Gani, Eds. Elsevier, 2021, pp. 1053–1058.

[26] A. M. Zidan, E. A. Saad, N. E. Ibrahim, A. Mahmoud, M. H. Hashem, and A. A. Hemeida, "PHARMIP: An insilico method to predict genetics that underpin adverse drug reactions," MethodsX, vol. 7, p. 100775, 2020, DOI: 10.1016/j.mex.2019.100775.

[27] S. Raschka and B. Kaufman, "Machine learning and AI-based approaches for bioactive ligand discovery and GPCR-ligand recognition," Methods, vol. 180, pp. 89–110, 2020, DOI: https://doi.org/10.1016/j.ymeth.2020.06.016.

[28] T. Ahmad et al., "Energetics Systems and artificial intelligence: Applications of industry 4.0," Energy Reports, vol. 8, pp. 334–361, 2022, DOI: https://doi.org/10.1016/j.egyr.2021.11.256.

[29] P. P. Senna, A. H. Almeida, A. C. Barros, R. J. Bessa, and A. L. Azevedo, "Architecture Model for a Holistic and Interoperable Digital Energy Management Platform," Procedia Manuf., vol. 51, pp. 1117–1124, 2020, DOI: https://doi.org/10.1016/j.promfg.2020.10.157.

[30] W. G. de Sousa, E. R. P. de Melo, P. H. D. S. Bermejo, R. A. S. Farias, and A. O. Gomes, "How and where is artificial intelligence in the public sector going? A literature review and research agenda," Gov. Inf. Q., vol. 36, no. 4, p. 101392, 2019, DOI: https://doi.org/10.1016/j.giq.2019.07.004.

[31] M. Abdallah, M. Abu Talib, S. Feroz, Q. Nasir, H. Abdalla, and B. Mahfood, "Artificial intelligence applications in solid waste management: A systematic research review," Waste Manag., vol. 109, pp. 231–246, 2020, DOI: https://doi.org/10.1016/j.wasman.2020.04.057.

[32] J. Chen, M. Zhang, B. Xu, J. Sun, and A. S. Mujumdar, "Artificial intelligence assisted technologies for controlling the drying of fruits and vegetables using physical fields: A review," Trends Food Sci. Technol., vol. 105, pp. 251–260, 2020, DOI: https://doi.org/10.1016/j.tifs.2020.08.015.

[33] N. A. Aziz, N. A. A. Adnan, D. A. Wahab, and A. H. Azman, "Component design optimisation based on artificial intelligence in support of additive manufacturing repair and restoration: Current status and future outlook for remanufacturing," J. Clean. Prod., vol. 296, p. 126401, 2021, DOI: https://doi.org/10.1016/j.jclepro.2021.126401.

[34] S. Fosso Wamba, R. E. Black, C. Guthrie, M. M. Queiroz, and K. D. A. Carillo, "Are we preparing for a good AI society? A bibliometric review and research agenda," Technol. Forecast. Soc. Change, vol. 164, p. 120482, 2021, DOI: https://doi.org/10.1016/j.techfore.2020.120482.

[35] E. Shin, S. Yoo, Y. Ju, and D. Shin, "Knowledge graph embedding and reasoning for real-time analytics support of chemical diagnosis from exposure symptoms," Process Saf. Environ. Prot., vol. 157, pp. 92–105, 2022, DOI: https://doi.org/10.1016/j.psep.2021.11.002.

[36] J. Jawad, A. H. Hawari, and S. Javaid Zaidi, "Artificial neural network modeling of wastewater treatment and desalination using membrane processes: A review," Chem. Eng. J., vol. 419, p. 129540, 2021, DOI: https://doi.org/10.1016/j.cej.2021.129540.

[37] T. Welz, R. Hischier, and L. M. Hilty, "Environmental impacts of lighting technologies - Life cycle assessment and sensitivity analysis," Environ. Impact Assess. Rev., vol. 31, no. 3, pp. 334–343, 2011, doi: 10.1016/j.eiar.2010.08.004.

[38] Z. Akkus et al., "A Survey of Deep-Learning Applications in Ultrasound: Artificial Intelligence–Powered Ultrasound for Improving Clinical Workflow," J. Am. Coll. Radiol., vol. 16, no. 9, Part B, pp. 1318–1328, 2019, doi: https://doi.org/10.1016/j.jacr.2019.06.004.

[39] M. Bagheri, A. Akbari, and S. A. Mirbagheri, "Advanced control of membrane fouling infiltration systems using artificial intelligence and machine learning techniques: A critical review," Process Saf. Environ. Prot., vol. 123, pp. 229–252, 2019, DOI: https://doi.org/10.1016/j.psep.2019.01.013.

[40] F. Sattari, L. Lefsrud, D. Kurian, and R. Macciotta, "A theoretical framework for data-driven artificial intelligence decision making for enhancing the asset integrity management system in the oil & gas sector," J. Loss Prev. Process Ind., vol. 74, p. 104648, 2022, DOI: https://doi.org/10.1016/j.jlp.2021.104648.

[41] P. Fantke et al., "Transition to sustainable chemistry through digitalization," Chem, vol. 7, no. 11, pp. 2866–2882, 2021, DOI: https://doi.org/10.1016/j.chempr.2021.09.012.

[42] Y. Xu et al., "Artificial intelligence: A powerful paradigm for scientific research," Innov., vol. 2, no. 4, p. 100179, 2021, DOI: https://doi.org/10.1016/j.xinn.2021.100179.

[43] G. Alam, I. Ihsanullah, M. Naushad, and M. Sillanpää, "Applications of artificial intelligence in water treatment for optimization and automation of adsorption processes: Recent advances and prospects," Chem. Eng. J., vol. 427, p. 130011, 2022, DOI: https://doi.org/10.1016/j.cej.2021.130011.

[44] R.-Q. Wang, "Chapter 13 - Artificial Intelligence for Flood Observation," in Earth Observation, G. J.-P. B. T.-E. O. for F. A. Schumann, Ed. Elsevier, 2021, pp. 295–304.

[45] Deloitte, "Winter 2022 Fortune/Deloitte CEO Survey," 2022.

[46] N. Ravi and D. P. Johnson, "Artificial intelligence-based monitoring system for onsite septic systems failure," Process Saf. Environ. Prot.,

vol. 148, pp. 1090–1097, 2021, DOI: https://doi.org/10.1016/j.psep.202
1.01.049.

[47] T. Ahmad et al., "Artificial intelligence in sustainable energy industry: Status Quo, challenges and opportunities," J. Clean. Prod., vol. 289, p. 125834, 2021, DOI: https://doi.org/10.1016/j.jclepro.2021.125834.

[48] I. Nogueira, C. Fontes, I. Sartori, K. Pontes, and M. Embiruçu, "A model-based approach to quality monitoring of a polymerization process without online measurement of product specifications," Comput. Ind. Eng., vol. 106, pp. 123–136, Apr. 2017, DOI: 10.1016/J.CIE.2017.01.030.

[49] E. Loh, "Medicine and the rise of the robots: a qualitative review of recent advances of artificial intelligence in health," BMJ Lead., vol. 2, pp. 59–63, 2018, DOI: 10.1136/leader-2018-000071.

[50] X. Ji, G. He, J. Xu, and Y. Guo, "Study on the mode of intelligent chemical industry based on cyber-physical system and its implementation," Adv. Eng. Softw., vol. 99, pp. 18–26, Sep. 2016, DOI: 10.1016/J. ADVENGSOFT.2016.04.010.

[51] I. Budiawan, P. Hidayah, E. M. I. Hidayat, and A. Syaichu, Design and Implementation of Cyber-Physical System-Based Automation on Plant Chemical Process: Study Case Mini Batch Distillation Column; Design and Implementation of Cyber-Physical System-Based Automation on Plant Chemical Process: Study Case Mini Batch. 2018.

[52] T. Gamer, M. Hoernicke, B. Kloepper, R. Bauer, and A. J. Isaksson, "The autonomous industrial plant – future of process engineering, operations and maintenance," J. Process Control, vol. 88, pp. 101–110, Apr. 2020, DOI: 10.1016/J. JPROCONT.2020.01.012.

[53] R. S.-V. Schätz, B.; Törngren, M.; Bensalem, S.; Cengarle, M. V.; Pfeifer, H.; McDermid, J.; Passerone, "A. Cyber-Physical European Roadmap & Strategy," 2015.

[54] M. Calzavara, D. Battini, D. Bogataj, F. Sgarbossa, and I. Zennaro, "International Journal of Production Research Ageing workforce management in manufacturing systems: state of the art and future research agenda," 2019, DOI: 10.1080/00207543.2019.1600759.

[55] R. L. Naylor, "Managing Food Production Systems for Resilience. In Principles National Resources Stewardship Resilience-Based Management a Change," in Springer: New York, NY, USA, 2008, pp. 1–46.

[56] A. McWilliams, A. Parhankangas, J. Coupet, E. Welch, and D. T. Barnum, "Strategic Decision Making for the Triple Bottom Line," Bus.

Strateg. Environ., vol. 25, no. 3, pp. 193–204, Mar. 2016, DOI: https://doi.org/10.1002/bse.1867.

[57] S. H. Bonilla, H. R. O. Silva, M. Terra Da Silva, R. F. Gonçalves, and J. B. Sacomano, "Industry 4.0 and Sustainability Implications: A Scenario-Based Analysis of the Impacts and Challenges," 2018, DOI: 10.3390/su10103740.

[58] B. t'Slusarczyk, "Industry 4.0: Are we ready?," Pol. J. Manag. Stud., vol. 17, pp. 232–248, 2018.

[59] K. Candle Haug, T. Kretschmer, and T. Strobel, "Cloud adaptiveness within industry sectors – Measurement and observations," Telecomm. Policy, vol. 40, no. 4, pp. 291–306, Apr. 2016, DOI: 10.1016/J. TELPOL.2015.08.003.

[60] S. Ford and M. Despeisse, "Additive manufacturing and sustainability: an exploratory study of the advantages and challenges," J. Clean. Prod., vol. 137, pp. 1573–1587, Nov. 2016, DOI: 10.1016/J. JCLEPRO.2016.04.150.

[61] M. Jelonek and M. Urbaniec, "Development of Sustainability Competencies for the Labour Market: An Exploratory Qualitative Study," doi: 10.3390/su11205716.

[62] G. Stock, T.; Seliger, "Opportunities of Sustainable Manufacturing in Industry 4.0.," in Procedia CIRP 40. In Proceedings of the 13th Global Conference on Sustainable Manufacturing-Decoupling Growth from Resource Use, Berlin, Germany, 2011, pp. 536–541.

[63] R. Burritt and K. Christ, "Industry 4.0 and environmental accounting: a new revolution?," doi: 10.1186/s41180-016-0007-y.

2

Managing the Transition Toward Industry 4.0: A Study on the Implementation of Digital Manufacturing Processes

Vijay Anant Athavale[1], Ankit Bansal[2], Akhilesh Kumar Mishra[3], and Anil Kumar Kapil[4]

[1]Walchand Institute of Technology, India
[2]Chitkara University Institute of Engineering and Technology, Chitkara University, India
[3]Panipat Institute of Engineering and Technology, India
[4]College of Technology, Surajmal University, India
E-mail: vijay.athavale@gmail.com; erankitbansal@gmail.com; akhilesh.mishra@gmail.com; anilkdk@gmail.com

Abstract

The rapid digitization and development of technologies has led to the fourth industrial revolution popularly known as Industry 4.0. To remain competitive, it is important for companies to assimilate Industry 4.0 and its technologies, especially within manufacturing. Industry 4.0 offers numerous opportunities, but the implementation also comes with some challenges that should be taken into account for a successful transition. The aim of this study is to create an understanding of what Industry 4.0 can offer for companies and their manufacturing processes, and what challenges and opportunities that may be involved. This study has been based on research of the phenomenon Industry 4.0 as well as data collected from semi-structured interviews with employees of an industrial manufacturing enterprise.

The result of the study suggests that firstly, the companies must have a customized strategy to embrace the implementation of Industry 4.0 into their existing structure. Secondly, it is important that all employees are informed about and involved in the new procedures to increase understanding as well as motivation. Thirdly, it is important that existing techniques and digital tools within the company are adapted for a transition to Industry 4.0, and they must also have a common standard to facilitate data management. Finally, if all these are taken into account, the implementation of Industry 4.0 can provide real-time information and understanding that contribute to a better overview of the manufacturing, the quality, and the efficiency. Industry 4.0 also allows businesses to create a more flexible production and provide good insight and better decision-making to have superior control over the business. Overall, it reduces energy consumption and offers companies an opportunity to potentially become leaders in the global market.

Keywords: Industry 4.0, digitization, transformation, IIoT, smart manufacturing.

2.1 Introduction

In today's society, the rapid development of digitalization and technology is a fact and organizations face the challenge of constantly keeping up to date to be competitive. The amount of computing power doubles every 18 months according to Moore's law, which gives an exponential curve for the development of technology. Today, society is in what is called the prelude to the fourth industrial revolution, which means that organizations are responsible for new technical and organizational changes. As a result of rapid technological development, it is expected to be more difficult to meet challenges, as well as take advantage of the benefits that the fourth industrial revolution makes possible, this in comparison with what has been experienced in previous industrial revolutions [1].

The first three industrial revolutions lasted for almost 200 years, including everything from the advent of steam engines, hydropower and mechanization, to mass production and use of computers, information technology, and automation in manufacturing processes.

Today, the fourth industrial revolution has taken shape, which means, among other things, that machines and technical equipment can communicate with each other and the environment with the help of sensors and cloud-based services, as well as collect and process data themselves [1].

ust as the development of technology, the phenomenon of internet of things (IoT) is also expected to grow exponentially as by 2020 there will be at least 20 billion connected objects. The IoT concept was first coined by Kevin Ashton in 1999 when he used it to name communication systems consisting of computers and sensors [2, 19]. IoT can be described as an umbrella concept and means that smart technology, such as sensors, computers, and cloud services, can be combined with and built into, for example, machines and household appliances. In this way, the collection, processing, and exchange of context-specific data in real time is made possible, as well as for physical objects to communicate with each other and be able to make joint decisions [2–4].

As machines and other equipment will be able to communicate and interact with each other and the environment through the exchange of data, they are expected to become active participants in social processes as well as in information and business processes. The goal of IoT is to enable objects to connect anywhere, anytime, and with anything or anyone using a network or service [1]. It is expected that IoT will transform the entire market and lead to increased innovation, productivity gains, and also economic growth. In order for organizations to be able to keep up with this revolution in the market, new business logic in combination with adapted strategies must be adopted so that the properties of the technology are utilized in an optimal way.

IoT is becoming increasingly integrated into operations and now proactive decisions are required so that industry and society can meet real future needs in its digitization. Within the fourth industrial revolution, the concept of IoT has evolved further and focus has ended up on the industry. The phenomenon is called industrial Internet of Things (IIoT), or Industry 4.0, and originates from Germany. Morrar *et al.* [1] describe Industry 4.0 as data and services that will change future production, logistics, and work processes. IIoT is thus based on the same principles as IoT, but instead focuses on connecting machines, robots, and other equipment in the industrial sector. The phenomenon is mainly about communication between machines, machine to machine (M2M) but also the autonomous processes that are based on the exchange of information between the machines in real time [2].

The IIoT phenomenon was introduced in 2011 and is believed to be a significant phenomenon among industries and organizations, whether they wanted it or not. Just as the internet created a great deal of uncertainty among consumers in the 1990s and later emerged as a dominant technical phenomenon, Industry 4.0 is a potential success rather than hype. Thus, all manufacturing companies must embark on this industrial revolution in order to remain a leader in the competitive market [3]. However, questions remain about what IIoT requires for technologies and methods, as well as what challenges and values it entails.

2.2 Main Text

2.2.1 Problem formulation

Although the development of digital tools and technologies is progressing fast, especially the development of IoT and Industry 4.0, organizations in the industry can still be considered relatively conservative and have not really adopted the tools and technologies provided in the market. It means there are challenges for organizations in this sector to keep up with developments.

Operations in the manufacturing industry face difficulties when it comes to understanding Industry 4.0 and identifying the steps required for the transition to the phenomenon [3]. The fourth industrial revolution places new demands on organizations in the industry and can therefore be seen as overwhelming to tackle.

2.2.2 Purpose and question

In accordance with the above problem formulation, the study aims to create an understanding of the requirements that Industry 4.0 places on organizations within the industry and what such a digital transformation can entail. To investigate this, the study is based on the following question:

What opportunities and challenges does the transition to Industry 4.0 entail for companies facing this digitalization?

2.2.3 Related research

Below is a description of the phenomenon Industry 4.0, then [3] model for strategic guidance, which has been used as a framework during the study to gain a better understanding of what steps businesses need to go through for successful implementation of IIoT. Values and challenges that this may entail are presented continuously throughout the section.

2.2.4 Industry 4.0

Industry 4.0 and the accessibility that the phenomenon brings in is no longer a future trend, many leading organizations have already introduced this as a central part of the strategy and the businesses that have not done so have the opportunity to take advantage of the competitive advantages organizations [3].

Industry 4.0 is often referred to as the industrial internet of things (IIoT) and deals with the industrial application of the IoT phenomenon. IIoT is not only about the network of physical objects but also includes digital representations of objects, processes, and infrastructure, including 3D models of machines. IIoT offers better visibility and insight into the business and its assets through the integration of sensors, software, cloud services, and storage systems. The phenomenon is also based on the philosophy that smart machines surpass humans through their ability to capture and process data accurately and stably.

Furthermore, industrial reports show that IIoT has great potential in terms of predicting service, sustainable manufacturing, and quality of the product and energy efficiency.

Various studies show that Industry 4.0 has the greatest impact on organizations' production and manufacturing. Sensors are built into machines, and monitor production processes and indicate deviations [4]. If a machine or robot needs to be maintained, repaired, or if a downtime occurs, the sensors can inform about this. Sensors can also track work pace and report if a break is detected, as well as they can be used to monitor and maintain the condition of the production line.

Ghobakhloo [3] describes this as a smart factory, which is characterized by a very productive manufacturing environment of connected and intelligent machines. He believes that manufacturing processes are streamlined through automation and self- optimization and that the smart factory is an integrated manufacturing system where the physical resources communicate with each other and people via IIoT. When collected data is then made available through network connection, comprehensive software can be used to optimize anything [2]. IIoT is also about digitization of entire production processes, which occurs when machines are part of a coherent network, preferably a well-adapted and flexible manufacturing system. Flexibility facilitates the implementation of new machines and in this way you can easily broaden the network.

Furthermore, IIoT also means that the organization uses technical tools to provide a clear and distinct overview of all production processes and steps.

Today, it is a fact that technology is an integral part of organizations and, as mentioned, technology and digitalization follows an exponential development curve. Many companies welcome the new technology with open arms. Saarikko *et al.* [5] believe that connected units have caught a lot of attention among companies. However, Ghobakhloo [1] believes that Industry 4.0 can often be seen as overwhelming and it is common for businesses to therefore avoid this journey toward digitalization for fear of not having the technical or organizational conditions required. Organizations thus need to develop a strategic roadmap to effectively visualize and understand each step and decision they need to go through to facilitate the transition to Industry 4.0.

2.2.5 Strategic guidance for IIoT

To further account for previous research in the subject area, [3] model for strategic guidance toward Industry 4.0 is used as a theoretical framework. The model in its entirety can be found in Appendix 1. His model strategic roadmap toward Industry 4.0 includes six steps whose purpose is to serve as a guide for businesses regarding the transition to Industry 4.0. On the other hand, he believes that there is no strategy that suits all organizations, which means that the roadmap should be designed based on the company's core competencies, priorities, budget, and goals. Due to the study's delimitation area, the focus has only been on the model's first (strategic management), third (human resource strategy), fourth (IT maturity strategy), and fifth (smart manufacturing strategy) steps, which the next four parts are structured from the outside.

The second (marketing strategy) and sixth (smart supply chain management strategy) step were excluded as these were not perceived as relevant for the delimitation area.

2.2.6 Strategy

According to Ghobakhloo [3], the first step in the strategic roadmap for the transition to Industry 4.0 is a strategic overview. The business should start by shaping both short-term and long-term goals with Industry 4.0. These should also be formed within a time-based plan and describe where the company is, where it is going and how to get there, and designed based on the already predetermined visions and plans that Industry 4.0 entails. Also Saarikko et al. [5] talk about the importance of having stated purposes with the phenomenon; Kaur et al. [20] say that many companies start IoT projects without knowing for what purpose the data is to be collected. Furthermore, they point out that a successful application requires that the business knows where it is aiming.

According to Ghobakhloo [3], digitization and Industry 4.0 require a committed management and basic resource allocation. Appointing a specific team that handles and leads the digital transformation toward Industry 4.0, as well as the integration of existing systems and infrastructure is therefore of great importance. Not all organizations have sufficient IT maturity to take on Industry 4.0, and not all manufacturing companies have sufficient IoT-based production to maintain their competitive position in the global market. The company and above all the appointed team should create a detailed project plan for the transition to Industry 4.0. They should look at each phase within it and map out its characteristics, as well as further conduct a comprehensive analysis of the costs and benefits associated with each phase.

Marr [4] points out the importance of an organization having a strategy, as well as goals and visions to get a clear picture of where you are going and where you want to go with the help of the organization's strategies. The tools used today to create strategies are not adapted for a digital transition as the technology is constantly evolving and involves greater uncertainty. A strategy where the goal is to become adaptable is thus most suitable for the unpredictable change that technology brings. The company should also decide whether they should be leaders or successors, and take risks and uncertainties into account.

2.2.7 Resources and competencies

Industry 4.0 results in a fundamental change in the division of labor between people and machines. This is something that is well agreed according to Ghobakhloo [3]. Experts believe that competent employees are among the most important success factors for digitization and as Industry 4.0 brings together the real and virtual worlds through modern technology trends, such as cyber-physical systems (CPS), IIoT, robots and simulation, good understanding and relevant technical qualifications are required on the part of employees. To succeed in the transition to Industry 4.0, the first step in resources and competence is therefore to make an assessment of the staff's competence for the phenomenon. The company must carefully assess the competence of the employees and identify the digital skills they have in order to then be able to identify which competencies the company currently lacks. Although the business is an expert in its field, data management requires expertise in things such as filtering and analysis of large amounts of data, thus Saarikko *et al.* [5] suggest that the business must think about what it strives for and who can help it get there.

Although current employees may not have the full skills required to operate a digitized factory, they are still well aware of the company's approach, standards, and workplace culture. The existing employees thus have a great advantage and you should start by training and educating them about the necessary skills required and adapting them professionally to the upcoming technology and routines. For example, all employees should undergo significant training in computers and IT, given that their real work will gradually be integrated with the virtual one. However, a certain aspect of Industry 4.0 requires advanced skills such as computer technology, and not all skills can be taught on site. Companies in the manufacturing industry thus need to perform a detailed cost and benefit analysis of various human resources, and place new talent where needed. The development of digitalization and technology is continuing at an exponential pace; this means that businesses should look for people who are highly qualified and flexible enough to adapt to all types of technology that can be developed with Industry 4.0 [3].

The employees' attitude to technology, motivation, and overall job satisfaction are important factors for successfully implementing new technology in a workplace. The aging of the workforce can be perceived as a problem as the elderly are often considered more difficult to learn, but this has proved to be a wrong assumption. The incessant technological development entails problems as older employees usually have a more negative attitude, which makes it more difficult to change their behavior. It is therefore important that managers inform, and motivate employees about upcoming changes. This can be facilitated by involving the employees in the work around the implementation, which can lead to them understanding the value and purpose of the change. Thus, the motivation for development and adaptation can increase. In order for this to be achieved, it is important, among other things, that the employees feel that they possess the right skills, have a good relationship and communication with managers and other employees, and an experience of control. In this way, negative attitudes and stress over the change can be reduced. Motivating your employees in this way can also mean that a perception that the new technology is something that leads to positive consequences [6–10].

2.2.8 IT maturity

Ghobakhloo [3] claims that IT governance is one of the weakest aspects within organizations. As the first step in IT maturity, businesses should therefore ensure that an IT management team is in place to draw up a strategy, as well as budget, implement, control, and report on IT projects based on the requirements for the transition to Industry 4.0. This is also something that Norfolk [11] believes is important as you do not have control over your company if you do not have control over your IT resources. The IT management team should therefore perform a detailed analysis of existing IT infrastructure, such as networks, hardware, software, sensors, and controllers, and identify the most meaningful approach to using the technology to support the transition to Industry 4.0. In this way, the IT management team can map different parts of the business where networking and integration are needed, as well as when existing IT infrastructure does not support digitization. Based on this, the group can then formulate and implement necessary development strategies for the areas that require change [3].

According to Wang *et al.* [12], the emerging information technology, such as IoT, big data, and cloud services, together with other technologies, helps to implement the smart factory in Industry 4.0. They believe that smart machines and information transfer technology can communicate and negotiate with each other through sensors, storage, and networks to adapt to a more flexible production. According to Ben-Daya *et al.* [15], there are mainly some technologies that should be applied for the use of IoT in industry—radio-frequency identification tags (RFID tags), wireless sensor network (WSN), industrial wireless network (IWN), cloud services, and IoT devices. RFID tags allow identification, tracking, and transmitting information from products or machines and with the support of industrial networks, these smart devices can be interconnected and also connected with the internet [15, 13]. In addition, Wang *et al.* [13] states that this together with cloud storage enables scalable data processing and storage space on demand for analysis of big data.

According to Ghobakhloo [3] and Athavale *et al.* [22], virtualization enables what is called a digital twin by combining sensor data collected from the physical world. The virtual twin in a smart factory could, among other things, enable process engineers and designers to improve existing processes or optimize production without disrupting the physical processes. The digital twin of a smart product can also enable manufacturers to have a complete digital footprint of their existing or upcoming products throughout their life cycle, from design and development to the end of the product. A digital twin provides access to real-time information about the physical objects and accesses historical data that can be used to prevent maintenance and thereby reduce maintenance costs and downtime. For this to be possible, however, Ghobakhloo [3] and Athavale *et al.* [22] claim that the business's potential for real-time is a crucial part, which in turn depends on a good network connection.

According to Ghobakhloo [3], the key to success with the transition to Industry 4.0 is the ability for all components within a smart factory to communicate with each other at field level, in real time, and with intelligent functionality that collects data, interprets it, and provides meaningful insight into the management system. But in reality, machines, equipment, and robots come from different suppliers; have different technical properties and different communicative working methods. Therefore, existing and newly added IT infrastructure needs to be harmonized and integrated so that all components in the smart factory are interconnected and interoperable [3].

The abovementioned technical aspects constitute for what are called cyber-physical systems (CPS); according to Boyes *et al.* [16], CPS is an essential part of Industry 4.0 and consists of the set of interactive physical and digital components within the IoT-based environment and provides functions such as reading, control, calculation, and networking to influence the results of, among other things, the physical processes of production. Ghobakhloo [3] believes that CPS is controlled and monitored by database algorithms and is tightly integrated via the internet with its users such as objects, people, and machines. For example, a smart production chain can be considered as a CPS, where machines, operators, materials, and even ongoing work can communicate with each other, as well as monitor and pass on information for calculation.

2.2.9 Smart manufacturing

As the penultimate part of its model, Ghobakhloo [3] mentions smart manufacturing systems, which are characterized by connectivity, integration, transparency, proactivity, and flexibility. He believes that Industry 4.0 enables a transition from traditional automation to fully connected and flexible manufacturing systems. According to Ghobakhloo [3], this development begins with IIoT to create smart connections across the factory's processes and enable interaction with the people who work with them. This includes the application of smart autonomous mobile devices and various types of production or process controllers to obtain machine and process data. These types of control systems communicate with each other and can be further developed and become an ERP system, which in turn provides access to information, connection to company data in real time, as well as an expanded understanding of the business. ERP combined with data mining can also create the conditions for a digital twin [3, 22].

Ben-Daya *et al.* [14] also talk about integrated management systems or ERP systems and that organizations today have a great tendency to use them. An ERP system is a company-wide system that integrates several software programs to be able to meet many functions within an organization and thus automate processes. In this way, all programs can be run in one and the same system, on a common database [14]. This integration has long been a problem in industry and with the help of an ERP system, all departments within an organization can in a smooth and efficient way share information and communicate with each other, especially between different management sectors within the organization [14]. Wang *et al.* [12] believes that ERP systems significantly improve an industry's productivity.

Based on the theoretical framework, collected data material is analyzed to investigate the current research question. The following method sections present the approaches used during the course of the study.

2.2.10 Method

This section describes the qualitative method as a basis for data collection and analysis. We present selected activities and how we have delimited the work and then give a deeper description of our approach based on grounded theory. The section also deals with parts such as literature review, interviews, and respondents, as well as how we have reasoned regarding selection and ethics. Furthermore, we describe how transcription and coding of the data has taken place and finally, criticism is presented about the choice of method.

When the chosen research area was use of IoT in operations, an IT company was contacted, which offers digitized solutions to other companies, which in turn helped to contact a company in the industry for further research. In this way, the research area came to focus on investigating IIoT solutions within the company's production and its processes. The company where the surveys were conducted is a global industrial company with a focus on mobile construction transport solutions.

Many industrial companies on the market today can be considered relatively conservative and may not have embraced all the digitized solutions that the market offers. The company that has been studied could, at present, be considered to be in the starting blocks for becoming an increasingly digitalized business. This was seen as an opportunity to investigate how the company could switch to using more digitized tools to enable new and additional value creation in production. In order for the company to have the opportunity to be at the forefront of digitized solutions in the industry, it was therefore examined how the company works with its manufacturing processes today to get an idea of which development areas and opportunities could be offered with the help of IIoT.

2.2.11 Delimitation

In order to gain a more in-depth view, the study has been delimited by only examining value creation in production and manufacturing processes at companies in the industry. Further delimitation was performed by analyzing the research questions based on a framework and common concepts within Industry 4.0, as well as previous research in the area.

2.2.12 Method selection

The basis for our study consists of a qualitative method where the emphasis is on words, how the respondents expressed themselves, rather than quantity when collecting data and in this way a deeper understanding is given about the chosen topic. Often when using a qualitative method, it is common for a theory to be generated based on the research that is done, as well as the data that is collected, i.e., an inductive approach. It is also common to start from the approach of constructivism where people themselves create and construct knowledge through interaction with the outside world [17]. This has been done when we have interviewed employees within the organization to get an idea of how the organization is structured and how it is currently run in order to be able to arrive at a theory based on our research.

In qualitative research methods, grounded theory has become the most widely used framework for data analysis [17]. The theory consists of systematic and flexible guidelines for collecting and analyzing qualitative data and then constructing theories based on them. In this way, the theory begins with inductive data which then leads to an iterative strategy of going back and forth between data and empirical data in the production of a report so that it eventually leads to a result. In this way, grounded theory, aims to make patterns visible to increase the understanding of collected data and can thus warn of aspects that can be misleading.

The grounded theory is not only an iterative comparison but also an interacting method that allows the researcher to interact with data. This is because the iterative process creates a deeper insight into data that can lead to new analytical questions as new connections can be created between them. Furthermore, this means that a comparison between different data helps to define differences and similarities that may occur in them, which can make implicit meanings visible. For us, this iterative and interacting approach means that we have worked between the different steps to interpret data, focus on concepts and the theoretical work, to then be able to further specify our research question and thus be able to collect more data directed to our chosen research area.

In qualitative approach, interviews are, according to [17], a common data collection method and large parts of the data material therefore consist of semi- structured interviews that aim to map the business's manufacturing processes. This is to gain a reality-based insight into what the research area can mean for organizations instead of only starting from research and scientific literature. In order to be able to analyze the material afterwards, we transcribed the recorded interviews. Based on the transcripts, we have then, coded, created a theme and categorized the collected data material through a well-founded approach. Then analyzed and compared this with related literature in the field of Industry 4.0.

Finally, we would like to inform you that we did not start from grounded theory in its entirety but instead took inspiration from this framework in the analysis of our material and have therefore chosen to instead call the phenomenon grounded approach.

2.2.13 Data collection

This section describes how we have proceeded in collecting data. First, it is described how we proceeded to collect relevant literature, then it is described how the interviews were conducted, how the selection of respondents went, and what ethical requirements we have related to.

2.2.14 Literature review

In order to be able to conduct a relevant analysis of the subject area, related research and scientific literature have been an essential component during the course of the study. In order to be able to form interview questions that deal with concepts and concepts within the area, the work began with an overall literature review of the area. To gain a deeper understanding of the phenomenon, we also chose to interview a professor of computer science at Umeå University. In doing so, we generated a basic knowledge of the relevant topics that could then be used to search for literature and verbalize significant follow-up questions in future interviews. Based on the interviews, further literature review was done, in accordance with the established approach, to further investigate relevant topics. When the subject area of the study finally took shape, relevant articles were used, primarily in value creation and Industry 4.0 (IIoT), which were collected via the search service for Umeå University Library and Google Scholar. In addition to scientific articles, books written by renowned authors in the field of research have also been used. With the enormous growth of digitalization, the literature was judged on the basis of how contemporary and reasonable it was perceived for the context.

2.2.15 Interview

As we have used a qualitative method, this has also meant that we have made an interview guide (see Appendix 2) adapted for the qualitative interviews that were conducted. In order to be able to make an interview guide with relevant themes and questions, we gathered information beforehand to have a stable foundation to start from. Bryman [17] believes that a qualitative interview tends to be flexible where the interviewer answers and follows the direction that the respondent's answer creates. After this, we have thus shaped our interview guide, which can be described as question areas or questions that a researcher wants to investigate in order to create a greater understanding of a phenomenon.

The interviews have in turn been semi-structured, which means that the questions have only been in a general form of an interview guide where the sequence of the questions can be changed and new questions can be asked that follow up the respondents' answers. In the case of semi-structured interview questions, the interviewer may also deviate significantly from the guide used and new question formulations may arise. During the interview itself, it can also be of positive importance if the respondent wanders off as this can provide insight into what he considers important [17].

To get a nuanced picture of the research area, seven interviews were conducted with seven respondents with appropriately selected people in different areas of responsibility. One of the interviews was a group interview where one of the participating respondents was also interviewed individually at a later time. During the interviews we have done, the adaptation to the respondents' answers is something we have worked with to a great extent, which we have also noticed afterwards that it has given us a clearer insight into how the surveyed organization is run and functions today, and what the respondents, who are from different parts of the organization attach importance. The interview guide was designed according to different areas in the form of organization, value creation and strategy, customers and customer relationships, technology and approaches, as well as the environment and sustainability. We have thus had a complete interview guide, but which during the course of the interviews has been adapted to and responded to what the respondent has said. We have also developed our interview guides as more and more interviews have been conducted. This is because other, relevant questions have emerged, and we have in some cases wanted to do a follow-up from a previous interview.

This meant that, based on previous interview responses, after transcribing certain interviews, we considered that there were certain gaps in the collected data that needed to be filled in. and that we have in some cases wanted to do a follow-up from a previous interview. This meant that, based on previous interview responses, after transcribing certain interviews, we considered that there were certain gaps in the collected data that needed to be filled in. and that we have in some cases wanted to do a follow-up from a previous interview. This meant that, based on previous interview responses, after transcribing certain interviews, we considered that there were certain gaps in the collected data that needed to be filled in.

There are different ways to register interviews for documentation and later analysis. It is not entirely uncommon to take notes during the interview, but one of the most common methods is sound recording when words, tones, pauses and the like are recorded in a permanent form. All interviews were recorded to facilitate transcription, and so that the focus would not be on taking notes, but instead could be placed on follow-up questions based on the respondent's answers. Below is a table of all interviewees, interview length, and interview form (see Table 2.1). Six of the interviews were conducted in work or conference rooms to have a calmer and secluded environment where we could talk undisturbed with the respondents and they could be confident with the realization that no outsider overheard their statements. Another advantage of holding the interviews in a calmer environment was that the sound quality of the recordings was much better. The seventh interview was conducted by telephone, due to technical problems (see Table 2.1). According to Bryman [17], the fact that we transcribed our data can also serve as a guarantee that we as researchers have not influenced the results and based this on preconceived opinions or values.

Table 2.1 Respondents from the industrial company

Respondent	Occupational title	Interview form	Length (min.)
1	Production manager	Physical	50:55
2	Production technician	Physical	50:55
			38:00
3	Production technician	Phone	28:20
4	IT technician	Physical	20:27
5	IT technician	Physical	07:41
6	Production coordinator	Physical	25:09
7	Production manager	Physical	14:02

2.2.16 Selection and respondents

In order to best answer the study's questions, respondents with good insights and experience in the organization's manufacturing processes were needed. To get a first contact, a first interview was booked in with the help of the IT company and then to establish further contact with relevant interview objects. Bryman [17] describes this as a snowball selection, which means that the researcher is looking for interviewees with knowledge and understanding of the phenomenon that are representative of the population. Of the seven respondents who participated, five worked in different areas of responsibility related to the organization's production and manufacturing processes.

2.2.17 Ethics

Before each interview, oral information was given about the research ethics principles that constituted an important starting point for all interview occasions. The main purpose of the research ethics principles is to create norms for what the relationship between researcher and respondent should look like so that the research is conducted in a correct, ethical manner in accordance with the individual protection requirement. The individual protection requirement, which means that each participating respondent must be treated in a good way and feel safe to participate in the research, can be divided into four general requirements, the information requirement, the consent requirement, the confidentiality requirement, and the use requirement (Swedish Research Council, 2002).

The information requirement means that all participants must be informed about the purpose of the research, their task at participation, and the conditions that apply. It is important that the respondent is aware that participation is voluntary and that they have the right to suspend their participation at any time during the interview. The consent requirement means that participating respondents must give consent to participate in the research, something we did before we started recording the interview as the participants themselves have the right to decide if, for how long, and on what conditions they want to participate. The respondents were also informed that it would not lead to any negative consequences if they chose to suspend their participation and in their decision to participate or suspend their participation, they were also not subjected to inappropriate pressure or influence. All respondents were also informed and assured that sensitive information about them is treated with great confidentiality so that no outsiders can access sensitive information and that they will be mentioned in such a way in the public report that identification is not possible. This requirement is called the confidentiality requirement. As a result, we were also careful to inform the participants the utilization requirement that collected data will not be used for nonscientific or commercial use and personal data will not be used for decisions that may affect the participant (Swedish Research Council, 2002).

2.2.18 Data analysis

This section describes how we have transcribed and coded collected data material. In connection with the coding, a table is also shown that shows how we coded the data material into different themes and categories.

2.2.19 Transcription

In order to be able to go back and look at our collected material afterwards, transcripts were made in connection with each interview. This is to be able to take out key parts that could then be used in the upcoming interviews. The recorded interviews were shared almost equally between us for transcription as it is a very time consuming process. The material was largely transcribed verbatim, but as it requires a lot of time, laughter, pauses, and other similar expressions that were not relevant to the analysis were sometimes excluded, in accordance with Bryman's [17] advice to include only what is considered important for the context. The transcript has since been used to be able to carefully examine what the respondents have expressed, and according to Bryman [17], transcribed data material also serves as an assurance that the researcher has not influenced the results based on their values and preconceived notions. The transcription was an essential part of the study in order to create an understanding and enable further processing of interview data in the form of coding and categorization, thus the transcription provided a comprehensive and relevant basis for further analysis.

2.2.20 Coding

The most common method of analyzing data is by coding and categorizing the respondents' statements. As we have started from a well-founded approach, we have carried out qualitative coding of collected data and we have separated, sorted, and synthesized our data. The coding involves sorting segments of data into different categories that are described with keywords in order to be able to get the most important of the data and later analyze it. In order to be able to develop a theory of our collected data, coding has been a basic prerequisite as this means that the content is defined and we then have the opportunity to get a good overview of our data. During our coding, we have interpreted words, sentences, and segments which we then categorized. The reason why we have coded our data is to enable a deep dive into the respondents' perceptions and experiences.

When all transcripts were completed, these were read through for coding of relevant segments. Based on the transcripts, 153 relevant text segments were coded. The coding took place initially through an inductive approach where both themes and categories were created freely based on collected empirical data and then switched to an iterative approach where we switched between research on Industry 4.0 and our developed categories. Through this approach, we considered [3] model strategic roadmap toward Industry 4.0 was appropriate to follow as the framework was in good agreement with what we obtained through our transcripts. The framework is divided into six different categories; however, we chose to focus only on four of these as they were considered to be more relevant for IIoT in production and manufacturing processes.

Thus, our five different themes are strategy, IT maturity, resources and skills, smart manufacturing, and Industry 4.0. The fifth theme, Industry 4.0, became a theme partly because we could link certain parts of the transcripts to something more comprehensive than the abovementioned frameworks but also because the very starting point for the framework is based on the technology that Industry 4.0 has brought to the industrial sector. Table 2.2 presents a summary of all the results' themes and categories.

2.2.21 Method discussion

Despite the fact that we used semi-structured questions during the interviews, there may still be a risk that the researcher controls the respondents with leading questions. We tried to avoid this carefully when we designed the interview guide in such a way that the respondent was given space to answer the interview questions freely. The interviews that were conducted were also recorded to make it easier for us when transcribing, this is something that Bryman [17] believes can make the respondents nervous and feel uncomfortable, which in turn can lead to incorrect and incomplete answers. This was something we did not experience as all respondents seemed to be relatively used to being interviewed. A problem that arose during the interviews was that the recording technique stopped working at one point can be a risk. The interview when the technology was complicated was redone, but then as a

Table 2.2 Summary of categorization and coding.

Theme	Category	Important comments
Industry 4.0		Map processes and prevent interruptions
		Savings in time and expenses
		Better information flow and collaboration between departments
		Efficient work with quick decisions and better quality
Strategy	Methods and working	Very analog work
		Inadequate information and communication
		Lots of visual work
		In need of integrated production
		E-mail and intranet for information dissemination
	Goals and visions	Want to enter more into the computer world, become more connected and accessible
		Be up to date on new technology
		Lacks a clear vision of IT and data collection
		Together drive improvement work and innovation
Resources and competences	Competence	Lack of competence for handling data
		Need for competence development and new profiles
	Attitude to IIoT	IIoT: a must for further development of manufacturing processes
		Aware of digital change
		Employees can feel monitored
		Can be complicated and demanding
IT mature	Data, standard and communication	Modern and technology-adapted equipment
		Free access to data but no analysis of it
		Various machine and robot suppliers
		No stated standard
		Computer-assisted simulation
		Inadequate integrated communication
	Storage	No storage of generated data
		Cloud service for Office 365 and SharePoint, other via server
	Network	Internet connection via 3G and network cable
Smart manufacturing		Outdated current business system
		The organization's stands still waiting for a new system

telephone interview. A problem with this may be those important aspects such as facial expressions and body language fall away and can affect how the answer is interpreted, but since our interviews emphasized approaches and methods rather than personal experiences, we did not see this as a problem.

Using interviews as a research method can be considered problematic as in too few interviews it can be difficult to generalize the results to other places, situations, and times. Bryman [17], on the other hand, believes that the most important thing is not how generalizable a result is, but rather how the researcher can theorize his result. It can be a challenge to let all respondents have their say during a group interview and that when transcribing the material it can be difficult for the researcher to perceive who is saying what. This was something we were well aware of before the group interview took place, but since only two respondents participated, we experienced that both were given the same amount of space, and that they could rather complement each other when needed. Using snowball selection can sometimes be a disadvantage as you can miss out on respondents who are more suitable to answer the interview questions and then give a different picture of the organization. However, we considered that the respondents who participated could provide a nuanced and broad picture of the organization, partly because they had good insight into the chosen research area but also when they worked in different departments within the organization. Backman (2008) believes that it is relatively common for researchers to look for literature that can strengthen their perception or collected data. As this tendency is something we have been aware of, we have taken a critical approach and carefully examined the selected related research in the area. Another aspect that we have had to address before we started our research was that our knowledge in the subject IoT was only very basic, which has meant that we have had to read more before we have been able to conduct interviews and start the report. In the next section we discuss the results of the study in relation to the five themes namely; Industry 4.0, strategy, resources and skills, IT maturity, and smart manufacturing which were previously identified.

2.3 Outcomes of the Study

2.3.1 Industry 4.0

Ghobakhloo [3] points out that many leading organizations have already adopted Industry 4.0 and its technologies, and that those organizations that have not yet implemented this technology can benefit from the competitive advantages that other companies have achieved. IIoT can offer the company good insight into the business via collected and processed data in real time. In this way, you can, among other things, predict maintenance and service;

have supervision of production, quality, and efficiency, as well as work pace and downtime. IIoT has the greatest impact on the organization's production as it is where sensors are built into existing machines and robots. Ghobakhloo [3] describes this as a smart factory and believes that the communicative machines are what enable efficiency. When the cohesive network of machines is well-adapted and flexible, it also facilitates the broadening of the network. Gierej [2] also mentions the importance of a good network to make data available and thus be able to optimize anything. Ghobakhloo [3] also believes that the implementation of IIoT is important for organizations to review their strategies, organizational models, and personnel issues, as a transition such as this places both new and different demands compared with the organization's previous approach.

In addition to the respondents being familiar with Industry 4.0, it is also these values that the respondents hope for if they were to adopt IIoT. They mention savings in the form of time and expenses, as well as more efficient work through a better flow of information, which in turn could result in better quality of parts and end product.

However, only the respondents working in production seem to be those who are familiar with the phenomenon. This indicates that the company strives to implement IIoT to enable new value creation, especially in their production. In order for such an implementation to be implemented, however, it is important that all employees are familiar with IIoT, and are aware of what such a change means, both for the organization as a whole but also for themselves and their tasks.

2.3.2 Strategy

Based on the interviews, it appears that the organization today to some extent works digitally in terms of information dissemination. In the main, however, a whiteboard is used to go through any events, as well as the day's work and goals. This is done at the beginning of the work shift so that all employees get a good overview. At the same board, all jobs are logged in a monitor to gain some control over the manufacturing processes. In addition to this, it is partly used by e-mail and the company's intranet, but as employees rarely has time to check these, the dissemination of information is deficient and the information may instead need to be communicated orally. A certain visual work is done, but this is not done digitally but manually by using notes in different color codes on the whiteboard.

The tools used today in the creation of strategies are not adapted for a digital transition as the technology is constantly evolving. Other research also indicates that not all organizations have the IT maturity required to tackle Industry 4.0 or large enough IoT-based production to maintain a competitive position [3]. As the organization today has not adopted a working method that is adapted to digitalization, this is therefore something that needs to be taken into account in further development and transition to Industry 4.0. Implementing digital tools and technologies that are adapted for a digital transition is necessary in order to continue to be a leader in the global market.

Ghobakhloo [3], Marr [4], and Saarikko *et al.* [5] talk about the importance of having a strategy, goals and visions to get a clear picture of where you are, where you want to go, and how to get there. Shaping short-term and long-term goals with Industry 4.0 is therefore of great importance and should be designed based on the predetermined visions and plans that IIoT has. As for the organization's goals and visions, it is not clear what these are, but all of them claim that they are well understood in them. The business works a lot with adaptation and innovations and has a requirement to stay up to date on new developments. All respondents know what they want with the technology and the organization strives to become more digital, but at present there is no clear vision about it.

The organization is aware that it is facing a digital change, but at the same time the respondents feel that it is standing still and that they do not really know how to proceed. Ghobakhloo [3] argues that businesses should appoint a team to manage and lead the transition to Industry 4.0, and create a project plan for the transition to Industry 4.0. The team should also review any phases of the transition and conduct a comprehensive analysis of costs and benefits enabling the identification of needs and priorities. There is uncertainty about whether the business should wait and see how other organizations do or whether they should be at the forefront. This is important for companies to decide on. Based on this, it can be noted that the organization needs to think about its strategy, as well as discuss any goals and visions so that these are stated and anchored in the employees. The business must also decide if they want to wait and observe other businesses or if they want to be at the forefront. A team that manages and leads the transition process can be an advantage as you currently do not know how to proceed.

2.3.3 Resources and competencies

Ghobakhloo [3] believes that employees within an organization will need training to be able to adapt to the new working methods that the implementation of new technology entails. Also, Saarikko *et al.* [5] believe that businesses may need to acquire new skills. This is something that the interviewed respondents are well aware of as a transition to Industry 4.0 will mean gaps in and a lack of competence among all employees. The majority of respondents also state that the company will be in need of new profiles in the workplace during a transition to IIoT. Even Ghobakhloo [3] claims this when he writes that Industry 4.0 requires advanced competence that the company needs to be mapped to see where it is in need of these cutting-edge competencies.

The respondents see the use of IIoT as a must to develop the business's production, and to be able to be at the forefront of their industry. Although the respondents have a positive attitude toward IIoT and show a great interest in the phenomenon, they also see some challenges in implementing this technology in production. This is both because they feel some uncertainty about what existing systems used today can offer, but mainly because employees in production risk having a more negative attitude toward such an implementation. The respondents mention both the risk that they will feel monitored in a negative way and also the fact that many employees are older and have a more negative attitude to new technology and changes.

This can be linked to what research says about the importance of employees having a positive attitude toward the implementation of new technology, feeling motivated, and realizing the purpose and value of this type of change. To facilitate changes in attitudes in the workplace, it is important to reduce the stress that can arise among employees when they experience a lack of skills and a lack of communication. Research also shows that age does not affect learning, but rather impedes change of attitude [6–10]. In other words, it is important that leaders in the workplace reduce the feeling of stress among employees by ensuring that they have the required skills, and have a good conversation. One should also be aware that age does not affect the ability to learn to the extent previously thought. By involving employees in the implementation, they can develop an understanding of the need for technology development and thus gain a better approach to it.

2.3.4 IT maturity

According to Ghobakhloo [3], businesses should appoint an IT management team to set up a strategy for future IT projects. He believes that the IT management team should map current resources in the form of, among other things, hardware, software, and sensors. Norfolk [11] also believes that this is important as you do not have control over your business if you do not have control over your IT resources. In this way, the business can map where there is a lack of networks and integration and when the existing IT infrastructure does not support digitization.

The respondents who work in the organization's production believe that they have relatively modern equipment that is well adapted to technology, however, not all machines and robots are from the same suppliers. The machines can signal about themselves when some kind of problem occurs but they cannot see history or statistics on exactly which part was manufactured when the problem arose as they do not use that type of sensor tags on the parts in the final product. This makes it difficult to subsequently evaluate what has gone well or badly. A new robot that they have enables computer-assisted simulation as the entire machine cell can be drawn in 3D in a virtual world. In this way, problems that could have arisen in reality, such as collisions with objects in its vicinity, can be avoided.

Due to the fact that the machines and robots used come from different suppliers, and that the older and newer machines are manufactured in different ways, the data generated is often of a different standard. You have different control systems with different standard interfaces for the older and newer robots where data is transferred. This makes data difficult to manage in the existing systems and it is difficult for the organization to get a good historical overview.

Marr [4] mentions big data as the heart of the smart revolution. He believes that companies have access to large amounts of data and that the value lies in being able to utilize it. Furthermore, he believes that big data enables organizations to utilize data for further analysis.

The respondents believe that they have the opportunity to generate data, as well as free access to it, but since they lack the technology required for collection and processing, it is not possible to evaluate equipment and work. In this case, not only are different platforms and systems missing, but also physical resources such as computers, among other things.

All respondents mention that they use cloud services, but only administratively in the form of Office 365 and SharePoint. On the other hand, no cloud is used for various manufacturing processes and for data collection and storage of information from technical equipment in production [21]. This information can be retrieved manually, if necessary, directly from the machines. This also clarifies that the machines do not have the ability to communicate with each other.

2.3.5 Smart manufacturing

Ghobakhloo [3] talks about smart manufacturing systems and how Industry 4.0 enables a transition from traditional systems to fully connected and flexible manufacturing systems. These connected and flexible manufacturing systems, such as ERP, provide access to information, connectivity and an expanded understanding of the business. According to Mexas *et al.* [14], being able to run several different programs in one and the same system has long been a problem in the industry. ERP systems enable businesses to efficiently share information and communicate between different departments. Wang *et al.* [12] argue that ERP systems significantly improve an industry's productivity. Caputo *et al.* [18] believe that IoT-based units create good conditions for ERP systems and that information flow and production efficiency can thus be improved.

The results show that the business today uses an ERP system, but this is considered ancient and you are therefore in the middle of switching to a new one. The current system is used for a number of different purposes, but as it is complicated to use, the respondents find it difficult to understand. The new system aims to be more user-friendly, provide access to more information, and that as much as possible can be coordinated in it. As the new ERP system has not been implemented, there is currently uncertainty about how the new ERP system will be able to be used and what opportunities it will offer. This means that at the moment you are standing still in the development toward a more digitalized work environment as you do not want to spend time and resources on something that you may already have access to.

2.4 Conclusion

The purpose of the study was to create an understanding of what Industry 4.0 can mean for organizations in the manufacturing industry and the focus was on what opportunities and challenges the phenomenon can result in for an organization's production. An organization's production, as well as its processes, were examined on the basis of [3] framework for strategic guidance. The survey showed that the business has some technology and thus partial potential for an implementation of Industry 4.0. The business is in the middle of a shift of ERP systems, and the upcoming system is expected to create access to more information, as well as enable connection to certain machines and robots. Following this, four general conclusions could be drawn regarding the implementation of Industry 4.0:

First, organizations must review their strategy, goals and visions, and ensure that all employees understand these. The fact that all employees strive for common goals and visions, as well as having a team that leads and manages the transition process, is also something that facilitates the transition to IIoT.

Second, it is also important that companies both inform and motivate all employees prior to the implementation of Industry 4.0. In order to motivate, it is above all important that a business carries out competence development in the workplace. In this way, you can minimize the risk of possible resistance. New cutting-edge skills in the workplace are also important to review in order to be able to handle new technology.

Third, businesses need to implement digital tools and technologies that are adapted after a transition to Industry 4.0 in order to take advantage of the opportunities offered by the phenomenon. It is important that these technologies have a common standard for easier handling of collected data. However, implementing this technology entails risks that it is very important that employees are highly aware of in order to succeed in avoiding ending up in vulnerable situations. As IIoT is a relatively new phenomenon, not all technology, which is specially adapted to this, is fully developed yet.

Finally, the implementation of Industry 4.0, if these three aspects are taken into account, can result in the following values: real-time information for effective decision-making and understanding that contributes to better supervision of manufacturing, quality, efficiency, work pace, problems, and downtime. Industry 4.0 also creates conditions for businesses to have flexible production, good insight and control over the business, as well as cuts in energy consumption and expenses. Thus, the phenomenon offers the opportunity to be a leader in the global market.

At present, many technologies and phenomena related to IIoT are not yet fully explored and available on the market, such as 5G. For future research, it would therefore be interesting to investigate further how these, once established, may affect the continued use and development of IIoT. For example, this study mentions the importance of a good and secure network connection, as well as the risks of applying IIoT technologies within organizations [22].

References

[1] Morrar, R., Arman, H., & Mousa, S. (2017). The Fourth Industrial Revolution (Industry 4.0): A Social Innovation Perspective. Technology Innovation Management Review, 7(11), 12–20. http://doi.org/10.22215/timreview/1117

[2] Gierej, S. (2017). The Framework of Business Model in the Context of Industrial Internet of Things. Procedia Engineering, Vol.182, pp.206-212. https://doi.org/10.1016/j.proeng.2017.03.166

[3] Ghobakhloo, M. (2018). The future of manufacturing industry: a strategic roadmap toward Industry 4.0. Journal Of Manufacturing Technology Management, 29(6), 910–936. https://doi.org/10.1108/JMTM-02-2018-0057

[4] Marr, B. (2015). Big Data: Using SMART Big Data; Analytics and Metrics To Make Better Decisions and Improve Performance. Wiley.

[5] Saarikko T., Westergren U. H. & Blomquist T. (2017a). The Internet of Things: Are you ready for what's coming? Business Horizons, 60(5), 667–676. https://doi.org/10.1016/j.bushor.2017.05.010

[6] Elias, S. M., Smith, W. L., & Barney, C. E. (2012). Age as a moderator of attitude towards technology in the workplace: work motivation and overall job satisfaction. Behaviour & Information Technology, 31(5), 453-467. https://psycnet.apa.org/doi/10.1080/0144929X.2010.513419

[7] Mankin, D. (2009). Human resource development. Oxford; New York: Oxford University Press. ISBN: 9780199283286

[8] Deci, E. L., Olafsen, A. H., & Ryan, R. M. (2017). Self-Determination Theory in Work Organizations: The State of a Science. Annual Review of Organizational Psychology and Organizational Behavior, 4(1), 19-43. https://doi.org/10.1146/annurev-orgpsych-032516-113108

[9] Clarke, S., & Cooper, C. (2004). Managing the Risk of Workplace Stress: Health and Safety Hazards. Routledge.

[10] Ajzen, I. (1991). The theory of planned behavior. Organizational Behavior and Human Decision Processes, 50(2), 179-211. https://doi.org/10.1016/0749-5978(91)90020-T

[11] Norfolk, D. (2005). IT governance: Managing Information Technology for Business. London: Thorogood. ISBN 1280233486

[12] Wang, S., Wan, J., Zhang, D., Li, D., & Zhang, C. (2016b). Towards smart factory for industry 4.0: a self-organized multi-agent system with big data based feedback and coordination. Computer Networks, 101, 158-168.

[13] Wang, S., Wan, J., Li, D., & Zhang, C. (2016a). Implementing Smart Factory of Industrie 4.0: An Outlook. International Journal of Distributed Sensor Networks, 2016(1), 10. https://doi.org/10.1155%2F 2016%2F3159805

[14] Mexas, M. P., Quelhas, O. L. G., & Costa, H. G. (2012). Prioritization of enterprise resource planning systems criteria: Focusing on construction industry. International Journal of Production Economics, 139(1), 340-350. http://dx.doi.org/10.1016/j.ijpe.2012.05.025

[15] Ben-Daya, M., Hassini, E., & Bahroun, Z. (2017). Internet of things and supply chain management: a literature review. International Journal of Production Research, 1–24. https://doi.org/10.1080/00207543.2017.14 02140

[16] Boyes, Hallaq, Cunningham, & Watson. (2018). The industrial internet of things (IIoT): An analysis framework. Computers in Industry, 101, 1–12. https://doi.org/10.1016/j.compind.2018.04.015

[17] Bryman, A. (2012). Social Research Methods. 4th Edition. Oxford University Press.

[18] Caputo, A., Marzi, G., & Pellegrini, M. M. (2016). The Internet of Things in manufacturing innovation processes. Business Process Management Journal, 22(2), 383–402. https://doi.org/10.1108/BPMJ-05-20 15-0072

[19] Athavale V. A., Bansal A., Nalajala S., Aurelia S. (2020). Integration of blockchain and IoT for data storage and management. Materials Today: Proceedings, https://doi.org/10.1016/j.matpr.2020.09.643.

[20] Kaur K., Verma S., Bansal A. (2021). IOT Big Data Analytics in Healthcare: Benefits and Challenges. 6th International Conference on Signal Processing, Computing and Control (ISPCC), 2021, pp. 176-181, doi: 10.1109/ISPCC53510.2021.9609501.

[21] Athavale, V. A. & Ameya A. (2021). Digital Twin-A Key Technology driver in Industry 4.0. Engineering Technology Open Access Journal, 4(1). https://juniperpublishers.com/etoaj/ETOAJ.MS.ID.5556 28.php

[22] Athavale, V. A., Arora, S., Athavale, A., Yadav, R. (2022). One-Way Cryptographic Hash Function Securing Networks. In: Gupta, G., Wang, L., Yadav, A., Rana, P., Wang, Z. (eds) Proceedings of Academia-Industry Consortium for Data Science. Advances in Intelligent Systems and Computing, vol 1411. Springer, Singapore.https://doi.org/10.1007/ 978-981-16-6887-6_10

3

Container as a Service in the Cloud: An Approach to Secure Hybrid Virtualization

Manoj Kumar Patra, Bibhudatta Sahoo, and Ashok Kumar Turuk

Department of Computer Science and Engineering, National Institute of
Technology, Rourkela, India
E-mail: manojpatracs@gmail.com; bibhudatta.sahoo@gmail.com;
akturuk@gmail.com

Abstract

Containerization is one of the most recent advancements in cloud computing.
Containers are a small package of software components that bundle an application, dependencies, and settings into a single image that may be operated
in isolated user environments on a physical machine or in a virtualized
environment. Containers as a service (CaaS) is a cloud service model that
allows users to execute their tasks in an isolated environment. In CaaS, containers are deployed in a virtual machine (VM), and in turn, virtual machines
are deployed in a physical machine (PM). Containers virtualize software,
i.e., operating system (OS), and the virtual machine is the mechanism of
hardware virtualization. Hence, CaaS is also termed hybrid virtualization.
This chapter discusses different virtualization techniques in the cloud, the
difference between hardware and software virtualization, what containers are
and why to use containers, and the architectural difference between virtual
machine and container. We discuss the CaaS cloud model, the architecture
and implementation of CaaS, and the advantage and disadvantages of CaaS.
Then a detailed discussion on the research challenges available with the
CaaS model is presented. Finally, the chapter is concluded with a few future
directions.

Keywords: Cloud computing, virtualization, hypervisor, container, container engine, container as a service, hybrid virtualization.

3.1 Introduction

Cloud computing refers to the provisioning of a set of Internet-based services to the end-user. Instead of maintaining a local server and private data center for data storage, remote servers are used for storing and accessing data. Someone just needs to connect to the Internet through a device to get access to the data stored in remote servers. Cloud computing offers access to different computing resources such as storage, computing, applications, and also provides a platform for software development. Resources in cloud computing can be provided instantaneously over the Internet, with the facility to scale up or scale down quickly as per the requirement. Cloud users need to pay the price for the amount of computing resources they use on a pay-as-per-use basis without worrying about hardware maintenance. Multiple servers in a cloud are placed in different geographical locations and connected through the internet. A cloud can be deployed in three ways: public cloud, private cloud, and hybrid cloud. The physical location of the cloud infrastructure, who and how the infrastructure is being controlled, and how different cloud services are provided to the users are the few factors that define the cloud deployment model. These cloud models are most suitable for enterprise space, and each satisfies various business demands.

The classic cloud paradigm is the public cloud, in which a cloud service provider (CSP) owns, maintains, and manages cloud infrastructure shared by the public. All the services in a public cloud are delivered to end-users over the internet. The major benefits of the public cloud include on-demand availability and scalability and pay-as-per-use pricing. File sharing, software development, and email are all common use of the public cloud. In a private cloud, the cloud infrastructure is accessed by a single organization only. The private cloud infrastructure is managed and configured as per the specific requirement of the organization. A private cloud runs behind a firewall on a company's intranet, and stored data in a private cloud are entirely under the company's control. The private cloud addresses an organization's security and privacy issues and is mainly used for secure applications. The hybrid cloud blends private and public cloud models, allowing enterprises to benefit from shared resources while maintaining crucial security needs on existing IT infrastructure. Companies can use the hybrid cloud architecture to store confidential data on their servers and access it through public cloud applications.

Hybrid cloud systems cannot have a single point of failure, making them an appealing alternative for businesses with highly unpredictable workloads.

Cloud computing service models are divided into three categories. You may choose the optimal set of services for business requirements based on the degree of security, adaptability, and maintenance offered by each form of cloud computing. The most flexible cloud platform type, infrastructure as a service (IaaS), allows you to lease hardware and other essential IT building blocks. The IaaS cloud service model relocates a company's complete data center to the cloud. It enables you to have complete control over the hardware, such as storage, virtual machines, servers, networks, etc., on which your program executes. The cloud service provider maintains the servers, storage, and networking infrastructure, removing the need for a resource-intensive local installation.

Platform as a service, often known as PaaS, is a paradigm of cloud computing that offers development tools already set up and ready for use. This allows programmers to concentrate on writing and running high-quality code in order to build customized applications. The PaaS cloud computing service model will enable businesses to develop software without worrying about the backend infrastructure. It allows you to quickly build an application without worrying about the underlying infrastructure. When leveraging PaaS to deploy a web application, you will not be required to install either an operating system or a server, nor will you need to do any system updates. This model of cloud computing is costlier than infrastructure as a service (IaaS), which is much less expensive than software as a service (SaaS). It makes the process of creating and deploying applications easier.

Usually, the software is installed on every device separately. With the development of software as a service (SaaS) cloud service model, pay-as-you-go pricing allows web applications to be hosted in the cloud to minimize cost. In SaaS, the software is hosted in the cloud and made available to the end-users with a subscription. End-users can access the cloud application via a web browser or a mobile app, and IT teams are not required to participate in management or maintenance. Some of the common examples of SaaS are Gmail, OneDrive, and Dropbox. Different cloud service models are depicted in Figure 3.1.

In addition to the conventional cloud service model such as infrastructure as a service (IaaS), platform as a service (PaaS), and software as a service (SaaS), another novel cloud service model has been introduced by Google called containers as a service (CaaS). A container is a package including libraries, binaries, and all other dependencies required to run an application.

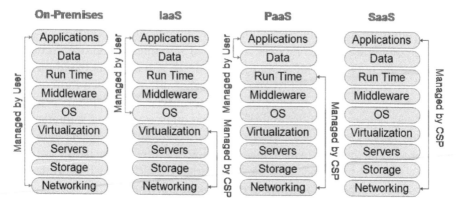

Figure 3.1   Conventional cloud service model.

Software teams can use CaaS to quickly deploy and scale containerized apps to high-availability cloud infrastructures. CaaS is especially useful for developers who want to create more secure and scalable containerized apps. Users can save money and increase efficiency by purchasing only the resources they need, such as scheduling capabilities, load distribution, etc. Enterprises interested in implementing a container service for their firm have the option of either purchasing a platform and managing the containers that are hosted on it themselves or selecting a container solution managed by vendors such as Google Cloud Platform. When using CaaS, the containers are placed in VMs, which are then placed in a physical system. A detailed discussion on CaaS architecture is presented in Section 3.3.

3.2 Virtualization in Cloud Computing

Creating a virtual version of any computing resources such as a server, a storage device, an OS, or network resources is interpreted as virtualization in cloud computing. It is a technique or procedure that enables several companies or individuals to share a single actual instance of a resource or program. Multiple operating systems and applications can work on the same machine and hardware using the virtualization technique. The host machine is the machine on which the virtual machine is built, while the virtual machine is known as the guest machine. Hypervisor-based virtualization is the most common type of virtualization. The hypervisor separates OS and applications

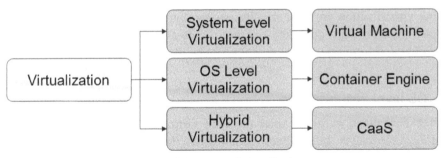

Figure 3.2 Types of virtualization.

from the underlying computer hardware, allowing the host machine to run several virtual instances as guests who share the physical computational resources.

Virtualization is crucial in cloud technology because it makes things more accessible and efficient. Virtualization can be done at the hardware and OS levels. Virtualization techniques can mainly be categorized into three categories, as shown in Figure 3.2; OS level, system level, and hybrid virtualization.

3.2.1 System level virtualization

Virtualization at the system level is also known as hardware virtualization. A virtual machine is the building block of system-level virtualization. A VM is a virtual system that acts as a computing system having all required computing resources such as memory, storage, CPU, and networking. The VMs are created on a physical machine, and more than one VM can be created in a single physical machine. Hypervisor is a software responsible for allocating resources to virtual machines. A hypervisor creates the VM by abstracting the computing resources from a physical machine, also called the host machine on which the VM or the guest machine runs [1]. On a single host machine, multiple VMs with different OS can coexist. VMs are entirely isolated from each other, and they can be moved from one host to another as per the demand to maximize resource utilization. This process of moving a VM fr one host to another is called VM migration [20]. There are two different types of hypervisors available; Type 1 and Type 2. Type 1 hypervisor is also called a bare-metal hypervisor, and it runs as a lightweight OS on the host machine's hardware. Type 2 hypervisor is also called hosted hypervisor, and it runs as a software application on the host OS. Some of the benefits of using virtualization are; that it partitions the hardware and is shared by multiple VMs, provides complete isolation, VMs are portable, so that they can be migrated from one host to another, and hardware independence.

3.2.2 OS level virtualization

OS-level virtualization, also called software virtualization, works on the operating system layer. In OS-level virtualization, the kernel of an operating system allows for the existence of several segregated user-space instances. Those instances are known as containers. A container in cloud computing is an approach to operating system virtualization. A container image is a small, standalone executable file that contains everything you need to run an application, including the code, runtime, and libraries. A container engine is the core component responsible for creating containers [8, 17]. The container engine creates a package of runtimes, libraries, and dependencies, including application code to run user applications. Containers provide an easier way of sharing computing resources such as CPU, memory, networks, etc., at the OS level.

They provide a kind of logical packaging for applications, allowing them to be abstracted from the actual environment in which they are executing. There are several benefits of containers; a few of them are portability of workload, isolation of application, and splitting of responsibility. A container can be moved among registry servers, i.e., a container image can be pushed or pulled into and from any registry in the same or different host [2, 10].

3.2.3 Container vs. virtual machine

Containers and VMs are quite similar technology used for resource virtualization in the cloud. Hardware resources such as RAM, memory, CPU, networking, etc., from a single host are shared and represented as multiple individual systems using the virtualization technique. In a container, software resource such as operating system is shared among multiple containers. The main difference between these two techniques is that VM virtualizes a system at the hardware level, and container virtualizes a system above the OS level. For hardware virtualization, a hypervisor is used, whereas, for software virtualization, a container engine is used. A virtual machine needs a full-fledged OS installed in it, but the container shares the same host operating system. The key differences between a VM and a container are presented in Table 3.1.

Both virtual machine and container have their advantage and disadvantages. Now, whether to use a virtual machine or a container completely

Table 3.1 Difference between virtual machine and container

Feature	Virtual machine	Container
Virtualization	VM provides hardware-level virtualization.	Container provides OS-level virtualization.
Operating system	Each virtual machine runs its own separate operating system.	All containers execute in same host operating system.
Isolation	VM provides better isolation because virtual machines are entirely self-contained systems that function in isolation. This implies that virtual machines are safe from vulnerabilities or interference by other VMs on the same host.	Containers are less isolated because they share the same OS and provide process-level isolation. So, other processes have a chance of interference and are possibly less secure than a virtual machine.
Startup time	Since each VM possesses a separate OS, booting up a system is time-consuming. So, the startup time of a VM is in minutes.	The containers only pull the required libs and bins to run an application and create a package. The startup time of a container is in milliseconds because it does not require booting a system.
Memory size	A separate OS is installed in each VM. Hence VM requires more memory space.	Container requires less memory space.
Security	Since VMs are fully isolated, they are more secure.	Container provides process-level isolation and is less secure.
Providers	VMware, Virtual Box, and Hyper–V are some VM providers.	LXC, LXD, CGManager, and Docker.
Portability	Virtual machines are heavily weighted, i.e., large in size, so moving a VM from one host to another is a costly task.	Containers are lightweight, so migrating a container from one host to another is comparatively easier than a virtual machine.

depends on users' specific requirements. Virtual machines are a preferable solution for applications that demand all of the OS's resources and operational capabilities. Containers are preferable when your primary goal is to execute as many apps or services as possible on a few servers while maintaining maximum scalability and portability. Containers provide increased portability and deployment simplicity in comparison to virtual machines. Choosing between a virtual machine and a container depends on the type of application you are using. Legacy systems typically rely on virtual machines, while cloud applications are typically built on containers. One can combine the two different approaches for the management benefits of virtual machines and the flexibility of containers. In the integrated approach, the virtual machines run on physical machines, and the containers run on virtual machines. This approach is called the container as a service model. This hybrid virtualization model is described in detail in Section 3.3.

3.2.4 Architectural difference between VM and container

The architecture of a virtual machine and a container is presented in Figures 3.3 and 3.4. In a virtual machine host OS is present on top of the infrastructure. The hypervisor responsible for creating and managing virtual machines runs on the host OS. Hypervisors create multiple virtual machines, and each virtual machine can have a different OS. VM #1 can run Windows while VM#2 runs a Linux OS. Each VM is completely isolated, and there is no interaction or interference between them. All types of communication between VM and host machine take place through the hypervisor.

On the other hand in containerization, the host OS is running on top of infrastructure similar to a virtual machine. Instead of a hypervisor, a container engine is running on top of the host OS. The functionality of the container engine is similar to the hypervisor. It creates and manages live instances of containers. As you can see in Figure 3.4, multiple containers are running on top of the container engine, but they do not have their own OS, unlike virtual machines. This makes a container lightweight and more portable than a virtual machine. All containers share a common host operating system. They only pull the required files such as libs and bins needed for executing an application. Since all containers share the same operating system, they are less isolated than virtual machines [6].

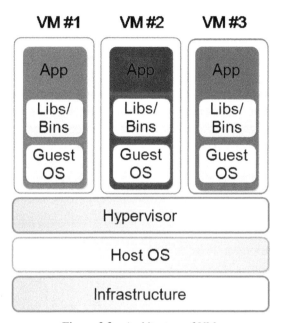

Figure 3.3 Architecture of VM.

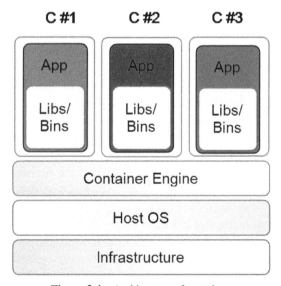

Figure 3.4 Architecture of container.

3.3 Container as a Service Model

Google and Amazon Web Services have introduced a new type of cloud service model called containers as a service (CaaS) [18, 14]. Containers are heading toward a new age in cloud computing because they are lightweight, easier to build and administer, and can significantly reduce startup time. Containers virtualize the operating system (OS), allowing multiple tasks to execute on a single instance of the OS. Because the containers are small in size, sometimes in the tens of megabytes, a single server can host several containers. CaaS can be considered a subset of infrastructure as a service (IaaS) and is positioned between IaaS and PaaS in cloud computing services. CaaS is a type of container-based virtualization technology. A cloud provider provides customers with container engines, orchestration, and the underlying computing resources as a service in which users can manage and execute containerized applications and clusters.

3.3.1 CaaS architecture

We have already discussed the architecture of a virtual machine and a container in Section 3.2.4. Now, moving a step ahead, the container as a service is a virtual machine and container combination. In a CaaS cloud model, containers are deployed in virtual machines, and virtual machines run on the host machine. The detailed architecture of a CaaS cloud model is presented in Figure 3.5.

In CaaS cloud model, the host OS runs on top of the infrastructure, and the hypervisor runs on top of host OS. The hypervisor creates several virtual machines, and each virtual machine has a guest OS. On top of the guest OS, the container engine is running, which is responsible for container management. The container engine creates and manages the live instances of the container. Each container creates a package of all the necessary files by pulling them from the guest OS. This is a hybrid approach of virtualization in cloud computing that uses both hardware and software virtualization. Now, what are the benefits of using this model? One of the main reasons for using this model is security. It provides better isolation to containers running in a virtual machine. As discussed, the virtual machine is more isolated and more secure than a container. Running many containers in a single host may lead to interference among the containers because they share the same operating system. Instead, if we distribute the containers and run them in a different virtual machine, there will be less chance of interference.

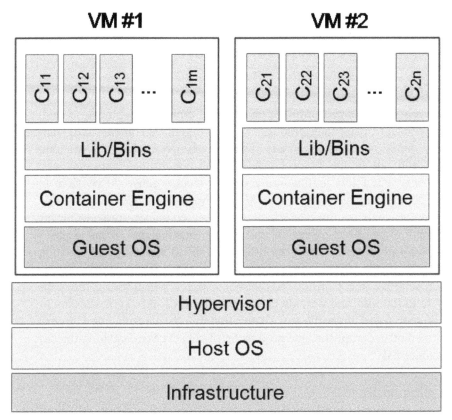

Figure 3.5 Container as a service (CaaS) model.

3.4 Containerization Techniques

Over the last few years, there has been a significant focus on researching container-based virtualization technologies. This section took a quick look at a few different container technologies, such as Docker, Singularity, and uDocker, among others. These technology solutions are typical examples of methods for virtualizing containers. At this time, they have been accepted by people devoted to scientific computing. Docker is an example of one of the container technologies that focus on a particular application in the business world. Others, such as Singularity, put more emphasis on the mobility of containers between different HPC platforms. Each of the solutions described above has its own unique set of procedures that must be followed in order to establish hardware and network isolation successfully. Isolation is the most fundamental and significant notion in terms of container security, and it is required to improve the security and privacy in all systems.

3.4.1 Docker

Docker is a software platform that enables the rapid development, testing, and deployment of software products. Docker organizes software into standardized units known as containers. These containers include the program together with all of the library resources, system utilities, program, and runtime that it requires to function properly. Docker makes it possible to rapidly deploy and grow apps into any environment while ensuring that your program will continue to execute. Docker containers provide means for achieving a desired degree of security by making use of namespaces and Cgroups procedures. These are used to establish hardware and isolation strategies for the process. There are several other sorts of namespaces, such as the user, net, PID, mnt, Cgroup, and time namespaces, which restrict the space available to the user and provide the container with Linux kernel resources that are segregated from the rest of the system. As a result, the Cgroups kernel method accomplishes the processing of subsets by mandating the use of resources [3].

3.4.2 Singularity

It is an open-source tool for generating and executing software containers built especially for high computation on shared computational clusters. It is possible to utilize Singularity containers to package whole scientific procedures, as well as software and library collections, and even data. Users are able to exercise complete command over their surroundings while using Singularity. This implies that you do not need to ask the administrator of your cluster to setup much for you; instead, you can just place the software in a Singularity container and execute it. The host system's Linux kernel can be accessed directly by Singularity containers. Therefore, there is no performance expense when utilizing containers instead of locally installed applications [22].

3.4.3 uDocker

uDocker is a simple software tool for cloud users that allows basic Docker containers to run in user space without root privileges. Downloading and running Docker containers is now feasible for nonprivileged clients on Linux machines when the Docker application itself is not installed. It is possible to pull and run Docker containers in Linux batch processing and engage clusters that are controlled by other units such as grid infrastructure or externally managed batch systems. These systems may be either managed internally or remotely. It is not necessary to have any special privileges in order to use uDocker, nor does it need the assistance of system administrators to install services. It is totally up to the end-user to download and run the program on their computer. The restricted root capability that a few of the uDocker execution modes make available is either emulated or made available via the usage of user namespaces. uDocker is a wrapper that includes a number of tools and libraries. Its purpose is to imitate a subset of Docker's characteristics, such as fetching images and executing containers with limited capability [7].

3.5 Research Challenges

Although containerization in cloud computing is considered one of the best strategies to achieve higher efficiency and resource utilization, several research challenges and issues still need to be addressed. Some of the current research challenges are discussed below.

- **Virtual machine sizing**: In CaaS, containers are executed on the virtual machine, and virtual machines are deployed on physical servers. So, determining the number of VMs to be instantiated and the amount of resources to be allocated to each VM is a challenging task. The exact allocation of resources for a VM in which a number of containers will run affects the overall resource utilization and energy consumption.
- **Container placement**: In CaaS, containers are placed in virtual machines, and virtual machines are deployed in physical machines. So, container placement involves two steps in CaaS. First, placing containers into the virtual machine and then the virtual machine into physical machines. Hence it can be considered a multi-objective optimization problem. Designing an optimal container placement algorithm that will assign containers into VMs and VMs into PMs is a significant challenge for container-based clouds [11].
- **Container consolidation**: Container consolidation in the CaaS cloud is used to balance different objectives and use of resources. There are three crucial decisions to take during container consolidation: when to initiate container migration, select a container to be migrated, and select a VM

to which the selected container to be migrated. Container consolidation can be achieved by migrating containers or virtual machines separately or through joint consolidation. Piraghaj *et al.* [19] proved that container consolidation is better than virtual machine consolidation for energy efficiency. Huang and Tsang [12] proposed a framework for automating virtual machine consolidation. It requires minimum configuration and makes a balance between power cost and network cost.

- **Comparison of VM vs container consolidation**: In CaaS, consolidation can be done at the container or virtual machine levels. For container consolidation, either we can move a container from one VM to another or move one VM containing multiple containers to another PM. The study of the performance of these two approaches for container consolidation is another research challenge. Designing and developing a consolidation algorithm that will consider both container and VM migration is another research challenge.

- **Workflow scheduling**: A workflow is a kind of workload comprising a number of distinct individual tasks with different types of association among them. The execution of such workflows is very challenging because the individual task should complete their execution before the deadline and the order of execution is also essential. The outcome of each individual task is combined to produce the output of the whole workflow. There are different approaches that various researchers have proposed to solve this problem in the virtual machine-based cloud [21]. Kang *et al.* [13] suggested a brokering system for scheduling jobs executed in containers to reduce energy usage while maintaining acceptable performance.

- **Load balancing**: The efficiency of a cloud system is greatly dependent on how the computing resources are being used. For better utilization and to maximize the efficiency of the cloud system, the total workload should be distributed evenly among all computing nodes, which is called load balancing. In CaaS, containerized applications should be distributed among all available VMs, so that all resources will be appropriately utilized. Several researchers propose several methods for solving this problem in the cloud. Paya and Marinescu [5] proposed a load balancing strategy that considers energy consumption and also exploits the mechanism of server consolidation.

- **Auto scaling of containers**: The orchestration facility in CaaS allows the deployment of applications from a single source to a collection of containers. The auto-scaling feature increases resource efficiency by automatically adding and deleting containers from a cluster as per the current requirement. This will improve resource utilization and improve service availability and performance.

- **Towards Serverless Cloud Computing**: Despite the growing popularity of container technology, cloud computing is still primarily centered on providing virtual machines (VMs), and cloud users are often charged based on the number of VMs provisioned. On the other hand, most of the applications fail to properly utilize the resources allotted to them, resulting in a waste of resources [16]. This issue can be addressed with a fine-grained cost model, which would present an attractive possibility for deploying apps inside containers. This is the major concept behind underlying serverless computing [4]. It is an event-driven execution approach in which the cloud user supplies the code, and the cloud service provider handles the code's execution environment throughout its lifecycle. Cloud customers are thus paid based on the amount of computing resources their application uses. Serverless computing might simplify cloud deployments by eliminating the requirement of deploying and maintaining many cloud instances. This cloud provides cost savings, particularly for the execution of small tasks and jobs [9]. Containers may play a vital role in the future of serverless computing since they can be deployed quickly and with little overhead [15].

3.6 Conclusion

Google and Amazon Web Services have introduced a new cloud service model called containers as a service (CaaS). Containers are heading toward a new age in cloud computing because they are lightweight, easier to build and administer, and can significantly reduce startup time. Containers virtualize the operating system (OS), allowing multiple tasks to execute on a single instance of the OS. Because the containers are small in size, sometimes in the tens of megabytes, a single server can host several containers. CaaS can be considered a subset of infrastructure as a service (IaaS) and is positioned between IaaS and PaaS in cloud computing services. The CaaS model provides better security and isolation for containers. Containers also lower administration costs. Because they all utilize the same operating system, just one needs to be cared for and fed in bug updates, patches, etc. Containers and VMs will both play significant roles in the future. Containers can run on virtual machines, allowing an organization to use its current automation, recovery, and analysis tools. CaaS is a strong new hosting architecture that can only be used if you are acquainted with containers. CaaS may be tremendously advantageous to exceedingly agile software engineers. It may be a huge help when doing continuous deployment on a business. Most current cloud hosting companies provide CaaS services at low pricing, so you will not have to hunt far for a suitable CaaS.

References

[1] https://www.redhat.com/en/topics/virtualization/what-is-a-virtual-machine

[2] https://www.redhat.com/en/blog/containers-understanding-difference-between-portability-compatibility-and-supportability

[3] Acharya, Jigna N., and Anil C. Suthar. "Docker Container Orchestration Management: A Review." In International Conference on Intelligent Vision and Computing, pp. 140-153. Springer, Cham, 2022.

[4] A. Eivy and J. Weinman, "Be Wary of the Economics of "Serverless" Cloud Computing," in IEEE Cloud Computing, vol. 4, no. 2, pp. 6-12, March-April 2017, doi: 10.1109/MCC.2017.32.

[5] A. Paya and D. C. Marinescu, "Energy-Aware Load Balancing and Application Scaling for the Cloud Ecosystem," in IEEE Transactions on Cloud Computing, vol. 5, no. 1, pp. 15-27, 1 Jan.-March 2017, doi: 10.1109/TCC.2015.2396059.

[6] Bentaleb, Ouafa, Adam SZ Belloum, Abderrazak Sebaa, and Aouaouche El-Maouhab. "Containerization technologies: Taxonomies, applications and challenges." The Journal of Supercomputing 78, no. 1 (2022): 1144-1181.

[7] Caballer, Miguel, Marica Antonacci, Zdeněk Šustr, Michele Perniola, and Germán Moltó. "Deployment of elastic virtual hybrid clusters across cloud sites." Journal of Grid Computing 19, no. 1 (2021): 1-16.

[8] D. Patel, M. K. Patra and B. Sahoo, "GWO Based Task Allocation for Load Balancing in Containerized Cloud," 2020 International Conference on Inventive Computation Technologies (ICICT), 2020, pp. 655-659, doi: 10.1109/ICICT48043.2020.9112525.

[9] H. Jin, X. Wang, S. Wu, S. Di and X. Shi, "Towards Optimized Fine-Grained Pricing of IaaS Cloud Platform," in IEEE Transactions on Cloud Computing, vol. 3, no. 4, pp. 436-448, 1 Oct.-Dec. 2015, doi: 10.1109/TCC.2014.2344680.

[10] H. Kang, M. Le and S. Tao, "Container and Microservice Driven Design for Cloud Infrastructure DevOps," 2016 IEEE International Conference on Cloud Engineering (IC2E), 2016, pp. 202-211, doi: 10.1109/IC2E.2016.26.

[11] Hussein, Mohamed K., Mohamed H. Mousa, and Mohamed A. Alqarni. "A placement architecture for a container as a service (CaaS) in a cloud environment." Journal of Cloud Computing 8, no. 1 (2019): 1-15.

[12] Huang, Zhe, and Danny HK Tsang. "M-convex VM consolidation: Towards a better VM workload consolidation." IEEE Transactions on Cloud Computing 4, no. 4 (2014): 415-428.

[13] Kang, Dong-Ki, Gyu-Beom Choi, Seong-Hwan Kim, Il-Sun Hwang, and Chan-Hyun Youn. "Workload-aware resource management for energy efficient heterogeneous docker containers." In 2016 IEEE Region 10 Conference (TENCON), pp. 2428-2431. IEEE, 2016.

[14] Kaur, Kuljeet, Tanya Dhand, Neeraj Kumar, and Sherali Zeadally. "Container-as-a-service at the edge: Trade-off between energy efficiency and service availability at fog nano data centers." IEEE wireless communications 24, no. 3 (2017): 48-56.

[15] Kumari, Anisha, Bibhudatta Sahoo, Ranjan Kumar Behera, Sanjay Misra, and Mayank Mohan Sharma. "Evaluation of Integrated Frameworks for Optimizing QoS in Serverless Computing." In *International Conference on Computational Science and Its Applications*, pp. 277-288. Springer, Cham, 2021.

[16] Maenhaut, Pieter-Jan, Bruno Volckaert, Veerle Ongenae, and Filip De Turck. "Resource management in a containerized cloud: Status and challenges." Journal of Network and Systems Management 28, no. 2 (2020): 197-246.

[17] M. K. Patra, D. Patel, B. Sahoo and A. K. Turuk, "A Randomized Algorithm for Load Balancing in Containerized Cloud," 2020 10th International Conference on Cloud Computing, Data Science \& Engineering (Confluence), 2020, pp. 410-414, doi: 10.1109/Confluence47617.2020.9058147.

[18] Piraghaj, Sareh Fotuhi, Amir Vahid Dastjerdi, Rodrigo N. Calheiros, and Rajkumar Buyya. "Efficient virtual machine sizing for hosting containers as a service (services 2015)." In 2015 IEEE World Congress on Services, pp. 31-38. IEEE, 2015.

[19] Piraghaj, Sareh Fotuhi, Amir Vahid Dastjerdi, Rodrigo N. Calheiros, and Rajkumar Buyya. "ContainerCloudSim: An environment for modeling

and simulation of containers in cloud data centers." Software: Practice and Experience 47, no. 4 (2017): 505-521.

[20] Verma, Garima. "Secure VM Migration in Cloud: Multi-Criteria Perspective with Improved Optimization Model." Wireless Personal Communications (2022): 1-28.

[21] Xu, Xiaolong, Wanchun Dou, Xuyun Zhang, and Jinjun Chen. "EnReal: An energy-aware resource allocation method for scientific workflow executions in cloud environment." IEEE transactions on cloud computing 4, no. 2 (2015): 166-179.

[22] Yang, Xu, and Masahiro Kasahara. "LPMX: a pure rootless composable container system." BMC bioinformatics 23, no. 1 (2022): 1-13.

4

Automated Framework for Detecting Unknown Activity in a Vehicular ad hoc Network

Atul B. Kathole[1], Dinesh N. Chaudhari[2], and Avinash P. Jadhav[3]

[1]Department of Computer Engineering, Dr. D. Y. Patil Institute of Technology, India
[2,3]Department of Computer Sci & Engineering JDIET, India
E-mail: atul.kathole1910@gmail.com; dnchaudhari2007@rediffmail.com; apjadhao@gmail.com

Abstract

This study aims to perceive abnormal behavior in vehicular ad hoc networks (VANETs) and recognize the relevant invaders or malfunctioning protuberances to prevent them from participating in the system's active communication and data exchange. By sharing cooperative awareness information and event-based messaging, vehicles and roadside equipment communicate ad hoc wirelessly in VANETs to improve traffic security and efficiency. Drivers can be instantly notified of impending potentially dangerous circumstances such as an abrupt braking action by an automobile driving in front of the tail termination of a traffic jam forward, or the hacking of shared information within a network by taking into account both the presence and position of vehicles moving within a definite range. VANET protuberances often broadcast mobility-related data (i.e., total values for location, time, direction, and speed) within a message range of numerous hundred meters to create collective alertness of single-hop neighbors. Low-latency traffic security applications become possible because of the ad hoc message between system nodes.

The suggested IHCMNDA (improve hybrid cooperative malicious node detection approach) approaches combined with automated predication offer security against external attackers in VANETs. Only registered VANET nodes have valid addresses that a reputable certificate authority has validated. Internal invaders who possess the necessary hardware, software, and legal certificates must be regarded as a severe hazard because of their ability to store data in a table using a clustering strategy. I explain how the processing of fabricated data might influence traffic's overall security and efficiency within the invaders' single- or multi-hop statement range. The majority of current techniques for detecting misbehavior in VANETs are data-centric in their approach and rely on plausibility and consistency checks.

I created a convincing proposal based on the information gathered from our actual tests inside the vehicular network to allow the secure and dependable long-term functioning of VANETs through an instruction detection technique. Attackers and malfunctioning nodes may be expelled from the network reactively once their misbehavior is recognized locally by independent network nodes and a central authority identifies offenders. This technique outperforms equivalent procedures solely implemented on VANET nodes regarding long-term attacker exclusion and false-positive detection reduction. As a result, the suggested notion will reduce prospective attackers' incentive to target VANETs to maximize throughput. Due to identifying anomalous node activity, this strategy should successfully counter even innovative attack methods that may appear in the future.

Keywords: VANET, IHCMNDA (improve hybrid cooperative malicious node detection approach), misbehaving node, attackers, security, receive signal strength (RSS).

4.1 Introduction

Internet and cyber technologies have permeated all spheres of social life at the moment. Globally, technological paradigms are shifting, and automation of devices is giving way to autonomy. The field of transportation communications is no different. A new kind of machine-to-machine (M2M) communication has emerged: a self-organizing wireless ad-hoc network with dynamic topologies, such as VANET, the intercar network; FANET, the aerial drone; and MARINET, the autonomous vessel network. Multiple viable connections between hosts are supported by the M2M network, which also enables dynamic route control at each site [1].

For instance, a vehicular ad hoc network (VANET) is a kind of mobile ad hoc network (MANETs) where it is generated using a wireless network connecting groups of moving or stationary vehicles [2].

As the European Transference Policy indicates, the adoption of Intelligent Transference Systems is critical for increasing the transport industry's safety, efficiency, and environmental friendliness. Intelligent Transference Systems are built on advanced communication systems that receive information from the many units that comprise the traffic scheme. The data is analyzed and transformed into valuable data and suggestions to aid transit passengers and authorities. The term "vehicular network" refers to this complex communication network [3].

Vehicle networks link cars and serve as a foundation for the future arrangement of large-scale, highly mobile applications. The applications are limitless: driver aid for faster, less jammed, and innocuous roads; more effective use of the transference system; more effective route forecasting and traffic flow control; more safe and environmentally friendly traffic through cardinal driver support; and improved planning and development of the scheme as a whole due to the accessibility of historical information based on traffic and utilization trends perceived via information mining methods and autonomous driving.

In next-generation mobile apps, deploying a dedicated structure for vehicular statements has advantages in generating shortcuts in the chart of connections and limiting ad hoc statements to limited areas and location-based applications [4]. Audi demonstrated the first commercial vehicle-to-infrastructure (V2I) communication structure in the United States at the end of 2016: car-to-traffic-light communication to determine the duration of red light. Shortly, these devices may aid in fuel conservation and pollution reduction and provide entertainment and commercial information. When messages travel a great distance, they may be tunneled to the destination location through the communication infrastructure, enhancing network connection [5]. On the other hand, deploying a large-scale setup would require substantial expenditure. As a result, the academic community's focus has shifted to efficiently deploying a distributed communication infrastructure.

4.1.1 VANET deployment challenges

Vehicular networks face numerous deployment challenges, including scalability (number of connected vehicles), quality of service (QoS) for connectivity (low latency, vehicle mobility), network conditions variability (rapidly changing topology, nonuniform network coverage based on vehicledensity), networking technology heterogeneity, and a lack of flexibility and programmability in network intelligence (resource allocation, prediction, and use), confidence and security [6].

With a rise in the number of cars on the road and the advancement of autonomous vehicles, road safety is becoming a more significant concern. VANET enables communication to disseminate safety, traffic control, navigation, and road service information. VANETs are deemed susceptible to various assaults, ranging from passive eavesdropping to active interference [7]. For instance, an attacker may look for and replay previous vehicle communications to access similar tools, such as toll services. An intruder may damage a targeted car by impersonating it and sending a fake alarm that might cause road traffic to stop [8]

Machine learning, abbreviated as ML, is a technology that utilizes artificial intelligence, or AI, to educate a computer about unknown concepts and to make exact judgments. ML is used in nearly every field, including manufacturing, robotics, arts, biotechnology, intelligent automated transportation systems, etc. It has gained widespread adoption due to its low cost and high capability (i.e., increased computing power and massive information storage) and the occurrence of large information volumes. It enables rapid and intelligent decision-making to optimize the system's presentation, including energy effectiveness, quality of service (QoS), and dependability [9]. Due to exponential population increase and automotive development, traffic congestion and public health have developed into dynamic and perplexing challenges in many urban regions. Around 1.25 million persons die each year in traffic accidents globally. They are the principal cause of mortality for individuals aged 15–29—congestion origins, expensive interruptions, excessive heat, pollutants, and fuel waste. Congestion in the United States cost $305 billion in 2017 [10]. An innovative and efficient transference network can reduce road accidents, promote environmental stewardship, and facilitate smooth traffic flow, contributing to increased performance. The VANET, or vehicular ad hoc network, improves road safety and traffic congestion, particularly during rush hours, to shorten passengers' travel time [11]. The rapid growth in demand for wireless policies has necessitated the necessity for a large spectrum that accommodates large volume capacity allocation over the market, which has hampered the implementation, adaptation, and scaling of next-generation switching technologies, which include smart cities, high-definition 3D video streaming services, augmented reality, the internet of things, or IoT, and VR, or virtual reality [12].

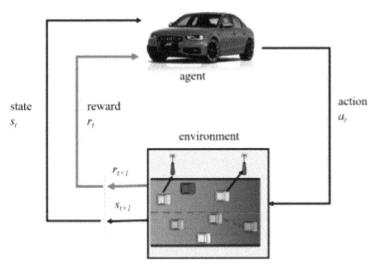

Figure 4.1 VANET architecture using intermediator.

The exponential expansion of wireless devices necessitates the development, implementation, and scaling of next-generation communication networks for high-definition 3D video streaming, smart cities, the internet of things (IoT), augmented reality, and cybernetic reality [13]

That is why this study presents a shared VANET IDS based on an automated technique that is focused on privacy preservation. To begin, ADMM is utilized to generate a distributed issue of empirical risk minimization on a VANET, allowing for decentralized training of the classifier to determine if an activity is regular or an assault [14]. At CIDS, we extend differential privacy to dynamic differential privacy and propose a privacy conserving technique called dual variable destruction to capture the concept of privacy in distributed machine learning. Additionally, we investigate the DVP's utility and define the DVP's fundamental interaction between security and privacy by constructing a convex optimization problem and executing data sets based on statistical tests to demonstrate the privacy measures' optimal existence [15].

4.1.2 VANET applications

While the primary motivation for VANETs was safety, several more uses have been suggested or implemented. Schoch *et al.* classify existing

VANET applications into four categories, as seen in Table: active safety, public service, driving improvement, and entertainment. As a result of the discussion above, it is clear that functional safety apps and moving performance enhancement applications may contribute to driving safety. There are coincidence prevention apps (such as rapidity warnings on curves or hilly roads, infrared visibility assistance, and blind-spot assistance) and accident cautioning applications for active safety (such as accident alertness warning) [16]. Emergency response (e.g., fire truck localization) and assistance to authorities are examples of public service applications (e.g., stolen vehicle tracking and finding). Enhanced driving applications (e.g., intelligent adaptive cruise control and a driving attention assister) and traffic proficiency apps are included in the improved driving applications (e.g., competent traffic flow supervision, digital map apprise.) Entertainment applications include internet admittance from any location, multimedia play, etc. Applications for public services and entertainment may enhance the driving experience.

To work correctly, most applications on VANETs need location information, if not all. Both vehicle and resource location must deliver a high-quality service in entertainment applications. Vehicles must exchange information about their whereabouts to prevent collisions. Vehicles operating in public service must be aware of emergency vehicles' positions to make way for them. We will now discuss several typical applications that make use of location data:

▶ Event management is a kind of application that facilitates processes and activities after an occurrence. It encompasses incident recognition, incident support, and post-event traffic retrieval. The incident recognition process establishes the special place of the issue and serves as the foundation for all subsequent incident management operations and actions [17].

▶ The most critical safety application is collision warning. This program alerts drivers of approaching collisions by detecting and calculating a safe distance between them and barriers like other cars, structures, or anything that might originate from a crash.

▶ Vehicle pursuing is a software program that enables automobile producers, logistics businesses, and other reliable third parties to display a vehicle's position and drive remotely. Position data is gathered, validated, and transferred to a central site server.

▶ Reserve vehicle evasion is a software program that warns motorists to yield to emergency vehicles such as fire engines, medical ambulances, police cars, and military vehicles. To prevent collisions with these vehicles, drivers must be aware of their position and own [18].

4.1.3 Possible attacks on vehicular network

Numerous attacks target the forwarding, control, and application layers. Man-in-the-middle attacks are possible between a switch and its controller due to a lack of transport layer security. Physical network security may help reduce such assaults. Attacks that cause a denial-of-service condition may overwhelm flow tables and buffers. These assaults result from reactive policies being implemented instead of a proactive strategy. Multiple controllers may be used to avoid them. Other potential dangers include dispersed multi-controllers, apps, unauthorized access, and conflicts between security policies or settings [19]. Regardless of current methods, increased mobility necessitates real-time authentication systems. Otherwise, the delay might result in traffic congestion, impeding SDVN implementation. This time-sensitive aspect complicates the task of enhancing security.

Application-level: Malicious apps may corrupt the SDN controller, resulting in permission violations, privilege escalation, resource exhaustion, service chain violations, or the injection of malicious control messages into the network, all of which can have a disastrous effect on the network's behavior (e.g., packet dropping, re-routing, and SDN controller termination). Third-party apps may potentially provide significant security risks due to vendor heterogeneity, lack of compatibility in security rules, and trust difficulties.

Control plane: Compromise switches may result in poisoning the SDN controller's perception of the network or its topology and the creation of bogus connections. Control messages may be abused to extract sensitive information or impersonate network resources. General-purpose attacks include breaking the SDN controller's authorizations, disrupting network isolation, or posing a danger to the controller's availability. Due to the control plane's reliance on the controller as the single decision-maker, it is susceptible to assaults and failures. Additionally, it may exploit its network knowledge to conduct new assaults. Vulnerabilities may also arise due to incompatibilities between several controllers [20].

Communication APIs: API vulnerability and a lack of standardization are significant issues. SDVN does not have a standardized, customized Open-Flow API or a Northbound API. Additionally, there is no standardization of eastbound/westbound APIs among controllers. Southbound APIs are often vulnerable to man-in-the-middle attacks, eavesdropping, and availability assaults.

The following is a list of several VN assaults [7],[10].

A. Phony data. Attackers square measure rational and aggressive insiders in this scenario, and they will broadcast incorrect data across the network, influencing the behavior of various drivers.

B. Cheating with sensing element data. This assault is being carried out by a UN official who is a business executive, reasonable, and proactive. He employs this assault to alter the observed location, speed, and direction of several nodes, evading accountability in the event of an accident.

C. ID speech act. A business executive is an adversary; he is passive and vicious. It will track a target vehicle's trajectory and utilize this data to determine its identity.

D. Denial of service (DOS). Throughout this example, the attacker is malevolent, present, and native. The attacker may want to put the network to a halt by flooding the guide with unnecessary messages. Direct electronic countermeasures and message injection are two examples of this kind of assault.

E. An adversary might drop replaying and dropping packets of unauthorized packets. For instance, an attacker may disable all alarm messages intended to warn drivers approaching the accident site. Similarly, an attacker will repeat packets the instant an incident occurs to create the appearance of an accident.

F. Hidden vehicle. This form of assault is possible in any circumstance when cars strive to reduce congestion on the wireless channel orderly. For example, consider a vehicle that has issued a warning message to its neighbor and is waiting for a response. Once the car receives an answer, it recognizes its neighbor is better positioned to convey the warning message and ceases propagating it to further nodes.

G. Worm hole attack. Detecting and preventing this assault is tough. A malicious node will capture packets at a single place on the network and tunnel them to another site through a shared personal network with other malicious nodes. If the malicious node uses the tunnel to deliver only management messages and not information packets, the attack's severity will grow [21].

H. Sybil attack. A car is used to impersonate many automobiles in this assault. As a result of this assault, many variants may be contending to control the network by faking the locations or identities of various nodes by using an automated framework capable of automatically detecting an assault. These identities are often used to launch multiple attacks on the system.

Additionally, these fabricated identities provide the sense of additional cars on the road.

4.1.4 ML for vehicular networking

Vehicle networks' dynamic nature and heterogeneous assembly have necessitated new network regulation and resource distribution methods. Network control comprises handover, directing, and offloading network traffic, while network supply allocation encompasses a spectrum, communication power, and computer properties. As conventional approaches cannot capture basic designs in vehicular networks, this section discusses the usage of machine learning approaches in this area and a combination of supporting technologies [22].

The research that uses machine learning techniques is classified in this overview by application, with more precise tasks listed in Table. The Table also summarizes the basic algorithm types presented in these studies (together with the associated significant machine learning types) and expected network-based task-specific problems. The 5G growth roadmap includes enabling skills (such as mobile edge/fog computation (MEC), net function virtualization (NFV), control/data plane parting, software-defined nets (SDNs), and network sharing [23–25]) that perform diverse functions in response to existing wireless network tests, such as network intricacy and spectrum supply efficiency. Current research in in-vehicle networks [26, 27] has also included similar know-how to address these issues, albeit not all have incorporated machine learning methodologies. Consider the following points while dealing with VANETs and machine learning.

4.1.5 Current challenges and opportunities

Several requirements for applying machine learning are discussed in this part, including the kind of issue to be addressed, the training data, the time required to complete the task, the implementation complexity, and the distinctions between machine learning approaches within the same category. These circumstances should be evaluated separately before deciding whether to implement machine learning methods and which machine learning techniques to apply.

4.1.5.1 Tests for vehicular networks

Despite advancements in directing and delivery in vehicular networking, the decision-making process for system formation/deformation has gotten little consideration from academic groups. Although some study on bio-inspired approaches has joint machine learning and routing conclusion optimization [58], vehicular networks' network node stirring speed is significantly faster than in outdated mobile networks, further constraining network formation and validation time. As a result, future research should emphasize efficient and robust network connectivity in automotive systems. Machine learning must cope with a high degree of network dynamics in in-vehicle networks, which requires a particular focus from academic groups. The growth of vehicular HetNets [59] enabled the introduction of numerous access skills for automobiles, RSUs, and BSs and the provision of differentiated facilities tailored to specific application needs while preserving a particular traffic load in a given situation. Although machine learning algorithms employed in many recommendation schemes may be used to understand individual node behavior and network traffic masses to match appropriate networks, this issue gets less consideration than the aspects discussed in this article [15]. MEC and NFV, on the other hand, have been actively researched as potential vehicle network technologies [23]. MEC provides dispersed explanations by relocating less computationally intensive jobs from cloud computing centers to system edges, reducing latency and facilitating information exploitation across several system nodes. By contrast, NFV adds elasticity to systems through hardware concepts, while machine learning techniques may be employed to provide a more generic, less hardware-dependent solution. Applying machine learning techniques to vehicle networks presents significant hurdles due to high network dynamics; this warrants special attention.

4.1.6 Research gap

Today, VANETs support many new services and protocols. Common VANETs are peculiar for their maximum decentralization where a network lacks a selected server, and their infrastructure and functionality are spread among hosts. This feature defines several problems caused by improving service provisions when using a new network and its low mobility capabilities and extended response amidst aggressive external influences.

The lack of connectivity and the consequent lack of VN authorization facilities hinder the standard procedure of forming a security line, separating nodes into trusted and untrusted ones. Such a distinction may have centered on a compliance policy, the necessary credentials, and node authentication capability. The VNs make security and privacy issues a real challenge because of their characteristics, particularly communication in open access. VNs are also made up of machines, and specific networks are resource-constrained nodes, thus reducing complexity as much as possible, and network security must be guaranteed.

The sources of problems related to information security in VANETs are as follows:

1. The absence of the tools of host protection from intruders.
2. The possibility of wiretapping the channels and substituting messages due to shared access to the communication environment.
3. The impossibility of using a usual security system due to the features of the VANET architecture.
4. The need to use complex routing algorithms that consider the probability of receiving incorrect information from compromised hosts due to changing the network's topology.
5. Any host that resides in the range of the signal source and knows the frequency of data transfer and other physical parameters (modulation, coding algorithm) can potentially intercept and decode the signal.
6. The impossibility of implementing the security policy due to the decentralization of the VANET and the lack of traditional security mechanisms working with dynamic topology.

4.1.7 Problem statement

In particular, VNs must be highly flexible to adapt to certain situations and use cases and have very low latency. It then has to be reconstituted to break into the rigid structure of the network and simplify hardware operation. For VN security, network nodes are susceptible to capture, hacking, and communication with attackers, and node communications may be eavesdropped on and fraudulent messages inserted or replayed into the network. Malicious nodes deliberately interrupt the network's routine operations to disrupt the network's standard functionality. These concerns would significantly impact the introduction of VPNs on a large scale. As a result, defending against VNs becomes more difficult.

These concerns include the lack of data communication due to the poor performance of wireless networks connecting various nodes and the absence of a source-destination link. As a result, VNs must be incredibly adaptable to any environment.

However, several challenges remain to be addressed. Some of those are as follows:

- Performance and flexibility
- Scalability
- Interoperability

Since VN comprises all kinds of devices and specific networks are resource-constrained nodes, the network's security must be assured, reducing overhead as much as possible.

4.1.8 Motivation

Among the many research topics, ad hoc networks, mobility management, internetworking in heterogeneous networks, and security continue to provide significant concerns. Due to vehicles' excellent mobility, their topologies are dynamic, and wireless channels are insecure. As a result, controllers have difficulty collecting real-time vehicle and network data. Additionally, the controllers' command distribution is slowed. Numerous security techniques have been created and suggested, but it remains challenging to assure that the whole network is immune to malicious assault. Security is complicated to address among the research challenges associated with ad hoc networks due to communication and the absence of infrastructure support. None of the network's components is committed to a single purpose, with routing most susceptible. Due to the dynamic nature of the SDN topology, it is impossible to depend on a single centralized certification authority (CA) to issue certificates and perform other administrative tasks.

4.2 Literature Work

This section will review prior efforts to secure VANETs. Numerous studies have investigated various intrusion detection system topologies that are well-suited for VANETs [3], and this section will summarize the primary options. Because VANETs are connected to sensor systems and mobile ad hoc networks (MANETs), it is necessary to examine security solutions in these sectors. This section will begin with the VANET problem. (1) Given that location availability is provided by secure data transmission through VANETs' networks, this addresses reliable communication by picking predictable and long-lasting routing connections in terms of link period and the likelihood of link period. As a result, routing algorithms are introduced. (2)This will address security-related issues. Digital signatures, resource-based algorithms, cryptographic algorithms, and algorithms based on radio signal intensity are among the solutions. (3) This outlines the algorithm used for encryption.

4.2.1 An overview

Routing systems may use numerous features of VANETs. They are connection, mobility, setup, geographic position, and the likelihood of particular actions occurring, such as the presence and longevity of links. These qualities enable us to categorize routing protocols in the literature. Connecting cars is accomplished by exchanging packets, that is, flooding messages to all nodes. MANETs and wireless sensor networks are the first to offer flooding-based protocols (WSNs). Specific flooding-based protocols, such as Ad-hoc On-demand Distance Vector (AODV) [32], DSR [33], and DSDV [34] that was initially suggested for VANETs and WSNs, have been prolonged to VANETs. Specific VANET procedures, such as Biswas [35], Murthy [36], Abedi [37], and DisjLi [38], are also based on flooding. As seen in Figure, we provide an overview of each directing mechanism.

Mobility encompasses distance, speed, acceleration, and the direction and movement patterns specified by maps. Compared to fixed networks such as Ethernet or ATM, mobility is unique. Even when compared to MANETs, the mobility scale of vehicles in VANETs is greater. They can try right and left, hasten and slow, and completely stop. Roads, too, impose restrictions on vehicles. As a result of this mobility information, it is possible to forecast the routing path's lifetime/duration. PBR [39], DisjLi [38], Taleb [40], Abedi [37], Wedde [41], and NiuDe [42] optimize message routing by adjusting mobility parameters.

4.2.2 Mobility-based routing in VANET with security

One of the primary distinctions between VANET and other net systems, including MANET, is mobility. Nodes are fixed in their locations in wired systems such as Ethernet and ATMs. Nodes in traditional wireless systems, such as MANETs, often utilized in congested areas such as airports, can have sluggish mobility. Although cellular networks may move quickly, infrastructure facilitates communication between them, that is, base stations. Nodes in a VANET usually exhibit excellent mobility, as seen by their rapid speed and frequent changes in direction and speed. Due to their high mobility, many of the current traditional routing techniques are inapplicable to VANET. Thus, mobility is a critical aspect in selecting and maintaining routing paths.

Quevedo *et al.* [20] note that vulnerability and weaknesses in VANETs are the main problems. In addition to traditional network attacks, VANETs are influenced by modern threats based on the disruption of authentication and false information dissemination, such as threats by Sybil. This paper proposed a system in this context for detecting Sybil attacks in VANETs, called SyDVELM. It is based on intensive learning tech techniques, providing more robustness, efficiency, and high precision to allow road protection, traffic congestion, digital entertainment, and other services. SyDVELM's proposed approach explains the mobility pattern of the vehicle nodes in urban scenarios. They are comparing absolute vehicle reliability in terms of inaccuracies in the relocation of Sybil nodes. They showed that the use of SyDVELM in VANETs guarantees a high detection rate with meager error rates and a versatile detection process. These features reflect the advantages of SyDVELM, as opposed to the existing Sybil attack detection system. They intend to validate the suggested approach in low density (sparse) scenarios and combine the ELM solution with other machine learning algorithms as possible implementation.

Mihai *et al.* [21] discuss the technological advancements that have resulted in the development of a connected, mobile, cooperative transportation system. They explore the most significant safety implications of VANETs and provide an in-depth review of current approaches for maintaining the privacy, stability, and confidentiality of vehicle network communications. However, widespread adoption is contingent upon resolving the remaining unresolved issues regarding safe automotive access and road networks. Additionally, the author discusses the need for coherent methodologies and governance to maintain flexibility and dependability while maintaining adequate levels of security and privacy. Further, network security must be bolstered to guarantee that the information is sent securely.

Goncÿalves *et al.* [29] discuss a detailed SLR on the usage of VANET smart IDs. Ns-2 with SUMO is the most common network-traffic-simulator combination employed in the studies. As for the most common ML algorithm, NN (its various variants) is the one chosen. For each analysis, the required datasets are usually generated, either from the simulation or the trace file of the network simulator. One of the SLR's purposes was to identify highly credible and publicly accessible datasets. Unfortunately, this does not appear to have been possible. Study assessment shows that most of them don't define how their databases and attacks are being developed.

Furthermore, neither of them makes their databases freely accessible for peer analysis. They use freely accessible, commonly reputed databases, such as the Kyoto dataset and the NSL-KDD. They also describe this engineering as an infrastructure for intelligent identification of attacks. One purpose of this work should be to create large enough datasets to allow the successful training of ML algorithms. Also, a thorough description should be made of how the dataset, attacks, and daily messages were created. Furthermore, this must be made openly accessible for peer review.

The author explaining machine learning (ML), Mohammad Asif Hossain *et al.* [4], one of the fastest developing computational methods, is widely used to resolve critical problems in many fields. Ad hoc vehicle network also referred to as VANET, is projected to play a crucial role in lowering congestion and road traffic accidents. To guarantee this place, an enormous data volume should be exchanged. Present connectivity assigned to VANET is therefore insufficient to accommodate such large volumes of data. Hence VANET goes through a problem of scarcity in the spectrum. The cognitive radio, i.e., CR, is a potential answer for solving issues of this type. VANET, based on CR or CR-VANET, could attain many steps for performance improvement, including connectivity with low latency and ultra-reliability. ML approaches can also be combined with CR-VANET to make CR-VANET very intelligent, achieve accelerated adaptability to environmental conditions, and increase service efficiency in an energy-efficient manner. They summarize CR, VANET, CR-VANET, and ML, including their architecture, features, problems, and open issues. The specification and roles of ML methods were evaluated in scenarios of CR-VANET. It also offers information on the use of ML in automated or driverless vehicles. They also define the implementations, and latest advances of methods in ML discussed in different areas of CR-VANETs, such as routing, spectrum sensing, security, and resource utilization. ML's functions have been extended to mitigate traffic congestion and road collisions, and many aspects of ML usage of AVs have been identified. They are using tools of ML to leverage the rewards of being researched because those fields are only in the early stages. He discussed some of these scopes in his thesis, unresolved questions, and future trends in the area.

4.2.3 Encryption and authentication

Asymmetric and symmetric algorithms are the two primary kinds of encryption techniques. Symmetric algorithms include sharing a secret key between communicating peers, and the private key must be secured because both encoding and decoding use the same undisclosed key. Each node has a communal and remote key in asymmetric algorithms. The collaborative and remote keys are unique in that a note encrypted using the public key of a node can only be decrypted with the node's private key and vice versa. A certification authority (CA) is answerable for verifying communal keys and issuing certificates required for verification in critical public infrastructure (PKI), a well-known technique for using and disseminating collective keys.

In VANETs, critical public infrastructure (PKI) and digital signatures have been extensively studied. A CA issues nodes with public and private keys. When node A transmits an encoded message M to node B, A encrypts the message using B's public key. Because only B has the remote key, he is the only one capable of decrypting the ciphertext. If B wishes to mark the note M digitally, B first encodes it using its remote key and then transmits both M and the engaged version of M to A. The signature is verified by A decrypting the signed version using B's public key. If the outcome is M, A will receive it as being transmitted by B since only B has the remote key necessary to produce the exclusive signature. Laberteaux *et al.* explored how a similar technique may be used to sign communications in VANETs. The digital signature enables the sender to be validated and authenticated. The encryption objective ensures that the message's content is only visible to nodes that possess the secret keys. PKI is an excellent approach for securing infrastructure, particularly roadside e-commerce, internet access points, and the like.

However, employing PKI in VANETs presents several challenges. The primary issue is that a trustworthy CA must distribute communal keys and certificates. All vehicles must trust the same CA to communicate with other cars, which is challenging when vehicles are built by diverse businesses in various nations. Additionally, certifications that have been revoked due to fraud or misuse must be revoked. All cars must then be informed of the list of revoked certifications. Additionally, asymmetric encryption/decryption often takes 1000 times as long as symmetric encryption/decryption. Additionally, VANET nodes may interact as part of a group. In this situation, public keys are not required for all nodes since cars in a group may share the message. By using symmetric methods in our work, we can accelerate encryption/decryption. As a result, the secret key must be designed. No additional fee is required for the undisclosed key, which depends on the automobile site. Additionally, the undisclosed key is a shared key used by a cluster of nodes.

4.3 Proposed Approach

A new algorithm called IHCMNDA is suggested with the help of CBD and K-means for improving the convergence behavior and enhancing the accurate results of activity recognition in autonomous vehicles. The IHCMNDA algorithm is used for "optimizing the number of hidden neurons and count of an epoch" to maximize the accuracy of recognition. This IHCMNDA algorithm is suggested by integrating the features of CBD and K-means machine learning algorithms. CBD is chosen here due to its advantages like robustness and optimization accuracy, good optimization outcomes, selection of the optimal features, classification efficiency, and superior searchability. However, it does not reduce the number of attributes; it faces complexities due to the complications in achieving optimal global solutions and slow convergence speed. Thus, the K-means algorithm maximizes the convergence speed because it offers convergence more quickly and requires fewer resources. Therefore, this new proposed algorithm helps in providing high-accuracy results with maximum throughput.

An IHCMNDA algorithm is implemented by considering the probability computation ρ and random number B utilized in the CBD algorithm. The position updating of the proposed approach is done through the following condition: if $\rho_{is} \geq B$ "the solutions are updated" based on the global leader phase of the K-means algorithm or else "the solutions are updated" by considering the CBD algorithm.

The global leader phase process assists in discovering the most appropriate positions using eqn (4.1).

$$X_{isk}^{new} = X_{isk} + B\,[0,1]\,(GL_{is} - X_{isk}) + B\,[-1,1]\,(X_{uis} - X_{isk}).\quad (4.1)$$

In eqn (4.1), "the random number is noted as B, the new position updating based on the global leader is termed as X_{isk}^{new}, the global leader position kth dimension is shown as GL_{is} and the uth spider monkey at kth dimension is termed as X_{uis}, the isth at kth dimension is derived as X_{isk}, where an arbitrarily chosen index" is specified as $is \in \{1, 2, \cdots, Is\}$.

Movement: The source is computed through the priority of fitness values, in which the highest value of roosters will be accessed first. This is mathematically formulated in eqn (4.2) and eqn (4.3).

$$X_{i,k}^{is+1} = X_{i,k}^{is} \times \left(1 + rn\left(0, \sigma^2\right)\right) \quad (4.2)$$

$$\sigma^2 = \begin{cases} 1, & if\ Fn_i \leq Fn_k, \\ \exp\left(\frac{(Fn_k - Fn_i)}{|Fn_i| + \varepsilon}\right), & otherwise, \end{cases} \quad k \in [1, Ns]\,, k \neq i. \quad (4.3)$$

Here, the smallest constant in the computer is ε employed to avoid the zero-division error, and the fitness value is denoted as Fn regarding cs.

A "rooster's index k is chosen randomly from the rooster's group, and a Gaussian distribution is noted as $rn\left(0, \sigma^2\right)$ with mean 0 and standard deviation σ^2."

Node movement: The node searches n neighbor based on their group-mate rooster and avoids the other error. Additionally, the other node can steal the different node positions. This behavior is formulated in eqn (4.4).

$$X_{i,k}^{is+1} = X_{i,k}^{is} + RN1 \times rdx \left(X_{z1,k}^{is+1} - X_{i,k}^{is}\right) + RN2 \times rdx \left(X_{z2,k}^{is+1} - X_{i,k}^{is}\right)$$
(4.4)

$$RN1 = \exp\left(\frac{(Fn_i - Fn_{z1})}{(abs\,(Fn_i) + \varepsilon)}\right)$$
(4.5)

$$RN2 = \exp\left((Fn_{z2} - Fn_i)\right).$$
(4.6)

Here, a uniform random number is mentioned as rd that lies in the range of $[0, 1]$. An "index of the rooster and an index of the node" is termed as $z1 \in [1, 2, .., Ns]$ at the ith hen's group-mate and $z2 \in [1, 2, .., Ns]$, respectively, where $z1 \neq z2$. In the same way $Fn_i > Fn_{z1}$, and $Fn_i > Fn_{z2}$, and so, $RN2 < 1 < RN1$.

Head node movement: The movement of "nodes around their cluster to search for neighbor node" is formulated in eqn (4.7).

$$X_{i,k}^{is+1} = X_{i,k}^{is} + FU \times \left(X_{Y,k}^{is} - X_{i,k}^{is}\right).$$
(4.7)

The term $X_{Y,k}^{is}$ is the "position of the ith head in the cluster" in the range of $(Y \in [1, Ns])$, and a new parameter is specified as FU that is utilized for following the head node based on their cluster that is randomly chosen among 0 and 2. Finally, the algorithm is terminated while reaching the last node in the cluster.

The pseudo-code of the proposed approach algorithm is given in Algorithm 1.

This section is organized as per the objective and corresponding architecture.

4.3.1 Methodology

The following procedure is followed to detect malicious nodes in the network and send the packet through the best path to the destination, as depicted in Figure 4.2.

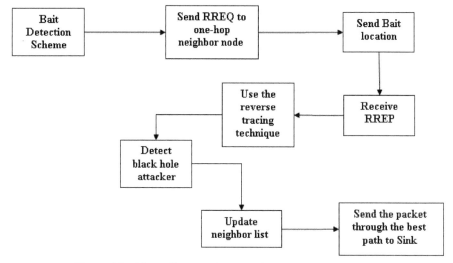

Figure 4.2 Flow of improved cooperative bait detection scheme.

The source node picks a neighbouring node stochastically because its address is utilized as a bait destination address to entice malicious nodes to send a reply message from RREP

- By using the reverse tracing approach, malicious nodes are identified and avoided.
- When a significant packet loss happens in PDR, an alert is expected to be sent back to the source.

4.3.2 Improve hybrid cooperative malicious node detection approach (IHCMNDA)

Objective 2: Development of ML-based system to increase intrusion detection system performance and prevent harmful warnings for false positive/negative factors.

$$(TLI)^X = ATL^X / q^X{}_{max}, \tag{4.8}$$

where $q^X{}_{max}$ is the length of interference of node X, ATL^X is the average traffic load at the node X, which is computed as follows

$$ATL^X = 1/Q \sum q_i, \tag{4.9}$$

Algorithm 1: Proposed

Initialize the number of nodes N_s and their parameters

Compute the positions of entire individuals

While $is < max. Generations$

    if $p_n \geq B$

        Update the solutions based on an algorithm

        Update the position using

$$X_{isk}^{new} = X_{isk} + B[0,1](GL_{is} - X_{isk}) + B[-1,1](X_{uis} - X_{isk})$$

    Else

        Update the solutions based on the CBD algorithm

        if $is == cluster\ node$

            Update solution using

$$X_{i,k}^{is+1} = X_{i,k}^{is} + FU * \left(X_{Y,k}^{is} - X_{i,k}^{is}\right)$$

        end if

        if $is == normal\ node$

            Update solution using

$$X_{i,k}^{is+1} = X_{i,k}^{is} + RN1 * rd * \left(X_{z1,k}^{is+1} - X_{i,k}^{is}\right) + RN2 * rd * \left(X_{z2,k}^{is+1} - X_{i,k}^{is}\right)$$

        end if

        if $is == malicous\ node$

            Update solution using

$$X_{i,k}^{is+1} = X_{i,k}^{is} * \left(1 + rn(0, \sigma^2)\right)$$

        end if

    End if

        Estimate new solutions

        Update best solutions

    end for

  end while

where Q is the total number of queue length samples, and qi is the ith queue length of the current time of forwarding node X. Based on the TLI value computed for forwarding node X [30], we estimate the packet forwarding probability of node X related to the queue congestion parameter as following:

$$P^{load} = 1 - (TLI)^X \tag{4.10}$$

We used P^{load} as the packet forwarding probability concerning the congestion load of node X. Higher the probability, the lesser the packet loss at node X.

$$(LCR)^X = (\sigma^X + \gamma^X)/(max(\sigma^X) + max(\gamma^X)), \qquad (4.11)$$

where σ^X is the link arrival rate, and γ^X is the link breakage rate of node X.

By using eqn (4.4), the probability of successful packet transmission concerning the mobility is computed as follows:

$$P^{mobility} = 1 - (LCR)^X. \qquad (4.12)$$

We can further compute the final probability for successful packet forwarding as follows.

$$P^{final} = (P^{load} + P^{mobility})/2. \qquad (4.13)$$

4.4 Result and Discussion

Since the network and layer are complicated to maintain an effective routing and connectivity mechanism, this work aims at ICBDS using clustering, which offers high data transfers between nodes and inexpensive genuine resources to the CU without any security issues in the network. The VANET middle of 1000 m to 1000 m with separate node numbers will be presented below. The CUs were mobile to flee or unite from their network at all times. With a transmission range of 30 m, the mobility rate of CU was set to 0–30 ms^{-1}. In comparison, the MAC standard underlining the layer was 802.11 for objects, with routers having a transmission range of 30 m. The malicious nodes or the CUs were incorporated using the probability distribution during transferal and connectivity to evaluate the protection.

4.4.1 Simulation

The investigation of the proposed gadget model is surveyed and appeared differently about the ICBD steering conventions dependent on the packet distribution ratio, to be specific, the improvement of energy proficient assignment of assets in an intellectual radio organization through clustering. Processing delay and throughput to verify and how efficiently the resources can allocate with the shortest path through the network in the presence of several nodes.

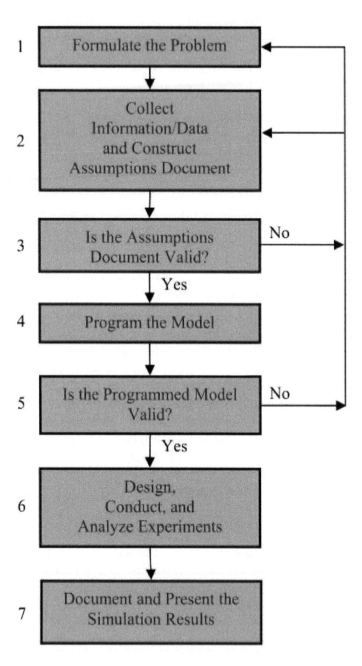

Figure 4.3   Steps in simulation result of DSR.

Table 4.1   Parameter used in ad hoc network for simulation

Parameters	Values
Simulation time	1000 s
Grid facet	1000 m × 1000 m
Ad hoc nodes	50
Number of malicious nodes	0, 5, 10, 15, 20
Transmission range	250 m
Data size	512 bytes
Number of source destination	20%
MAC protocol	IEEE 802.11a

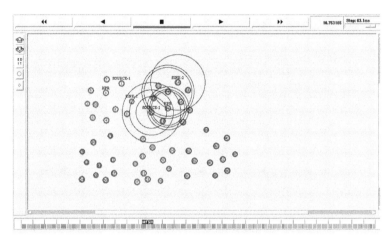

Screenshot 1   Simulation of improved hybrid cooperative malicious node detection approach using clustering.

Network simulation using NS$_2$

As proposed algorithm works iteratively for improvement in accuracy with maximum throughput. We are going to consider for simulation purposes and measure the performance of the system in Screenshot 1,

The above screenshot shows the cluster-based approach for securely sharing the data to improve performance. As resources are allocated from the specific cluster, we know how much resources are free to distribute and how to manage with less power utilization with secure data sharing. That is, allocation of data securely to get the maximum outcome with less energy utilization.

In the above Screenshot 2, we share the data between different clusters using a proper resource-sharing approach with the proposed methodology.

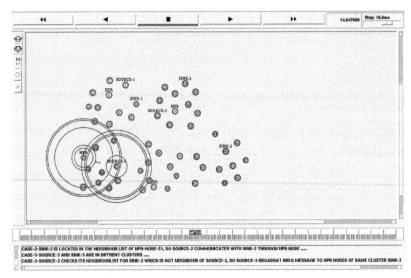

Screenshot 2   Sharing of data using the proposed method using the cluster-to-cluster sharing approach.

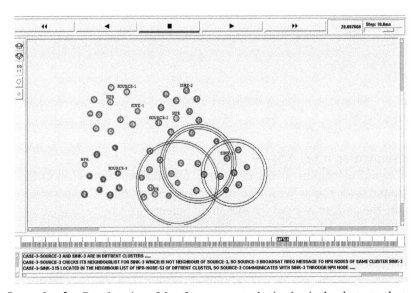

Screenshot 3   Board casting of data from source to destination in the cluster to cluster.

Screenshots 2–6 show the proposed (IHCMNDA) routing scheme, and each node is allowed to have its proper location by using that forward the

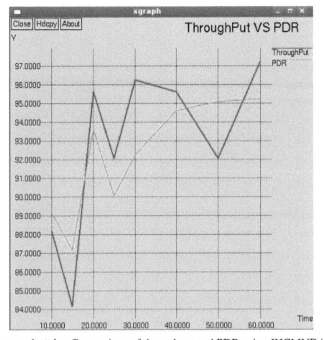

Screenshot 4 Comparison of throughput and PDR using IHCMNDA.

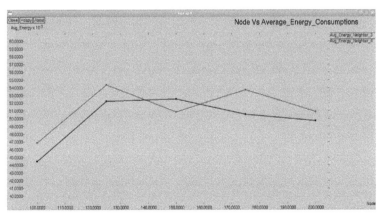

Screenshot 5 Comparison of energy utilization using IHCMNDA.

RREP and RREQ messages. In other words, only the shortest path nodes can be used to create a route between S and D. Provided that usual nodes will not retransmit RREQ or forward RREP messages, the cluster's overhead and transmitting capacity will be decreased. Since each node can transfer messages to longer distances, the cumulative path length can be reduced.

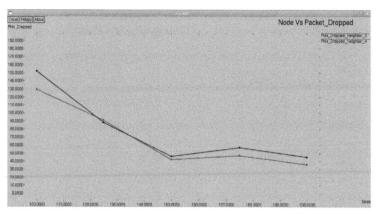

Screenshot 6   Comparison of packet dropped using IHCMNDA.

```
File  Edit  View  Terminal  Tabs  Help
using backward compatible Agent/CBR; use Application/Traffic/CBR instead
using backward compatible Agent/CBR; use Application/Traffic/CBR instead
using backward compatible Agent/CBR; use Application/Traffic/CBR instead
using backward compatible Agent/CBR; use Application/Traffic/CBR instead
using backward compatible Agent/CBR; use Application/Traffic/CBR instead
using backward compatible Agent/CBR; use Application/Traffic/CBR instead
using backward compatible Agent/CBR; use Application/Traffic/CBR instead
using backward compatible Agent/CBR; use Application/Traffic/CBR instead
using backward compatible Agent/CBR; use Application/Traffic/CBR instead
using backward compatible Agent/CBR; use Application/Traffic/CBR instead
using backward compatible Agent/CBR; use Application/Traffic/CBR instead
using backward compatible Agent/CBR; use Application/Traffic/CBR instead
using backward compatible Agent/CBR; use Application/Traffic/CBR instead
 energy = 399.26163079182692
 energy = 417.91060306965863
 energy = 365.51475448790694
 energy = 470.01196675468793
 energy = 497.72503343304851
Loading scenario file...
Loading connection pattern...
channel.cc:sendUp - Calc highestAntennaZ_ and distCST_
highestAntennaZ_ = 1.5,  distCST_ = 550.0
SORTING LISTS ...DONE!
```

Screenshot 7   Comparison of energy utilization using IHCMNDA.

Reducing the gap would minimize the delay from end to end. In addition, typical nodes need only be transmitted to the next closest node, where the power (tx) is reduced according to the distance, representing the overall electricity consumption and facilitating the overhead routing shown in Screenshot 7 and 8.

```
File  Edit  View  Terminal  Help
[root@localhost script]# awk -f wireless.awk out.tr
No of pkts send              1570
No of pkts recv              1543
Pkt_delivery_ratio:          98.2803
Control_overhead:            6609
Normalized_routing_overheads:  4.28321
Delay:                       0.0381882
Throughput:                  40518.2
Jitter:                      0.101109
Pkts_Dropped                 27
Dropping_Ratio:              1.71975
Total_Energy_Consumption:    6.37221
Avg_Energy_Consumption:      0.0637221
Overall Residual Energy:     9993.63
Avg_Residual_Energy:         99.9363
[root@localhost script]#
```

Screenshot 8   Simulation output.

4.4.2 Improved cooperative bait detection (ICBDS)

Graph 1 shows that PDR with increasing malicious nodes in proposed ICBDS is better than existing systems DSR and CBDS.

The proposed ICBDS outperformed DSR CBDS by producing over an average throughput as shown in Graph 2. The DSR is the least, and its throughput had numerous fluctuations under the Sybil attack.

The results conclude that the projected model is flooding a minimum number of delays as shown in Graph 3.

From the Graphs 4 and 5, it is clear that the rate of false positives decreases to a far more significant degree in the proposed ICBDS system compared to the other schemes like DSR and CBDS.

The detection rate is the number of malicious nodes found in the scheme relative to the total number of malicious nodes. Graphs 6 and 7 show a better detection rate for the proposed method.

Our projected ICBDS scheme does not increase the number of exchanged messages. Instead, it uses existing routing packets to exchange information like queue and connection status (already required according to routing protocol standards). In addition, the data path continues to convey actual

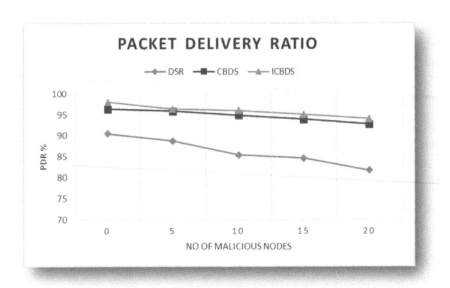

Graph 1 Packet delivery ratio vs. number of malicious nodes.

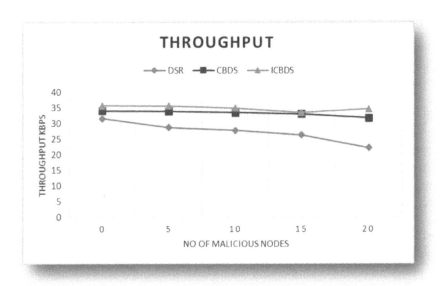

Graph 2 Throughput vs. no malicious nodes.

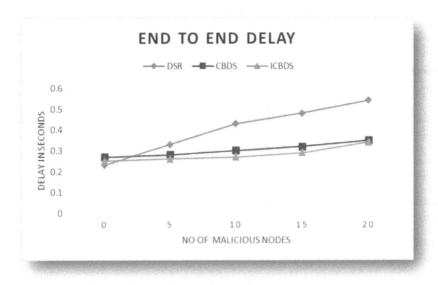

Graph 3   End-to-end delay vs. number of malicious nodes.

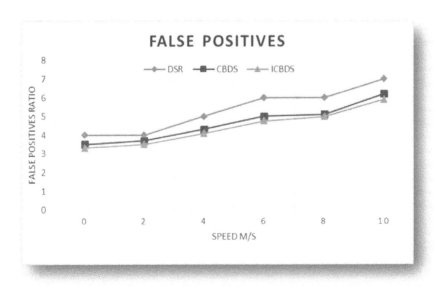

Graph 4   False positive vs. node speed.

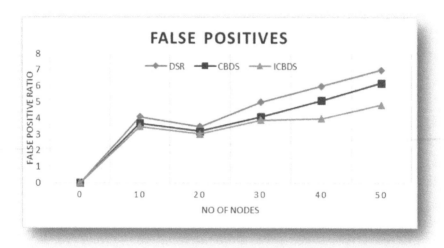

Graph 5   False positive vs. number of nodes.

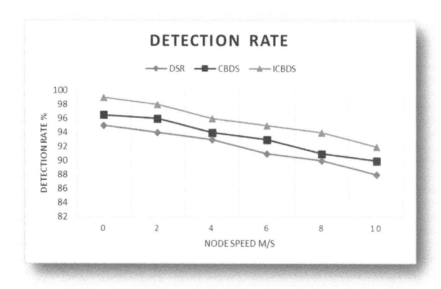

Graph 6   Detection rate vs. node speed.

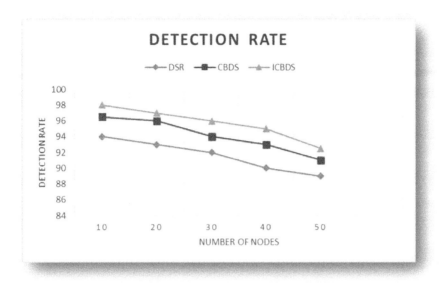

Graph 7 Detection rate vs. number of nodes.

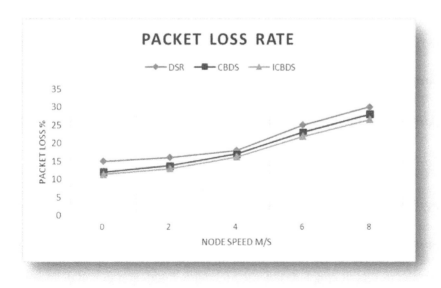

Graph 8 Packet loss rate vs. node speed.

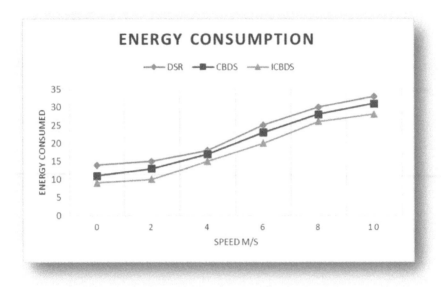

Graph 9   Energy consumption vs. node speed.

malicious nodes with minimum energy consumption, as shown in Graphs 8 and 9.

4.4.3 Improve hybrid cooperative malicious node detection approach (IHCMNDA)

Graph 10 shows PDR with increasing malicious nodes in the proposed IHCMNDA is better than existing systems. The percentage of data loss in DSR, CBDS, and FGA under Sybil attack is increased more than the IHCMNDA routing protocol in all scenarios.

The proposed IHCMNDA outperformed DSR, CBDS, and FGA by producing over 38.37 average throughputs as shown in Graph 11. The DSR is the least, and its throughput had numerous fluctuations under the Sybil attack.

The end-to-end delay performance of the conventional DSR, CBDS, FGA, and proposed IHCMNDA are shown in Graph 12.

From the results, it concludes that the model is flooding minimal delays as compared to others along with prevention to Sybil attack.

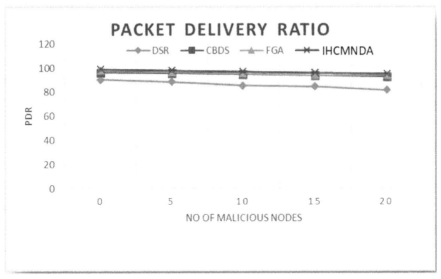

Graph 10   Packet delivery ratio vs. number of malicious nodes.

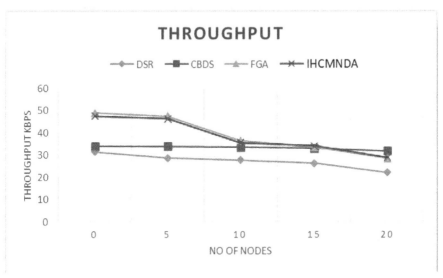

Graph 11   Throughput vs. number of malicious nodes.

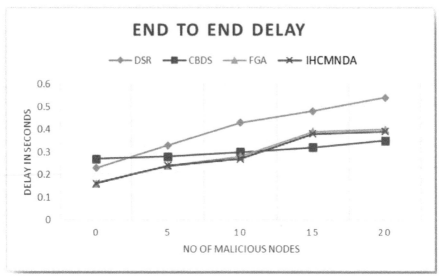

Graph 12   End-to-end delay vs. number of malicious nodes.

4.5 Conclusion and Future Work

In the proposed work, I am attempting to build an architecture for a collaborative intrusion detection system that utilizes distributable IHCMNDA (improve hybrid cooperative malicious node detection approach) to preserve user privacy. Private collaboration requires a privacy-preserving approach for centralized collaborative learning. Otherwise, the dispersed understanding of the software would result in the leaking of the personal instruction data. It protects user privacy using a shared intrusion prevention mechanism based on machine learning. The alternate direction approach of multipliers is used to circumvent the critical issue of empirical risk reduction while designing collaborative learning through a well-suited hierarchical structure to construct VANET systems.

Numerous studies have been conducted to examine the relationship between privacy and security. Dynamic differential privacy must be implemented to maintain the confidentiality of training data, and dual variable perturbation must be created by manipulating the double variable. I am attempting to quantify the performance of a low-error classifier in a real-world situation using the fewest amount of training data possible.

The enhanced cooperative bait detection scheme surpasses dynamic source routing and the old cooperative bait detection scheme. These techniques are ineffective in determining the precise source of the performance degradation. It has been found that the current strategy results in a drop in network performance owing to the reduction in the number of normal nodes. The IHCMNDA (improve hybrid cooperative malicious node detection approach) scheme is a two-step approach for detecting malicious nodes accurately. The first step involves the cooperative detection of malicious nodes, verifying discovered dangerous nodes through receive signal. It is performed utilizing characteristics related to packet losses, such as mobility and congestion. Combining a cluster-based certificate blocking method with an enhanced false accusation algorithm increases the number of regular nodes, resulting in improved outcomes. In terms of throughput, PDR, and detection rate, the simulation results indicated that the IHCMNDA technique outperformed the benchmark methods DSR, CBDS, and FGA.

I have proposed a design that selects the optimal value for the data protection parameter by solving an optimization problem while simultaneously optimizing privacy and security. The experiments examined the influence of various VANET sizes and VANET architecture on the effect of cooperative learning. I want to investigate both supervised and unsupervised virtual IDS and extend the differential privacy dynamic to other machine learning approaches in future research.

I aim to examine the following in future work:

1) Evaluation of varying densities in a clustered network.
2) Additional packet loss parameters, such as MAC layer information.
3) Our method's performance in the presence of various security concerns in a mobile ad hoc network.

References

[1] Shunyuan Xiao; Xiaohua Ge; Qing-Long Han; Yijun Zhang, "Secure Distributed Adaptive Platooning Control of Automated Vehicles Over Vehicular Ad-Hoc Networks Under Denial-of-Service Attacks," IEEE Transactions on Cybernetics, 25 May 2021.
[2] Megha V Kadam, "Recent Security Solutions For VANET Communications: A Systematic Review," Turkish Journal of Computer and Mathematics Education Vol.12No.7 (2021), 674-683.

[3] Atul R. Deshmukh; Pranjali Nirmal; S. S. Dorle, "A New Approach for Position-Based Routing Protocols Based on Ant Colony Optimization (ACO) Technique in Vehicular Ad Hoc Network (VANET)," IEEE Xplore: 04 August 2021.

[4] C R Komala; Srenivas Shetty; G R Smitha, "Junction Based Angular Routing in Vehicular Ad Hoc Networks (VANET)," IEEE Xplore: 22 June 2021.

[5] Atul B Kathole, Dr. Dinesh N. Chaudhari, "Performance analysis of Secure Hybrid Approach for Data Transmission in vehicular adhoc network" in IEEE International Conference on Mobile Networks and Wireless Communications (ICMNWC-2021) during 03 – 04 December 2021 in Pune.

[6] Atul B Kathole, Dr. Dinesh N. Chaudhari, "Secure Hybrid Approach for Sharing Data Securely in VANET" in springer International Conference on Computational Science and Applications (ICCSA21) during 10 – 11 December 2021 in Pune.

[7] Surjeet, Priyanka Bhardwaj, Raghavendra Pal, Nishu Gupta, "An intelligent scheme for slot reservation in vehicular ad hoc networks," China Communications (Volume: 18, Issue: 7, July 2021).

[8] Pooja Badukale, Samrat Thorat; Dinesh Rojatkar, "Sum up Work on Intrusion Detection System in Vehicular Ad-hoc Networks," 2021 5th International Conference on Trends in Electronics and Informatics (ICOEI).

[9] Atul B Kathole, Dr. Dinesh N. Chaudhari, "Pros & Cons of Machine learning and Security Methods, "2019. http://gujaratresearchsociety.in/index.php/JGRS, ISSN: 0374-8588, Volume 21 Issue 4.

[10] Atul B Kathole, Dr. Prasad S Halgaonkar, Ashvini Nikhade, " Machine Learning & its Classification Techniques," International Journal of Innovative Technology and Exploring Engineering (IJITEE) ISSN: 2278-3075, Volume-8 Issue-9S3, July 2019.

[11] Atul B Kathole, Dr. Dinesh N. Chaudhari, "Fuel Analysis and Distance Prediction using Machine learning," 2019, International Journal on Future Revolution in Computer Science & Communication Engineering, Volume: 5 Issue: 6.

[12] Ammara Anjum Khan; Mehran Abolhasan; Wei Ni; Justin Lipman; Abbas Jamalip, "An End-to-End (E2E) Network Slicing Framework for 5G Vehicular Ad-Hoc Networks," IEEE Transactions on Vehicular Technology (Volume: 70, Issue: 7, July 2021).

[13] Leticia Lemus Cárdenas; Ahmad Mohamad Mezher; Pablo Andrés Barbecho Bautista, "A Multimetric Predictive ANN-Based Routing Protocol for Vehicular Ad Hoc Networks," IEEE Access (Volume: 9), 11 June 2021.

[14] Xiao Zheng; Mingchu Li; Yuanfang Chen; Jun Guo; Muhammad Alam; Weitong Hu, "Blockchain-Based Secure Computation Offloading in Vehicular Networks," IEEE Transactions on Intelligent Transportation Systems (Volume: 22, Issue: 7, July 2021).

[15] Jitendra Bhatia; Jashvant Dave; Madhuri Bhavsar; Sudeep Tanwar, "SDN-Enabled Adaptive Broadcast Timer for Data Dissemination in Vehicular Ad Hoc Networks," IEEE Transactions on Vehicular Technology (Volume: 70, Issue: 8, Aug. 2021).

[16] Rasheed Hussain; Fatima Hussain; Sherali Zeadally; JooYoung Lee, "On the Adequacy of 5G Security for Vehicular Ad Hoc Networks," IEEE Communications Standards Magazine (Volume: 5, Issue: 1, March 2021).

[17] S. M. Farooq; S. M. Suhail Hussain; Taha Selim Ustun, "A Survey of Authentication Techniques in Vehicular Ad-Hoc Networks," IEEE Intelligent Transportation Systems Magazine (Volume: 13, Issue: 2, Summer 2021).

[18] Yuhan Yang; Lijun Wei; Jing Wu; Chengnian Long; Bo Li, "A Blockchain-based Multi-domain Authentication Scheme for Conditional Privacy-Preserving in Vehicular Ad-hoc Network," IEEE Internet of Things Journal (Early Access), 24 August 2021.

[19] Majed Al-Qutwani; Xingwei Wang; Bo Yi, "Request/Advertise-Based Content Forwarding in Vehicular Named Data Networking," IEEE Access (Volume: 9), 23 December 2020.

[20] Carlos H. O. O. Quevedo, Ana M. B. C. Quevedo, Ahmed Serhrouchni, "An Intelligent Mechanism for Sybil Attacks Detection in VANETs," 978-1-7281-5089-5/20/$31.00 ©2020 IEEE.

[21] Stefan Mihai, Nedzhmi Dokuz, Meer Saqib Ali, Purav Shah, and Ramona Tristian, "Security Aspects of Communications in VANETs," 978-1-7281-5611-8/20/$31.00 c 2020 IEEE.

[22] Organic U, Soydas M, Sertel E. Comparative research on deep learning approaches for airplane detection from very high-resolution satellite images. Remote Sensing. 2020;12(3):458.

[23] Zhao ZQ, Zheng P, Xu ST, Wu X. Object detection with deep learning: a review. IEEE Trans Neural Netw Learn Syst. 2019;30(11):32123232.

[24] Xu D, Wu Y. Improved YOLO-V3 with DenseNet for multiscale remote sensing target detection. Sensors. 2020;20(15):4276.

[25] Butt UA, Mehmood M, Shah SBH, Amin R, Shaukat MW, Raza SM, Piran M. A review of machine learning algorithms for cloud computing security. Electronics. 2020;9(9):1379.

[26] Chen W, Huang H, Peng S, Zhou C, Zhang C. YOLO-face: a real-time face detector. The Visual Computer 2020:1–9.

[27] Fan D, Liu D, Chi W, Liu X, Li Y. Improved SSD-based multiscale pedestrian detection algorithm. In: Advances in 3D image and graphics representation, analysis, computing and information technology. Springer, Singapore; 2020, p. 109–118.

[28] Mittal P, Sharma A, Singh R. Deep learning-based object detection in low-altitude UAV datasets: a survey. Image and Vision Computing 2020:104046.

[29] Fabio Goncÿalves, Bruno Ribeiro, Oscar Gama, "A Systematic Review on Intelligent Intrusion Detection Systems for VANETs," 978-1-7281-5764-1/19/$31.00 ©2019 IEEE.

[30] WANG TONG, AZHAR HUSSAIN, WANG XI BO, AND SABITA MAHARJAN, "Artificial Intelligence for Vehicle-to-Everything: a Survey," 2169-3536 (c) 2019 IEEE.

[31] C. Chembe, D. Kunda, I. Ahmedy, R. Md Noor, A. Q. Md Sabri, and M. A. Ngadi, "Infrastructure based spectrum sensing scheme in VANET using reinforcement learning," Veh. Commun., vol. 18, p. 100161, 2019.

[32] Dimitrios Kosmanos, Apostolos Pappas, Francisco J. Aparicio-Navarro, "Intrusion Detection System for Platooning Connected Autonomous Vehicles," 978-1-7281-4757-4/19/$31.00 c 2019 IEEE.

[33] W. Tong, A. Hussain, W. X. Bo, and S. Maharjan, "Artificial Intelligence for Vehicle-to-Everything: A Survey," IEEE Access, vol. 7, pp. 10823–10843, 2019.

[34] R. Boutaba et al., "A comprehensive survey on machine learning for networking: evolution, applications and research opportunities," J. Internet Serv. Appl., vol. 9, no. 1, p. 16, 2018.

[35] S. Pouyanfar et al., "A Survey on Deep Learning: Algorithms, Techniques, and Applications," ACM Comput. Surv., vol. 51, no. 5, pp. 92:1–92:36, Sep. 2018.

[36] Y. Gordienko et al., "Deep learning with lung segmentation and bone shadow exclusion techniques for chest x-ray analysis of lung cancer," in International Conference on Theory and Applications of Fuzzy Systems and Soft Computing, 2018, pp. 638-647: Springer.

[37] L. Liang, H. Ye, and G. Y. Li, "Towards Intelligent Vehicular Networks: A Machine Learning Framework," IEEE Internet Things J., vol. PP, no. c, p. 1, 2018.

[38] H. Ye, L. Liang, G. Y. Li, J. Kim, L. Lu, and M. Wu, "Machine Learning for Vehicular Networks: Recent Advances and Application Examples," IEEE Veh. Technol. Mag., vol. 13, no. 2, pp. 94–101, 2018.

[39] B. Khalfi, A. Zaid, and B. Hamdaoui, "When machine learning meets compressive sampling for wideband spectrum sensing," in 2017 13th International Wireless Communications and Mobile Computing Conference (IWCMC), 2017, pp. 1120–1125.

5

Control of Mobile Manipulator with Object Detection for EOD Applications

Mukul Kumar Gupta[1], Vinay Chowdary[2], and C. S. Meera [3]

[1,2]School of Engineering, University of Petroleum and Energy Studies, India
[3]Advanced Remanufacturing and Technology Centre, A* STAR, Singapore

Abstract

This book chapter highlights the control of mobile manipulators with object detection for explosive ordnance disposal (EOD) application. The proposed design of a remotely operated mobile manipulator can be useful in finding suspicious activity and environment. The robot prototype is an unmanned ground vehicle (UGV) with a multi-terrain tracked wheel chassis and a 6-DOF robotic arm mounted on the top along with an FPV camera for visual feedback. The robot operation is performed wirelessly from a remote computer or a mobile device to assist bomb disposal teams in safely detecting the type of explosive device and disposing of it remotely with help of the manipulator. Deep learning methods are used for real-time object identification and detection using an open-source framework called TensorFlow object detection API, using the SSD Mobilenet V2 trained on the MS-COCO dataset. A custom dataset for detecting explosive devices and suspicious objects has been prepared by using photographs of various explosive devices, IEDs, bombs, and wrapped packages that are likely to create suspicion. The custom images are annotated and labeled into a pre-trained SSD model for custom object detection. The bomb or an explosive device that has been identified is reported to the operator for manual interaction using the robotic arm. The detected bomb then can be tactically diffused, contained, or disposed of at a remote location. The prototype is tested by performing remote teleoperation on a rough terrain also the custom object detection model developed using the SSD Mobilenet V2 model is tested for a random package detected as an unidentified package as test results.

Keywords: Explosive ordnance disposal (EOD), UGV, visual feedback, object detection, TensorFlow's object detection API.

5.1 Introduction

Mobile manipulator is an articulated arm mounted over a mobile base to improve mobility and achieve greater degrees of freedom (DOF). In such mechanical structures, the DOF of the robotic arm is added with the DOF of the mobile base. They also improve the workspace of the robotic manipulator when mounted over a mobile base than with a fixed base. Any object in free space is represented with a degree of 6 DOF, 3 to represent the object in the Cartesian plane, and 3 degrees to denote the object orientation. Therefore, to fully manipulate an object in a free space 6 joints are needed to achieve complete movement.

Mobile robots can be classified into various categories depending on their locomotion systems, design aspects, and medium for the robot movement and other technical aspects such as terrain conditions, stability, controllability, maneuverability, etc. Mobile manipulators are either autonomously operated or teleoperated remotely [1, 2].

A mobile manipulator has the advantages of mobility of a mobile base and dexterity of a manipulator. However, it also leads to greater challenges in operation and control as any system with degrees of freedom of more than 6 is highly maneuverable in different directions and requires each joint to actuate individually. A mobile manipulation system comprises a mobile platform, robot manipulation, and other vision systems and tools depending on the applications. Mobile manipulators have been the subject of research in the field of robotics as they have widespread applications such as:

1. Space exploration
2. Underwater robots
3. Robotic construction
4. Automation and manufacturing industries
5. Home and healthcare robots
6. Mining and excavation
7. Military operations

Extensive research is ongoing in the development of space robots and concepts of programming for trajectory planning of manipulators in free-floating space have been introduced to research on space robots [2].

Robotic manipulators are considered most suitable for performing subsea and marine operations. Researchers with concepts of design features, prototypes, and analysis of electrically and hydraulically actuated manipulators in the area of underwater robots for different subsea intervention applications [3]. Robotic machinery and advanced equipment have widespread applications in the field of construction. The use of robots on construction sites helps in the reduction of cost and labor and improves material efficiency during construction [4]. Articulated robots or robotic manipulators are used in the field of manufacturing in various applications such as arc welding, spot welding, material handling, painting, picking, and assembly operations in industries.

Robotic manipulators have wide applications in the field of healthcare and medical sciences as research is undergoing in the development of robotic equipment and robotic manipulators to improve the quality of care and achieve surgical precision. The popular robots in the field of healthcare are the da Vinci surgical robot, Xenex Germ-Zapping robot, PARO therapeutic robot, and the development of frameworks to improve healthcare with the use of robotic technology [5]. Mobile robot manipulators are being used for excavation and mining in hazardous locations by autonomous or remote operation. Autonomous robot excavators have been developed to add autonomy to produce robot excavators by adapting to different soil types without human intervention [6].

Robots in the field of military applications will be able to substitute soldiers on the battlefield by performing complex and tactical tasks by the use of vision systems and robotic manipulators to interact with the weapons and explosives on the battlefield. Military robots designed can be used for fighting, reconnaissance, and suicide attacks under critical circumstances and perform complex military missions without compromising human lives.

5.1.1 Explosive ordnance disposal robots

Bomb disposal or explosive ordnance disposal robots are unarmed machines capable of disposing of explosive devices tactically by the explosive ordnance disposal teams remotely from a safe distance. "Robots can go where humans fear to tread." Explosive ordnance disposal robots serve as the remote presence for a bomb disposal expert to allow them to examine the explosive and render tactical disposal operations without putting themselves in harm's way. It has been said that distance is your friend when it comes to bomb disposal and the first bomb disposal robot Wheelbarrow Mark1 was created in 1972, with the use of an electrically powered chassis to move suspicious devices to detonate in a remote location. Robotic unmanned vehicles and drones have been used ever since for over 40 years for bomb disposal applications worldwide with the use of advancing technology and innovative control and tracking systems.

Indian Defence Research and Development Organisation has developed a remotely operated robot with the primary objective of bomb recovery named Daksh as shown in Figure 5.1 has been an inspiration for the selection of this project. Daksh uses varied bomb detection features such as X-ray vision, robotized arm to lift and water jets to disarm, and other complex weaponry. It can navigate through staircases, steep slopes, and narrow corridors and tow vehicles, and can reach and pick hazardous materials. Daksh is equipped with water jet disrupters to diffuse explosive devices and is equipped with a shotgun for short-range firing [8]. Object detection algorithms in Daksh can scan cars and vehicles for explosive devices using X-ray vision.

Bomb disposal robots had been in service since the 1970s. Researchers across the world have developed unmanned ground vehicles for explosive ordnance disposal for antiterrorism and military purpose.

Reduanur Rehman, 2019 [9] developed a joystick-controlled industrial robot arm that can perform different movements and pick and place operations using a wireless transmitter PS2 controller interfaced with an Arduino Mega Controller.

Hongfu *et al.* [10] proposed a research on DC servomotor control system for bomb disposal robots based on PID control and PWM output for the DC motor drive.

Motaleb *et al.* [11] developed a bomb disposal robot for discarding explosives through a wireless control method by using a 6-DOF articulated arm mounted on a movable base to help the bomb squad dispose of the bomb safely from a distance.

Albert *et al.* developed a robot to assist in landmine detection with the incorporation of inductive sensors and video cameras to detect metallic landmines in the area and controlled wirelessly using RF communication [16]. Meera *et al.* [18] proposed disturbance observer-assisted hybrid control for an autonomous robot.

Daksh by DRDO

Figure 5.1 Explosive ordnance disposal robots [8].

5.2 Object Detection

Object detection is widely used in various applications such as suspicious objects, vehicle and face detection, and obstacle avoidance in autonomous vehicles and people. Object detection API is a tool used to deploy an image recognition algorithm by classifying and recognizing objects by constructing bounding boxes around the image based on the class defined in the dataset.

Krishna Sai *et al.* [14] proposed a research article for detecting threatening objects using the CNN algorithm. An intelligent video surveillance system is widely used in the military. Due to changes in the environment and dynamic obstacles, achieving a stable video surveillance system is a complex problem [20]. Brian Day *et al.* [21] proposed a paper that describes the concept of visual display to provide depth of the objects grasped to use with a bomb disposal robot.

Huang *et al.* [17] developed a real-time object detection model using neural network and deep learning techniques using YOLO—YOLO-LITE: object detection algorithm optimized for non-GPU computers by Lite [22].

5.2.1 Computer-aided design of the proposed design

Fusion 360 is an Autodesk product software tool designed for 3D modeling. It has a cloud-based development procedure to carry out work on any platform. To design the assembly or geometry of a design several constraints such as Dimensions of lengths, diameter, tangent, chamfers, shells and other features that are not sketch-based are examples of operation-based features. Relations are used to define features such as tangency, parallel, perpendicular, and concentricity. The sketch's dimensions can be modified with factors either individually or in a group that is within or outside of the sketch.

In creating an assembly of the design Mates is analogous to sketch relations and define joints required to define the motions of each component in the assembly. Assembly mates define equivalent relations concerning specific parts or components and define their joint properties.

5.2.2 System architecture

The section discusses the design and development of a semi-automatic mobile manipulator for EOD applications by using a robotic arm manipulator over a mobile robot base along with an FPV camera to facilitate visual feedback and perform object detection in the remote location. The proposed model as shown in Figure 5.2 uses task space control methods to implement a control strategy in the system. The system comprises three individual blocks that are separated, controlled, and integrated to form a single robot control system. The proposed model is teleoperated with a remote computer using internet protocol (IP) communications.

5.3 Hardware Architecture

The designed prototype shown in Figure 5.3 of the mobile robot with a tracked wheel chassis is controlled remotely with the control signals provided through the controller through an H bridge motor driver using a Raspberry Pi

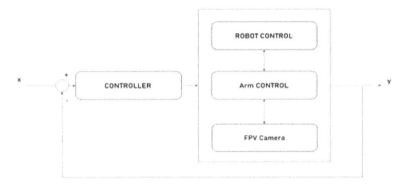

Figure 5.2 Block diagram of control system.

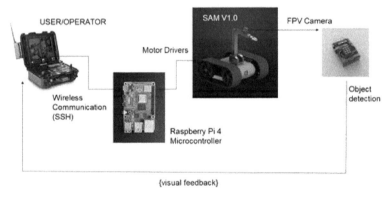

Figure 5.3 Hardware architecture of the proposed model.

computer. The robot's motion is controlled by controlling the control inputs to the DC motor in forward, reverse, left, and right directions.

A 5-DOF robotic arm with a gripper as an end effector is mounted over the chassis for pick and place applications and interaction with the objects in the robot's environment. An FPV camera mounted on the facade of the robot is used to track, detect, and identify the objects in the robot's field of view and report to the user as visual feedback through the controller.

Raspberry Pi, DC Motor, Servomotor, 5-DoF robotic arm, USB Camera has been used as hardware specifications.

5.3.1 Robot-arm contol and actuation

The 5-DOF robotic arm mounted over the mobile base is used to pick and place the objects in the robot's environment. For a robotic arm to interact with any object in Cartesian space, the position of the object must be derived concerning the joint angles using forward kinematics given by the D-H convention. The extent of the actuation required by the joints calculated as the joint angles from the end-effector position is obtained by using inverse kinematics.

A 5-DOF robotic arm is simulated using a MathWorks open-source GUI using a robotic system toolbox to derive forward and inverse kinematics. The joint angles are used to determine the Cartesian coordinates of the end-effector and vice-versa. The robotic arm when placed on the movable mobile robot base, the end effector coordinates alter concerning the position of the robot base.

5.4 Object Detection

Object detection is a computer vision approach for identifying and locating objects. In object detection, the number of objects is counted in a scene, using identification and localization. Object detection distinguishes itself by identifying the type of object and its coordinates based on its location. Object detection can take several forms, one of which is face detection [23].

In most cases, object detection algorithms are pre-trained model weights, which one can custom tune according to our applications.

Object detection is carried out in three steps:

Step1: Small segments are generated in the input image. A large set of bounding boxes is spanned over a full image.

Step2: Feature extraction is carried out for each segmented region to predict whether the boxed region contains a valid object.

Step3: Boxes that are overlapping are combined into a single bounding rectangle to label the image class.

5.4.1 TensorFlow

TensorFlow (TF) is an open-source framework used for machine learning that simplifies the process of acquiring data, training models, and offering predictions and refining results. It is built on both machine learning and deep learning models and algorithms. On the front end, it uses python for making it convenient for the user, whereas it runs on C++ in the backend. It allows developers to create computational graphs in which each node represents mathematical operations, and each connection denotes the data allowing the user to focus on the logic of the application. TensorFlow is used by Deep Face, Facebook's image recognition system.

5.4.2 TensorFlow object detection API

The TF object detection API contains a collection of pre-trained models that have been trained on a variety of datasets, such as Common Objects in Context, KITTI, and open images dataset.

5.4.2.1 Object detection API using tensorFlow 2.0

The major challenges occur when creating accurate models for the localization and identification of multiple objects in a single image. TF object detection API is an open-source framework built on top of TF to train and deploy object detection models. TFOD API (TensorFlow Object Detection) is compatible with both TF1 and TF2.

a. **Installation:** Install the TensorFlow object detection API with python package installer (PIP) to run using the Google Cloud.

1. Install python package: Install python packages by navigating into the directory and compiling protos using the command:

   Protoc object_detection/protos/*.proto –python_out=.

2. Install TensorFlow object detection API: Install the TFOD API using the following commands:

   "cp object_detection/packages/tf2/setup.py"

   "python -m pip install –use-feature=2020-resolver"

b. **Training:** Train the TFOD API with a novel class based on the application by tuning an architecture on novel examples after initializing from a pre-trained COCO dataset.

1. Import libraries:
   Clone the TensorFlow model's repositories.
   Install the object detection API

2. Utilities:
   Add utilities to load images from a file to an array to feed into the TensorFlow graph with features such as shape, height, width, and channels.

3. Explosive devices data: The proposed project uses an object detection algorithm for the detection of explosive devices in an environment using TFOD object detection API. To perform the object detection algorithm to detect explosives and a new class of images containing the images of explosives is added to the COCO dataset by a technique called data augmentation.

Data augmentation: is used to expand the training dataset to improve the performance and ability of the model to generalize.

4. Annotate images with bounding boxes: The required class of objects is annotated by drawing bounding boxes around the objects in each image as shown in Figure 5.4.

c. **Prepare data for training:** Before adding class annotations, convert everything to the format for the training loop to convert into tensors and classes to representations. Each class of data is predicted by assigning it with a class id:

Example:
explosive_class_id =1
num_classes = 1

Figure 5.4   Annotating image.

category_index
= {duck_class_id: {'id': duck_class_id, 'name':Explosive_Device}}
label_id_offset = 1
Visualize the class explosives as a sanity check. Then create the model and
restore the weights of all classification layers.

5.5 Results and Discussion

The model for the manipulator is designed using Fusion 360 software and
a hardware model is implemented based on the design specifications. The
manipulator developed is redundant which means it has extra DOFs to what is
required. The kinematic model for the mobile manipulator is developed using
the D-H convention as it provides a better method for writing the kinematics
of the serial link manipulator as compared to the analytical approach.

5.5.1 Hardware implementation

The proposed hardware is shown in Figure 5.5 developed using a Raspberry
Pi 4 Model B computer and it is teleoperated with the help of internet protocol
using SSH on a remote laptop or mobile device. The Raspberry Pi computer

Figure 5.5   (a) Robot prototype in indoor environment, (b) Robot prototype in outdoor
environment.

is programmed using python scripts to teleoperate the mobile robot using arrow keys and custom-designated keys to operate the robotic arm. The object detection algorithm is performed on the USB camera connected to the Raspberry Pi to detect the objects in the robot's field of view.

The following commands are used to run the python script for teleoperation of the robot in the terminal after connecting to the Raspberry Pi using a remote SSH connection or Raspberry Pi desktop using a VNC server/viewer.

$ python keyboard_control.py

To run the object detection algorithm as shown in Figure 5.6 with the USB camera, navigate to the directory/TensorFlow/object detection and then the following command launches the python script that runs the object detection algorithm:

$python Pi-Camera-OD.py

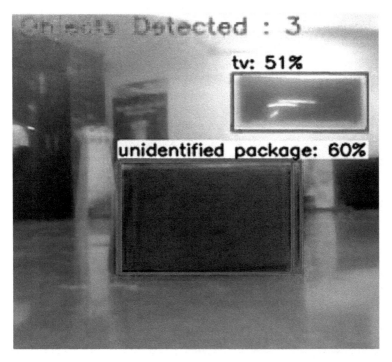

Figure 5.6   Object detection results in a laboratory environment.

5.5.2 Robotic arm simulation using MATLAB GUI

A 5-DOF robotic arm is simulated using a MathWorks open-source GUI using a robotic system toolbox to derive forward and inverse kinematics as shown in Figure 5.7 The joint angles are used to determine the Cartesian coordinates of the end-effector and vice-versa. The robotic arm when placed on the movable mobile robot base, the end-effector coordinates alter concerning the position of the robot base.

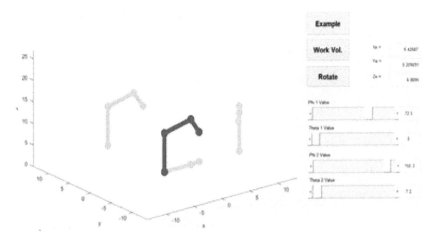

Figure 5.7   Robotic arm simulation using MATLAB GUI.

5.5.3 Simulation in gazebo using ROS

ROS provides the services that one expects from an operating system, including hardware abstraction, low-level device control, and implementation of commonly used functionality. It also provides tools and libraries for obtaining, building, writing, and running code across multiple computers. The CAD model of each component of the mobile robot is designed using Solidworks or Fusion 360 software, assembled, and then exported to URDF format using the sw_urdf_exporter or plugins as shown in Figure 5.8.

URDF is a universal robot description format model that consists of a robot's physical description to ROS. The robot model and environment are then imported into the gazebo and the simulation of the robot is controlled with the ROS python node as shown in Figure 5.9.

The following commands are used to launch the gazebo in ROS.

$roscore
$rosrun gazebo_ros gazebo

The mobile robot is teleoperated remotely with the keyboard using ROS python controller node "teleop_twist_keyboard.py."
The following command is used to run the teleoperation script:

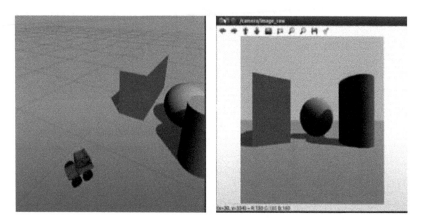

Figure 5.8   (a) Mobile robot simulation in gazebo, (b) Robot perspective.

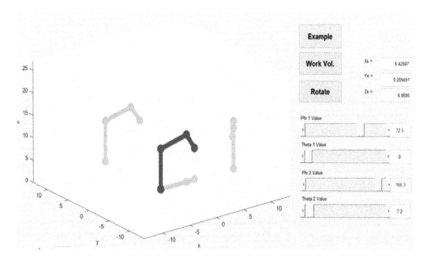

Figure 5.9   Three-link manipulator MATLAB GUI simulation.

$rosrun teleop_twist_keyboard teleop_twist_keyboard.p

The environment office_world is obtained from a pre-built launch file from ROS packages.

5.6 Conclusion

The book chapter describes the research that has been carried out to develop a working prototype of a mobile manipulator for EOD applications. This feature facilitates the robot to navigate conveniently on all platforms irrespective of the terrain as the mechanism is designed to climb rocks and navigate in rugged environments. A 5-DOF robotic manipulator is mounted over the mobile base to improve the workspace of the robotic arm for pick and place applications. The arm equation is developed for the considered 5-axis manipulator using forward kinematics. The camera connected to the front of the robot serves as an FPV camera to display the robot's perspective vision and also to perform object detection in the robot's environment.

For the future scope of this work, the development of a fully autonomous bomb disposal robot with improved object detection algorithms and the addition of an extra manipulator making it a dual-arm mobile manipulator for added dexterity can be implemented. Virtual gazebo worlds inspired by real-time situations can be developed in the gazebo and the control scheme can be tested. Optimal path planning of the mobile base is another scope where the study can be focused on keeping the obstacle avoidance schemes in mind and autonomous navigation in complex environments by clearing objects interfering with the robot's path.

Acknowledgments

The author likes to thank the Bosch center of Automation, at UPES Dehradun for providing the necessary support for developing the hardware facility.

References

[1] F. Rubio, F. Valero, and, C. Llopis-Albert 'A review of mobile robots: Concepts, methods, theoretical framework, and applications' International Journal of Advanced Robotic Systems, 16(2), 2019.

[2] R. B. AS, Z. Hao, U. Montanaro, S. Dixit, A. Rathinam, Y. Gao, G. Neumann, S. Fallah 'Autonomous Robots for Space: Trajectory Learning and Adaptation Using Imitation'. Frontiers in Robotics and AI. 2021.

[3] S. Sivčev, J. Coleman, E. Omerdić, G. Dooly, and D. Toal, Underwater manipulators: A review. Ocean Engineering, 163, pp.431-450., 2018.

[4] D. Luo, L. Yu 'From Factory to Site—Designing for Industrial Robots Used in On-Site Construction. InAutomating Cities' 2021 (pp. 87-109). Springer, Singapore.

[5] A. Wynsberghe, 'Healthcare robots: Ethics, design, and implementation. Routledge', 2016.

[6] D. Seward, F. Margrave, I. Sommerville, and R. Morrey, LUCIE is a robot excavator designed for system safety. In Proceedings of IEEE International Conference on Robotics and Automation (Vol. 1, pp. 963-968). IEEE, April 1996.

[7] S. Naskar, S. Das, A. K. Seth, and A. Nath, Application of radio frequency controlled intelligent military robot in defense. In 2011 International Conference on Communication Systems and Network Technologies (pp. 396-401). IEEE, June, 2011.

[8] Asian Defence "Bomb Disposal Robot Daksh". Theasiandefence.blogspot.com. Retrieved 2010-08-31, 2009.

[9] R. Rahman, M. S. Rahman, and J. R. Bhuiyan, Joystick controlled industrial robotic system with robotic arm. In 2019 IEEE International Conference on Robotics, Automation, Artificial-intelligence and Internet-of-Things (RAAICON) (pp. 31-34).

[10] H. Zhou, 2008, October. DC servo motor PID control in mobile robots with embedded DSP. In 2008 International Conference on Intelligent Computation Technology and Automation (ICICTA) (Vol. 1, pp. 332-336). IEEE.

[11] F. Y. C. Albert, C. H. S. Mason, C. K. J. Kiing, K. S. Ee, and K. W. Chan, 2014. Remotely operated solar-powered mobile metal detector robot. Procedia computer science, 42, pp.232-239.

[12] A. K. B. Motaleb, M. B. Hoque, and M. A. Hoque, 'Bomb disposal robot' In International Conference on Innovations in Science, Engineering, and Technology (ICISET) (pp. 1-5). IEEE, October, 2016.

[13] S. Masunaga, 'Controlled Metal Detector Mounted on Mine Detection Robot' International Journal of Advanced Robotic Systems, 4(2), p.26, June 2007.

[14] B. K. Sai and T. Sasikala, 2019, November. Object Detection and Count of Objects in Image using Tensor Flow Object Detection API. In 2019 International Conference on Smart Systems and Inventive Technology (ICSSIT) (pp. 542-546). IEEE.

[15] T. Kaur, D. Kumar, 'Wireless multifunctional robot for military applications' 2nd international conference on recent advances in engineering & computational sciences (RAECS) (pp. 1-5). IEEE, Dec 2015.

[16] J. W. Kim, B. D. Choi, S. H. Park, K. K. Kim, and S. J. Ko, 'Remote control system using real-time mpeg-4 streaming technology for mobile robot' Digest of Technical Papers. International Conference on Consumer Electronics (IEEE Cat. No. 02CH37300) (pp. 200-201). IEEE., June 2002.

[17] R. Huang, J. Pedoeem, and C. Chen, 2018, December. YOLO-LITE: a real-time object detection algorithm optimized for non-GPU computers. In 2018 IEEE International Conference on Big Data (Big Data) (pp. 2503-2510), IEEE.

[18] M. C. S., M. K. Gupta, S. Mohan 'Disturbance observer-assisted hybrid control for autonomous manipulation in a robotic backhoe' Archive of Mechanical Engineering. 2019:153-69.

[19] B. Day, C. Bethel, R. Murphy, and J. Burke, 'A depth sensing display for bomb disposal robots' IEEE International Workshop on Safety, Security and Rescue Robotics (pp. 146-151). October 2008.

[20] R. Huang, J. Pedoeem, and C. Chen, YOLO-LITE: 'a real-time object detection algorithm optimized for non-GPU computers', IEEE International Conference on Big Data (pp. 2503-2510), December 2018.

[21] B. Padmaja, M. B. Myneni, and E. K. R. Patro, 'A comparison on visual prediction models for MAMO (multi activity-multi object) recognition using deep learning' Journal of Big Data, 7(1), pp.1-15, 2020.

[22] Q. She, F. Feng, X. Hao, Q. Yang, C. Lan, V. Lomonaco, X. Shi, Z. Wang, Y. Guo, Y. Zhang, and F. Qiao, 'OpenLORIS-Object: A robotic vision dataset and benchmark for lifelong deep learning' IEEE International Conference on Robotics and Automation (ICRA) (pp. 4767-4773). May 2020.

[23] Y. H. Lee and Y. Kim, 'Comparison of CNN and YOLO for Object Detection' Journal of the semiconductor & display technology, 19(1), pp.85-92, 2020.

6

Smart Agriculture: Emerging and Future Farming Technologies

S. Sathiya [1], Cecil Antony[2], and Praveen Kumar Ghodke[3]

[1]Department of Electrical Engineering, National Institute of Technology, Calicut, India.
[2]School of Biotechnology, National Institute of Technology Calicut, India.
[3]Department of Chemical Engineering, National Institute of Technology Calicut, India.
E-mail: sathiyas@nitc.ac.in; cecil@nitc.ac.in; praveenkg@nitc.ac.in

Abstract

The advent of smart agriculture has revolutionized agricultural management. It is an emerging technology that uses modern sensing and communication technologies such as internet of things (IoT), unmanned aerial vehicles, robotics and automation and artificial intelligence. The technology recognizes the factors responsible for diminished agricultural yields and helps to track them in a real-time mode. Simultaneous monitoring of qualitative and quantitative improvement in the agricultural production, optimization of parameters such as environmental conditions, soil conditions, plant growth and its health can be done on a real-time basis. Interestingly, the data collected from various sensors with wireless sensor networks (WSNs) can be utilized to measure parameters that ensures better plant growth for enhancing agricultural management. On the contrary, manual intervention can be reduced to the maximum as the wireless monitoring of plant parameters provides accurate status of plant nutrition and health. Furthermore, the technology focuses on development of devices and tools to monitor, display, and inform the farmers regarding agricultural parameters. The recent development of IoT-based robotics and automation lead to dynamic, fast, and accurate farming with increase in agricultural yields. Additionally, geological and meteorological information existing as numerical, or image data could be analyzed promptly using artificial intelligence. This would further help to make an effective decision on pest control, fertilizer usage, and water management, etc. This chapter reviews the existing and emerging sensors along with wireless network, IoT, artificial intelligence, and other communication technologies, focused on agricultural industry. It also discusses the existing practical difficulties and implementation challenges of the latest technologies; also extends discussion on the future direction of smart agriculture.

Keywords: Artificial intelligence (AI), internet of things (IoT), smart agriculture, wireless sensor network (WSN).

6.1 Introduction

The expanding global human population is likely to reach about 9.6 billion by 2050 and this might create a high demand for food and agricultural products. In order to make sure the availability of food for the entire human population, it is mandatory to double the agricultural productivity. As the expanding global human population needs a consistent supply of high-quality agricultural produce that must be met through the globalization of markets and demands the implementation of modern technologies in agriculture. Undoubtedly, the recent AI-based technologies in farming have been successfully implemented in the developed world. Among global nations, the United States, Canada, and Australia recognized and implemented smart farming, whereas France is to be the first from Europe [1]. Smart agriculture is an advanced technology that revolutionized agriculture in 1900s with improved mechanization and genetic modifications to increase the productivity as illustrated in Figure 6.1. However, it has been considered as a scientific way to improve the agricultural and food management by implementing recent technologies to monitor and control agricultural parameters thereby increasing the yield. Therefore, it precisely helps to identify and analyze the soil quality, nutrients requirements, various diseases, and water management to enhance agricultural yields.

Smart agriculture enhances productivity or yield by closely monitoring spatiotemporal variations of various parameters in agriculture without harming the ecosystem. The accurate and precise evaluation of spatiotemporal variability and its management has been considered the critical achievement of smart agriculture. For instance, crop management of sugarcane, beetroot,

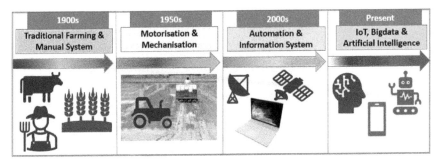

Figure 6.1 Evolution of agricultural technology.

coffee beans, and tea has been successfully implemented and evaluated by many nations. Conversely, the ability to enhance the economy, ecosystem quality, and social benefits are yet not realized. One of the main reasons for this might be the lack of information addressing the spatiotemporal variability of the crop parameters. Moreover, the quality production of farming is nothing but a yield of the complicated interaction between quality seeds with soil, environment, fertilizers, pesticides, and water management system, as depicted in Figure 6.2. Therefore, monitoring and controlling such complicated agricultural ecosystem parameters seems essential. It is also important to consider the ecological impact when improving agricultural productivity, which reduces environmental effects. In recent decades, the production of crops has been increasing according to the demand created by the rapidly expanding global human population. The increased use of agrochemicals has facilitated the rise in food production globally. Conversely, the productivity of the major crops is still below the expected level in meeting the huge demand [2]. Unfortunately, the increased usage of agrochemicals to enhance productivity might increase the negative impacts on the ecological system. Hence, the degradation of land, water erosion, wind erosion, salinization, and loss of nutrients directly or indirectly affect the ecosystem and human wellbeing.

In general, smart agriculture is more specifically a management technique that depends on the observation, measurement, and response to the variations in agricultural parameters. Explicitly, the target of precision agriculture is preparing a system to support decision-making in agricultural management. As a result, achieving optimized returns from utilizing minimal resources [3]. Among various techniques of smart agriculture, the phyto-geomorphological approach describes the crop growth characteristics for multi-year and dictates

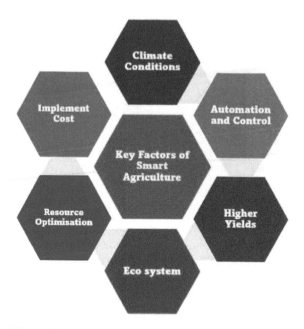

Figure 6.2   Key factors of smart agriculture implementation.

the hydrology of agricultural fields [4, 5]. One of the essential steps in smart agriculture is identifying the spatiotemporal variations in the field for particular crop production. At the same time, it helps to increase the crop yield without negatively impacting our ecosystem. The spatial mapping for large fields comprises of two components: spatial mapping within the field and spatial mapping between fields. The former is the site-specific crop management (SSCM) system which defines the agricultural management system for a specific crop based on the soil conditions and climate. It involves data collection on the required parameters and helps in decision-making [6]. The accurate selection of suitable crops, precise dispense of pesticides and fertilizers, and appropriate irrigation methods help to achieve optimum production and increase the yield with reduced labor cost. Furthermore, the equipment and technologies used in smart farming provide accurate information and precise decision to the farmers, for instance, preparation of agricultural fields, sowing seeds, monitoring the soil nutrients, application of agrochemicals, and post-production procedures to improve economic growth. More importantly, the precise dispense of agrochemicals at an appropriate time avoids residues in the soil and water, reducing the negative impact on the environment.

Over the last few decades, smart agriculture has gained attention from researchers that started with automation in the drilling of seeds. The recent information technology revolution in developing countries like India and China has been influential in implementing smart technologies in agriculture. Especially the aerial images of fields, climate prediction, precise fertilizer application, and health monitoring of crops are collected initially in smart agriculture. Next, the data containing information about plant health, soil parameters, and climatic conditions will be tracked and trained through machine learning models. Smart agriculture improves the economy with efficient management of water, fertilizer, and other related parameters while supporting farmers with sufficient spatiotemporal information about agricultural fields. In addition, the field mapping with spatial variability using global positioning system (GPS) measures various parameters such as crop yield, organic matter, soil nutrients, and plant health status. Different monitors and sensors mounted on GPS-equipped farming systems or vehicles collect these data. Variable rate technology (VRT) could further use this data for seed sowing, agrochemical dispensing, and water supply system [3]. Unmanned aerial vehicles (UAVs), commonly called as drones having hyperspectral RGB imaging cameras, can collect the data required for decision-making in smart agriculture. Subsequently, the images obtained from UAVs upon further processing through photogrammetric methods, help to create field mapping [7]. However, various developed and developing countries are currently testing this technology, while many have already implemented in their fields. Therefore, the knowledge of the existing technologies in smart agriculture helps to overcome future challenges in agricultural management. This chapter provides an overview of current technologies and developments in smart farming and its future directions.

6.2 Role of Smart Agriculture in Yield Enhancement

Introduction of recent technological developments, such as advanced sensing methods and the internet of things (IoT) has revolutionized agriculture by changing its fundamental procedures. As a result of integrating IoT and wireless sensor technology, smart farming methods emerged as an alternative to conventional farming. It immediately provides improvised solutions for various issues of conventional farming, such as water management, optimum yields, selection of appropriate fields, irrigation, and disease management, as illustrated in Figure 6.3. Undoubtedly, these advanced technologies can help

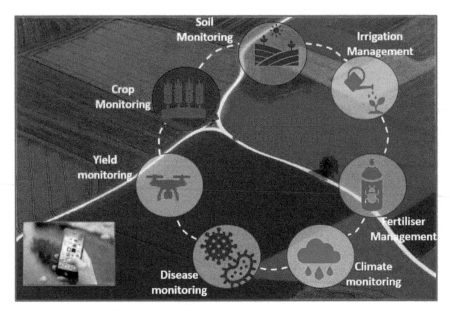

Figure 6.3   IoT-based agricultural monitoring.

improve agricultural yield efficiency at various stages, as discussed in the following subsections.

6.2.1 Soil samples and its mapping

Soil is the major factor of plant growth, and the primary step of soil inspection is soil sampling, which helps to identify and characterize the fields. Although further information is required to choose the right crop, the main aim of the soil sampling is to evaluate the nutrients needed for efficient plant growth. Frequent testing or sampling of the soil should be done to ascertain the soil characteristics under different weather conditions [8]. More importantly, the soil type, history of cropping, application of agrochemicals, irrigation, and geographical requirements of land are critical factors in the soil analysis. The factors mentioned above give a deep insight into the soil's physiochemical and biological characteristics, which supports choosing the appropriate crop and identifying its growth-limiting factors. Similarly, soil characterization and mapping allow the sowing of various crops in a selected field, efficiently managing seed suitability, sowing time, and the maximum depth of the deep-rooted plants. Moreover, planting multiple crops in the same field helps in agriculture with optimum utilization of resources. The advent of a different range of sensing systems or monitoring tool kits has simplified agricultural management. Farmers can easily track the soil nutrients and take appropriate measures for better plant growth without compromising the quality. For instance, the sensors and wireless networks can monitor soil characteristics like texture, water retaining capacity, and water absorption rate. As a result, soil erosion, salinization, and pollution due to over usage of agrochemicals can be reduced. However, many manufacturers have developed portable soil testing tool kits with a complete setup for soil characterization [9]. Therefore, portable kits facilitate farmers to carry out up to 100 soil sampling tests at any time without prior laboratory knowledge.

In particular, drought is the primary concern in agriculture, restricting crop growth and yield. All over the world, several countries and regions are facing drought issues at different intensities. A remote sensing system can measure the soil moisture from time to time, which can be further used for drought analysis in remote regions, especially in rural areas. For the data on global mapping of soil moisture, the Soil Moisture and Ocean Salinity (SMOS) was launched in 2009, which provides soil moisture mapping at the global level for every one or two days [10]. Few researchers used a moderate resolution imaging spectroradiometer (MODIS) sensor to map the soil characteristics for risk analysis of land degradation in sub-Saharan Africa. The soil and survey information mapping of farming lands covered almost all climate zones of Africa, which further helped develop prediction models [11]. Various sensors and imaging techniques determine the depth for sowing the crop seeds. Interestingly, the vision-based sensors mounted on an autonomous robot called "Agribot" perform seed sowing. The mapping of fields, either locally or globally, using the global positioning system assisted by the computerized vision-based device, supports the placement of robots in fields to perform their actions efficiently [12]. Several non-contact sensors equipped with LEDs for transmission of infrared (IR), visible light, and LED lasers along with receivers have been developed [13]. However, the output voltage depends on the seed movement passing through the sensor and light rays. Therefore, shades falling on the receiver element help determine the seeds' flow rate. Notably, using UAVs or drones to collect precise data for analyzing soil before sowing the seeds of a specific crop, helps to assess the crop suitability, type of seeds, and planting procedures. Recently, a fixed-wing aircraft equipped with Panasonic Lumix GFI digital camera has been developed to monitor Morocco's soil erosion [14].

6.2.2 Smart irrigation systems

About 97% of water sources on earth are from oceans and seas, which are salty, and only 3% are fresh water, among which 70% are in the form of glaciers and ice. The remaining freshwater, about 0.55%, is from the water-bodies, underground water, or evaporated in the air [15]. In particular, it is essential to preserve the fresh water in rivers, lakes, and other water reservoirs to maintain the ecosystem and agricultural/domestic needs. Although agricultural procedures consume around 70% of freshwater from the reservoirs, the quantity varies based on regions and climatic conditions [16]–[18]. UN Convention Combat Desertification (UNCCD) has reported that 168 countries will be affected by drought and desertification in 2030 [19]. In parallel to the increase in drought and desertification, food demands are also increasing faster due to rapid global population growth. More importantly, agricultural water management is essential to supply water in precise quantities at required locations. In addition, self-awareness is required among the farming communities regarding water usage and preserving freshwater through smart irrigation systems. At the same time, excessive or insufficient water supply to the crops would significantly affect the yield. However, few conventional methods, like drip and sprinkler irrigation systems, can significantly deal with the water shortage. Therefore, the estimation of accurate water demand for a specific crop depends on the type of crop, irrigation system, climatic conditions, and soil properties.

In order to optimally use water and maintain crop health, an accurate and precise moisture control system for soil and air is mandatory. Moisture control systems with sensors and wireless networks could suffice those mentioned above. It is necessary to adopt IoT technologies in irrigation systems, to monitor and control efficient water usage by crop water stress index (CWSI) and maintain the highest crop yields that increase crop efficiency [20]. To precisely calculate CWSI, measuring the atmospheric air temperature and humidity is essential. Installing appropriate sensors at various field locations and retrieving data from different sensors through wireless sensor networks can easily access soil and air moisture content. Finally, the collected data is transmitted to the data processing center for further analysis using in silico techniques. The information from the satellites and weather reports can be included along with the data collected at fields to calculate CWSI for water requirement and site-based irrigation index value [21]. However, UAVs with various sensors and digital cameras can help identify the locations under water stress and implement appropriate irrigation systems in affected locations. Recently, a fixed-wing drone has been developed to retrieve multispectral image data to identify and determine the physiological changes in citrus crops [22]. Therefore, it is used to determine the water-stressed locations and to provide precise quantities of water to those areas, thereby saving water efficiently.

6.2.3 Smart fertilizer system

A fertilizer is a substance that provides essential nutrients to the plants for better growth and yields, which may be naturally available, or chemically synthesized. The most important nutrients for plant growth are nitrogen (N), phosphorous (P), and potassium (K), which support leaf growth, development of roots, flowers and fruit, and stem growth, respectively [23]. On the contrary, deficiency or improper application of these nutrients negatively impacts plant health. Indeed, the overuse of these agrochemicals could lead to economic loss and also threaten the soil quality and the environment. As a result of soil and environmental degradation, the climatic conditions of the specific location might be affected due to the agrochemicals emitted into the environment. The crops can absorb only half the amount of nitrogen-enriched fertilizer while the rest remains as residues in soil or get emitted into the atmosphere. Around 80% of global deforestation is because of the improper usage of agrochemicals in agricultural practices resulting in climate change [24]. Of course, the technology involved in smart farming accurately determines the required quantity of nutrients for enhanced plant growth without affecting the ecosystem negatively. For accurate soil nutrient measurement, it is essential to measure the soil and crop type, nutrient absorption ability of soil, type of fertilizer and its utilization rate, and weather conditions. The traditional soil nutrient estimation is expensive and time-consuming, as it is required to make a site-specific measurement.

Eventually, the incorporation of IoT in smart fertilization has simplified the spatial mapping of soil nutrients with greater accuracy and minimum labor involvement [25, 26]. The system monitors and controls soil nutrient status by automatically computing the normalised difference vegetation index (NDVI) obtained from the image data reported by satellites or drones. The images are obtained from the reflection of visible light or near IR radiation from the crop for evaluating the crop health status, vitality, and density for analyzing the soil nutrient. Undoubtedly, the accurate determination of soil nutrients in smart agriculture help to assess the precise usage of fertilizers, thereby preventing soil and ecosystem degradation [27, 28]. Indeed, the efficiency of smart fertilization is improving with the incorporation of emerging technologies such as high-performing GPS systems, geo-mapping, VRTs, and UAVs that support IoT for data collection [29]. Conventionally, irrigation systems are used for the dispersal of water-soluble fertilizers and pesticides. Fertigation and chemigation are the methods through which farmers apply fertilizer and pesticides before the emergence of smart fertilization techniques. IoTs outsmarting these conventional methods, have been evident in recent field evaluations of smart fertilization systems [30]. Furthermore, fertigation integrated with IoT system is a best practice for fertilizer and pesticide management as its effectiveness and outcomes increase agricultural yields. However, UAVs spraying fertilizers on crops are more critical and efficient than conventional procedures. Therefore, fertilizer dispersal or application using UAVs is highly targeted with computerized control, reducing the overall cost.

6.2.4 Smart disease control and pest management

In 1950s, "potato blight" a crop disease caused Irish potato famine, leading to the death of one million people [31]. It was also reported that there has been a huge economic loss faced by farmers of USA and southern Canada due to "corn leaf blight" disease [32]. The recent estimations by Food and Agriculture Organization (FAO) confirms about 20–40% of global yield loss is due to the inefficient management of pest and diseases [33]. Conversely, pesticides and other agrochemicals play a vital role in controlling the pests and pathogens to avoid the yield loss. Unfortunately, about half a million tons of pesticides are in use only in United states and around two million tons of pesticides are used worldwide [34]. As a result of pesticide overusage, human and animal health is under threat as well as causing irreparable loss to the ecosystem [35]. However, the emerging technologies such as IoT-based devices, wireless sensor networks, autonomous robots, and unmanned aerial vehicles facilitate the farmers to accurately spot the pests or pathogens and apply required quantities of agrochemicals. Therefore, smart pest control in precision farming involves continuous monitoring of pest, prediction models, disease forecast, which help to implement efficient pest management [36].

In general, the efficient pest control and disease monitoring is mainly relying on the accurate sensing, determining the pesticide requirement and precise application. The recent advancements in the image processing methods have been incorporated for acquiring field and crop images using imaging sensors, drones, and remote satellites. Undoubtedly, the satellites can cover a wide area providing greater efficiency with less expensive monitoring. The sensors mounted on the fields helps in data collection for environmental sampling, crop health status and pest status, even in the smaller areas time to time. The recent advancement of IoT- based automatic traps are capable of capturing, counting, and characterizing the pest type. Further, the data could be uploaded in cloud for pest analysis, that is not possible with remote sensing satellites [37]. More importantly, the precise sprayers using vehicles and automated VRT chemigation are used for efficient pest control and disease management. Furthermore, the autonomous robot equipped with hyperspec-tral sensing devices and precise sprayers can easily locate and control the pest more efficiently than the remote sensing systems. Similarly, the pest management with inclusion of IoT technology offer several benefits such as reducing overall cost, and negative impacts to the environment. Especially, UAVs mounted with visible, and infrared (IR) light sensors can detect the plants affected by bacteria or fungal diseases frequently with better flexibility. However, the early-stage detection of such diseases helps to control the further spread of diseases in the crop areas. Therefore, evaluation of the plant health status using the data collected from various sensors for instance, chlorophyll meter, water potential meter, and spectroradiometer efficiently supports crop management.

6.2.5 Smart monitoring of crop yields and climatic conditions

Indeed, monitoring the crop yields by analyzing various parameters such as mass flow of grains, moisture content and its harvested quantity helps to eval-uate the crop performance. The ease of collecting the required data anytime at a given spatiotemporal conditions could further predict the necessary reme-dial actions upon crop damage. In contrast to yield monitoring during harvest, it is also important to measure the yield quality that relies on various param-eters. For instance, such as selection of good-quality pollens for pollination while predicting the yields of seeds for dynamic weather conditions [38, 39]. In present scenario, the global market consumers are more concerned about the quality confirming that the farmers should attain an effective yield at the right time with best quality. On the other hand, crop forecasting plays an important role in smart agriculture which predicts the crop production even before the harvest. This information is helpful to the farmers for planning and decision- making. Essentially, the analysis of crop quality and maturity are other important factors to decide the right time for harvesting. The important factors that dictate yield monitoring involves certain phenotypic determinants such as crop or fruit size, color, etc., Moreover, deciding the right time for harvesting using yield monitoring, not only helps to enhance the quality, also the quantity. Furthermore, a yield monitor integrated with mobile app has been mounted on the harvester that captures live harvesting data and transfers it automatically to the cloud-based platform. Interestingly, the yield maps generated by the yield monitoring system are shared with the manufacturers, farmers, and the agricultural experts for further analysis. In order to assess the yield qualitatively and quantitatively, the maturity of crop and the fruit is supposed to be measured. Recently, a fruit growth monitoring system has been developed for this purpose [40] and a satellite Sentinel-1A interferometric images are used for rice crop monitoring in Myanmar. Likewise, the fruit size is measured to determine its maturity using RGB color images for mangoes [41], and multi-sensors of optical type are used for papaya for shrink measurement [42]. Alternatively, covering large areas for crop monitoring is quite challenging. In such cases, the introduction of drones offers real-time monitoring of large farms and turnout to be more economic than satellite images. Significantly, the nutrient levels, moisture level, and other related parameters can be observed using recently developed microdrones +m. However, for assessing the relation between the fertilizer availability and crop spectral characteristics, a UAV with digital camera has been developed. Therefore, smart agriculture has eased impossible and labor-intensive procedures involved in monitoring large areas for crop management. A novel procedure has been implemented in crop monitoring that relies on computation and mapping of 3-dimensional characterization of trees. This procedure facilitates to access the growth of trees and other field-related parameters for efficient crop management.

6.3 Technologies Involved in Precision Agriculture

In contrast to traditional agriculture, most of the equipment used in modern agricultural practices are large and heavy. For instance, tractors, harvesters, and agricultural robots mounted with remote sensor system and associated wireless technologies. Whereas, in smart agriculture, the seed sowing, fertilization, water irrigation, and crop harvesting are performed by vehicles integrated with sensing technology with GPS and GIS facilities. Consequently integration of technologies help to perform regular agricultural practices accurately and autonomously within specific location. Notably, the performance enhancement in smart agriculture greatly depends on accuracy and precise data collection. The data collection can be done with devices such as remote satellites, drones, balloons, and the other methods by mounting different type of sensors at several locations of the farming field. As the collected data is identified by its location with the help of GPS system, the necessary steps and precautionary measures can be taken efficiently and effectively as illustrated in Figure 6.4. Agriculture has been highly commercialized due to the demand and creation of global markets, that attracts many leading corporations and investors. Thus, the yields with good quality and quantity have become an important factor to sustain in the market. The well-organized monitoring and control of various parameters involved in agricultural practices helps the manufactures with sustainable market. Similarly, automation and control of every procedure in agriculture can be supported by recent developments in the IoT technologies. Of course, integration of these latest technologies in agriculture provides more sophisticated equipment such as autonomous multifunctional robots that performs sowing, irrigation, fertilization, chemigation, picking, and transportation. Smart agriculture is revolutionized not only due to emerging of latest technologies but the high demand for food production in the global market due to degradation of resources. Recent data suggests that the development of smart agriculture at global level will increase the growth rate by 19.3% every year [43]. In addition to that, the drones support the smart agricultural system to achieve highest revenue among the different equipment. The highest crop yields and market growth can be achieved by the inclusion of latest automation and communication technologies in the agricultural practices. However, the performance of sensing and communication technology is being updated often by the manufacturers for its utilization in better farming. Therefore, the recent and most widely used technologies available in the market and their implementation in smart agriculture are discussed in the following subsections.

6.3.1 Wireless sensing technology in agriculture

In particular, the crop conditions are monitored from the remote locations through the wireless technologies that play an important role in the data collection from different sensors in crop monitoring as depicted in Figure 6.5. Based on the requirement, the sensors can be placed as stand-alone devices or can be mounted on the tools or equipment such as tractors and autonomous robots, etc. The commonly used sensor systems in the smart agriculture are discussed here.

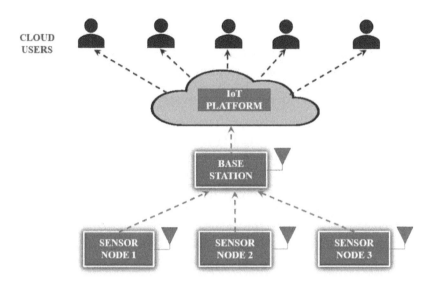

Figure 6.4   Smart agriculture based on wireless technology.

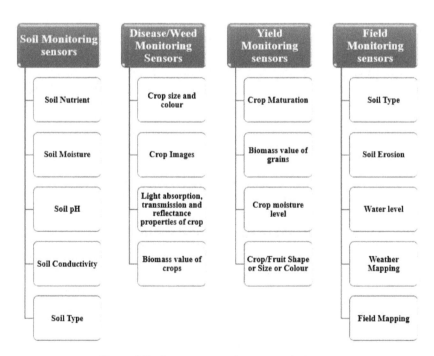

Figure 6.5   Sensor systems in smart agriculture.

6.3.1.1 Soil monitoring sensors

Generally, the sensors of mechanical type measure the soil mechanical resistance with the use of pressure unit that determines the soil compaction. For instance, the density variation of soil due to the applied stress that displaces air gaps between the soil grains. It basically measures the force need to cut into the soil using a tool, and the force is assessed with strain gauges or load cells [44]. On the other hand, the electrochemical types of sensors are used to analyze the soil properties based on its pH and other nutrient levels. In comparison to other sensors for measuring pH and soil nutrients, electromechanical sensors are simple to operate, and cheaper. In order to accurately measure soil salination, the micro and macro nutrients and pH, various electromechanical sensors are available [45]. Moreover, optical sensors are used to assess the organic matter present in the soil by using light reflectance principle. It measures the soil moisture level, soil type, color, mineral composition, etc., The basic principle of measurement is to verify the reflecting ability of soil for different regions of electromagnetic spectrum. Based on the reflection of specific region of spectrum and intensity, various parameters of soil are measured [46]. Additionally, the air flow sensors can measure the permeability of the soil, its moisture level and structure to identify the soil type. Sensors can measure the parameters in a fixed location or during motions with mobile mode. The output is the required pressure to push the known volume of air into the ground at a fixed depth. More importantly, this helps to determine the soil type with its structure, density, and moisture level [47]. The ability of soil for conducting or accumulating electricity is measured using contact or non-contact type of electromagnetic sensors. However, it not only measures the conductivity but also record electromagnetic response and electrical response dynamically in real-time scenario. Therefore, these kind of sensors are used to measure excessive usage of fertilizers and soil organic matter [48].

6.3.1.2 Yield monitoring sensors

Acoustic sensors are economic, fast in response, and portable have wide application in yield monitoring. It can be used to measure parameters during cultivation, management, and harvesting. It works based on the noise level when the tool or equipment interacts with the materials such as soil particles [49]. Field- programmable gate array (FPGA) sensors are also used to monitor plant irrigation and environmental parameters. Since, the high-performing FPGA sensors are bigger in size, expensive, and consumes more power, it is not suitable for continuous monitoring in the smart agriculture [50]. Notably, the optical sensors-based fluorescence principle is used to monitor the fruit maturity, and further characterization of the olive canopies or similar kind of crops by integrating with microwave scattering [51]. Conversely, the mass flow sensors are most widely utilized for yield monitoring by measuring the mass flow rate of the grains. Furthermore, the mass flow sensor is equipped with the harvester along with other modules such as sensors for measuring grain moisture level, devices for data storage, and analyzing software. However, the sensors system and its associated interfacing devices are mounted on tractor during the time of harvest. Therefore, the sensor could measure the necessary parameters while passing through the field real-time [52].

6.3.1.3 Weeds, pest, and disease monitoring sensors

Acoustic sensors can also be used for the pest detection based on the principle of sound absorption spectra. The ultrasonic sensors are used to detect the weeds by measuring the plant heights with the integration of cameras [53]. On the other hand, optoelectronic sensors can detect weeds in the large farm area by differentiating the plant type from the other plants and herbicides. Upon combining the sensor with location data, the weed spread in the field can be mapped, which can differentiate soil and the vegetation using the reflectance of light [54]. In general, the photographic RGB images with digital cameras are significant tools to assess the plant health status. Alternatively, video scanners or cameras are the other methods to identify plant diseases using digital images. However, angle of imaging and resolution of imaging sensors play an important role in accurate identification of diseases. Therefore, multi- and hyperspectral based on reflection principle are used to detect the affected crops by bacterial and fungal diseases. The hyperspectral image data is observed with spatial matrices in x, y, and z axes, and the spatial resolution is strongly depending on the distance between sensor and the crop. Furthermore, the optical properties of leaf such as transmission, reflection, and absorption of light are used to differentiate the healthy and disease-affected plants. Interestingly, the thermal sensors are used to detect the affected plants by determining the plant temperature using infrared thermography (IRT). However, the temperature of plant is correlated with water content in the leaves, microclimate in crop strands, and transpiration that changes due to early stage of diseases. Therefore, various fluorescence sensors are utilized to identify the diseases by assessing its photosynthetic process. Mainly, the fluorescence imaging sensors measure chlorophyll content in the plants using LED or laser light. Undoubtedly, the sensor differentiates the photosynthetic activity of healthy plants and the plants affected by pathogens. Recently, some non-invasive sensors are used to measure the biomass values of crop to identify the plant diseases. However, the density of the crop is one of the significant parameters to assess the health status and for the application of fungicides. Therefore, stereo cameras and 3D scanners are used to obtain data of plant diseases based on reflectance intensity [55].

6.3.1.4 Field monitoring sensors

Light detection and ranging (LIDAR) sensors are widely used to measure parameters such as field characteristics, type of soil, field mapping, soil erosion, and forecasting the yield for harvesting. By integrating with GPS, it can also be used to monitor and map the different parts of plant time to time and the biomass value of crops and trees [56]. The eddy covariance-based sensors are used to determine the exchange rate of CO_2, water vapor, methane, and various gases. By measuring the difference between the earth surface energy and the atmospheric energy the trace of gas fluxes around the fields can be estimated. As these types of sensors are capable of continuously measuring the gas fluxes precisely, it is widely used for large field areas [57]. Soft water level sensors are used for characterizing and mapping water level and flow dynamically, which is determined by rain fall measurement, flow rate of streams, and other water resources, that helps for irrigation management [58]. Telematics sensors are used for data collection with location information from the remote areas and enables communication between machines at remote locations or vehicles. This helps to record all kind of information about the farming field automatically so that the resources can be effectively utilized and also enhances the security of the yields [59]. Argos sensor is used for collection of environmental data worldwide with the use of satellite communication. For reporting telemetry data, automatic packet reporting system is being integrated with this satellite-based sensor [60]. Remote sensing system can also be used for capturing and storing the geographical data for weather and spatial mapping which can be used further for analysis and decision-making in smart agriculture. The ultrasonic sensors are also widely used in smart agriculture, as it can measure various parameters, easy to operate, inexpensive, and adaptable in several applications. The ultrasonic sensor also measures storage tank level, spray distance of agrochemicals, crop maturation, collision detection, etc. [61]. LIDAR sensors also measure various parameters such as crop growth status, in yield modeling and monitoring, weed and pest detection, field mapping, etc. [62]. The working of various sensors and its application in smart agriculture is listed in Table 6.1.

Table 6.1   List of various sensors used in smart agriculture(continued).

Work	Sensor type	Working principle	Pros and cons
[44]–[48]	Soil monitoring sensors: i. Soil mechanical resistance sensor ii. Soil pH and nutrient sensor iii. Multi-parameter sensor iv. Soil permeability sensor v. Soil conductivity sensor	i. Mechanical type works with pressure unit such as strain gage or load cell. ii. To measure the soil salination and nutrients, electrochemical type sensor is used. iii. Optical type sensor based on reflecting ability of soil for different regions of electromagnetic spectrum. iv. The air flow sensors with the output as required pressure to push the known volume of air into the ground at a fixed depth of the soil. v. Electromagnetic sensors can measure not only the conductivity and record electromagnetic response and electrical response dynamically in the real-time scenario.	i. Cheaper, easy to operate, but not accurate. ii. Simple to operate, cheaper and macro and micronutrients are accurately measured. iii. Can measure soil type, moisture level, color, mineral composition, etc. iv. Can measure the parameters in a fixed location or during motions. v. Can also measure residues due to excessive usage of fertilizers and organic matter contained in the soil.
[49]–[52]	Yield monitoring sensors: i. Cultivation monitoring sensor ii. Irrigation monitoring sensor iii. Crop maturity monitoring sensor iv. Crop harvesting sensor	i. Acoustic sensors work based on the noise level when the tool or equipment interacts with the materials such as soil particles. ii. FPGA sensors, monitors plant irrigation and environmental parameters. iii. Optical sensors-based fluorescence principle is used to monitor the fruit maturity. iv. Mass flow sensors measuring the mass flow rate of the grains.	i. Less cost, faster response, and portable. ii. Advanced FPGA sensors are bigger in size, expensive, and consumes more power; not suitable for continuous monitoring. iii. Accurate, and also used to characterize the olive canopies and similar kind of crops by integrating with microwave scattering. iv. Along with other modules, mounted on tractor during harvesting which measure the necessary parameters while passing through the field.

Table 6.1 List of various sensors used in smart agriculture

[53]–[55	Weeds, pest, and disease monitoring sensors: i. Pest detection sensors ii. Weed detection sensors iii. Plant health monitoring and disease detection sensors	i. Acoustic sensors based on the principle of sound absorption spectra. ii. Optoelectronic sensors differentiate the plant type from the other plants and herbicides. iii. The optical and imaging sensors use properties of leaf such as transmission, reflection, and absorption of light; thermal sensors determine the plant temperature using infrared thermography (IRT); fluorescence sensors assess the chlorophyll content in the plants; non-invasive sensors measure the biomass values to differentiate healthy and diseased plants.	i. Simple to operate, cheaper. ii. Accurate but expensive. iii. Accurate, but expensive and interfacing modules required.
[56]–[59]	Field monitoring sensors: i. Field parameter monitoring sensors ii. Gas monitoring sensor iii. Water level monitoring sensor iv. Field security sensors	i. Light detection and ranging (LIDAR) sensors measure parameters type of soil, field mapping, soil and to forecast the yield for harvesting by integrating with GPS. ii. The eddy covariance-based sensors determine the exchange rate of CO_2, water vapor, methane, and various gases. iii. Soft water level sensors characterize and mapping water level and flow dynamically iv. Telematics sensors are used for data collection with location information from the remote areas and enables communication between machines at remote locations or vehicles.	i. Accurate measurements, but expensive.

6.3.2 Communication methods and latest technologies in smart agriculture

The primary support of smart agriculture for data collection and analysis is provided by the recent advancements in the communication technologies. It is also essential to make communication methods as application specific, reliable, and highly secured among various objects involved. The telecom operators play a vital role for the security of data communication in the smart agriculture. When it is applied for large-scale farming with IoT technology, it is necessary to frame the suitable architecture. The communication methods must be wisely chosen by considering the factors such as cost, area coverage, consumption of energy, etc. As the low-energy communication system cannot connect the sensing system from remote locations, based on the requirement and cost, the communication methods are implemented as discussed below.

6.3.2.1 Mobile communication systems

Communication with different generations of mobile networks from 2G to 4G are suitable in smart agriculture, based on suitability, bandwidth value, reliability, and area coverage, especially for the data collection from the remote locations. Other than mobile communications, the same data from remote locations can be collected by satellite communication, but it is comparatively expensive and not suitable for small area farms. The communication method is also chosen based on its requirement, applicability, number of data required, and battery life. Low-power wide area network (LPWAN) provides longer battery life with affordable cost and suitable for crop management [63]. The data communication for shorter and medium area coverage is done by mesh networks where the different sensor nodes collect data from its corresponding locations and transmits to a gateway located in the same area. The gateway or base station sends the data further to the farm monitoring system through wide area network (WAN). The wireless communication between sensor nodes within the mesh network can be done by Zigbee or Bluetooth technologies. Further, the data transmitted to the farm monitoring system helps to predict the weather conditions, product quality, resource utilization, etc. The wireless sensor system and its communication method are decided mainly based on the specific application requirement and power consumption.

6.3.2.2 ZigBee wireless technology

ZigBee is introduced to replace the conventional non-standard technology and based on the requirement of application it is categorized into three types such as coordinator, router, and end user. The three protocols are supported by different network topologies such as star, tree, mesh, and cluster networks [18]. For agricultural application, the ZigBee technology plays a crucial role to transmit the data from the sensor nodes to the end server for smaller range of communications. For irrigation monitoring and precise dispensing of fertilizers, the ZigBee-based communication network is most widely used. For long-distance communication and for transmission of information about field data to the farmers, the GSM module is used and for any shorter-distance communication, Bluetooth technology is used.

6.3.2.3 Bluetooth wireless technology

Bluetooth is one type of wireless communication method between devices at shorter distances. It is most widely applied in smart agriculture systems as it consumes less power, easy to operate, and inexpensive. The study on Bluetooth Low Energy in IoT systems is tested with programmable logic controllers for smart irrigation system by controlling the timer and soil moisture level. This technology helps in optimization of water and energy for different field applications. The moisture sensor based on Bluetooth Low Energy is designed for assessing weather conditions in the fields as it supports accessibility of smart phones [64], [65]. A similar kind of sensor is designed to measure the atmospheric light intensity and temperature using Bluetooth Low Energy technique for IoT-based agriculture management system. The remote monitoring system in smart agriculture is done with Wi-Fi modules where the deployed sensor nodes send data wirelessly to central monitoring system or base station which can be further stored, analyzed, and displayed [66], [67].

6.3.2.4 LoRa and sigFox technology

LoRa is a wireless technology suitable for long-distance communication in IoT applications with less power consumption. As it provides WAN connectivity with less power between the wireless remote sensors and the cloud-based server, it is more effective and reliable than Bluetooth and Wi-Fi technologies. It can penetrate thicker concrete buildings and shielded devices with wide area coverage, and it requires less maintenance with better lifespan [68]. It is also used to measure temperature and humidity conditions in the warehouse, and in grains transportation to ensure food quality [69]. SigFox is another narrow or ultra-narrow band wireless communication network technology based on narrow chunks or narrow spectrum that encodes the data by changing the phase of carrier radio frequency wave. It has the ability to transmit the data from more than 100 sensors simultaneously with high performance [70].

6.3.2.5 Smartphones-based communication

Due to the wide usage of mobile communication even in the rural areas, smart phones can be used as primary communication technology for data acquisition of various parameters in the agriculture. As the recently advanced microfabrication technology resulted in price decrease of smart phones, made the devices more attractive, even the small-scale farmers afford the technology. As the smart phones has multiple functionality such as image capturing with camera, location identification with GPS, inbuilt micro accelerometer, gyroscope sensors, and proximity sensors, it is more flexible and suitable for development of application specifically for smart agriculture [71]. As the smart phones are easy to operate, handy, multifunctional with affordable cost, it helps to adopt the smart agricultural systems even in the developing countries. While developing the mobile applications, it is necessary to consider various associated factors such as farmer needs, ease of operation, cost, market prices, and also the application should support various languages. In addition to that, it is also essential to include the issues related to wide community, not only the farming community while developing agriculture-based applications.

6.3.2.6 Cloud computing

The performance of smart agriculture can be improved with better data communication and decision-making. Thus, improvement in the tools and technology with affordable cost is required so that the decision-making in agriculture will be effective based on the efficient data received from the different wireless sensor systems. The farmers can utilize the database provided by the agricultural experts using cloud services so that they can make an appropriate decision for their crop requirements. As IoT has the drawback of data storage and processing, the features of cloud computing support IoT to overcome the limitations in data communication. The integration of Cloud and IoT technology helps to implement a high-speed communication between the farm monitoring system and the sensor nodes located in remote areas [72]. The database repositories in the cloud contain information regarding agricultural practices, previous experience, and the details of tools and equipment available in the market. The main challenges in the implementation of cloud computing in agriculture are the data format of various sensor systems and it must be application-specific. Thus, the cloud services should support diversity of data and suitable for different kind of applications. CLAY-MIST is a cloud-based service to monitor various parameters related to specific crop, developed to support the farmers for effective decision-making toward better yields [73]. Agri-info is also a service for smart agriculture that combines IoT and cloud computing, is capable of processing data collected from various IoT-based devices deployed at different locations [74]. It also provides suggestions and decision to the farmers to automatically troubleshoot and diagnose the issues in farming. Another cloud-based service called AgJunction is developed to accumulate and analyze the data from different sensor and monitoring systems to take preventive measures for reducing the cost and negative impacts for the ecosystem. To focus on the food and agricultural production with the help of communication technology "Akisai" cloud service was developed and it offers prominent relation between the farming community and commercial markets. A technology called "eService Everywhere" cloud-based service was developed by considering the distance of farming regions and its coverage [75].

6.3.2.7 Fog/edge computing

Fog or edge computing basically offers virtual platform with traditional cloud services between the sensor nodes and the data centers, located on the edge of the communication network. The fog computing can process the data at the edge of the network that gathers data from the IoT-based devices and sensors regarding geographical distribution and hierarchical organizations. It processes the data with any device with storage computing and the network accessibility and supports low-energy wide area networks. It is also important to consider the network security, traffic, and other legal-based challenges while implementing the cloud-based services for agriculture. A three-layer edge computing based-IoT platform was developed to support the greenhouse farming with low-cost tools and devices. A fog computing–based intelligent cyber-physical system is introduced for food tracing [76–79].

6.3.3 Latest technologies for large data processing

6.3.3.1 Big data

As there are several factors to be measured and controlled in smart agricultural systems, IoT technology must deal with huge amount of data

Table 6.2   List of communication technologies used in smart agriculture(continued).

Work	Type of communication	Characteristics	Application range	Pros and cons
[63]	Mobile communication systems	Mobile networks from 2G to 4G, chosen based on its requirement, applicability, number of data required, and battery life.	Long range for remote locations	Low cost, but reliability is the major concern.
[18]	ZigBee wireless technology	Supported by different network topologies such as star, tree, mesh, and cluster networks to transmit the data from the sensor nodes to the end server for smaller range of communications.	Smaller range of communication	For irrigation monitoring and precise dispensing of fertilizers, the ZigBee-based communication network is most widely used.
[64], [65]	Bluetooth wireless technology	Bluetooth Low Energy in IoT systems is used with programmable logic controllers for smart irrigation system by controlling the timer and soil moisture level.	Shorter- distance communication	Consumes less power, easy to operate, and inexpensive.
[68], [70]	LoRa and SigFox technology	LoRa provides WAN connectivity with less power between the wireless remote sensors and the cloud-based server Narrow or ultra-narrow band wireless communication network technology	Long-distance communication Long- distance communication	Less power consumption, more effective and reliable, requires less maintenance with better lifespan Able to transmit the data from more than 100 sensors simultaneously with high performance.
[71]	Smartphones-based communication	Smart mobile phones are more attractive, has multiple functionality such as image capturing with camera, location identification with GPS, inbuilt micro accelerometer, gyroscope sensors, and proximity sensors.	Long-distance communication	Easy to operate, handy, multifunctional with affordable cost and more flexible.

Table 6.2   List of communication technologies used in smart agriculture

Work	Type of communication	Characteristics	Application range	Pros and cons
[72]	Cloud computing	The integration of Cloud and IoT technology helps to implement a high-speed communication between the farm monitoring system and the sensor nodes located in remote areas.	Long-distance communication	Affordable cost, application specific but the cloud services should support diversity of data.
[76]–[79]	Fog/edge computing	Offers virtual platform with traditional cloud services between the sensor nodes and the data centers.	Long-distance communication	Processes the data with any device with storage computing and the network accessibility and supports low-energy wide area networks.

from various sensors, devices, and autonomous vehicles etc. To process, analyzed and for decision making, data analytics can be used to create useful information from these data for food safety and its sustainability. An IoT framework was developed for the efficient working of smart farming model. It analyzes sample data from 10 sensors with less computation time and memory utilization and better accuracy [80]. Analyzing huge data to retrieve useful information for decision-making is a quite difficult process in big data analytics. A system is introduced to monitor environmental data from any farming field received from IoT devices, and further the useful information is received using data mining to increase the productivity. It helps to find the suitable temperature and humidity conditions for a specific crop. With the help of developed IoT system, the suitable humidity conditions for high productivity of lemons was found within 72% to 81%. [81]. Another intelligent IoT-based surveillance system was introduced for environmental monitoring in agriculture. The collected data from IoT was analyzed using 3D cluster method [82]. For intelligent agriculture, a decision support system was designed to make effective decisions in agricultural practices based on the data obtained from sensors and meteorological database [83].

Predictive analysis is also useful in smart farming as it predicts the future outcomes using previous data and statistical models. A framework AgriPredition was developed with IoT technology and predictive engine which estimates the malfunctions in the crop growth at an early stage and sends the information to farmers for effective solution without any delay [84]. An AI-based prediction engine was introduced for processing large-scale data, where the results shows that the efficiency of computation has been decreased while improving the accuracy [85]. An IoT compatible frost prediction engine was established to gather environmental data for the prediction of frost events. The algorithm was evaluated with training regression and classification patterns by using multiple machine learning algorithms [86]. Visualization is the most important tool in analysis of datasets. Compared to numerical data, the visual image dataset provides better impression of results, which enables the user to verify its correctness and display in a smart manner.

6.3.3.2 Artificial intelligence

Artificial intelligence, inspired from the working of biological neural networks, helps to enhance the efficiency of various sectors including agriculture. AI supports the farming community, to find effective solutions for complicated issues in agriculture such as disease detection in crops, crop storage, management of agrochemicals, weed control, and water management [87]. An intelligent semiconductor manufacturing based on computational scalability and AI algorithms were analyzed in detail. A dynamic algorithm was also proposed for acquiring valuable information on the manufacturing process of semiconductors which can help to address various challenges involved [88]. A distributed clustering protocol based on fuzzy logic was proposed for intelligent wireless sensor network, used for monitoring various conditions in agricultural systems. The main aim of the system is to enhance the energy efficiency and coverage area in wireless sensor networks. The protocol efficiently balances the energy usage among the sensor nodes [89].

6.3.3.3 Deep learning

Deep learning has been successfully applied in various fields including agriculture. A deep learning model, PestNet was developed to identify and categorize the distribution and type of pests. It is done by evaluating the set of pest image data using suitable image processing technique [90]. A reinforcement learning-based smart farming system was introduced for effective decision- making in water management system to monitor the environment for better crop growth [91]. An expert system was proposed for farmer's assistance to assess the field conditions for cultivation based on the information collected by various sensor systems along with artificial neural networks and multilayer perceptron with an accuracy of 99% [92]. Apple leaf disease detection was proposed with convolutional neural network (CNN)-based deep learning model. It showed a detection accuracy of 78.80% [93]. A CNN-based plant seedling classification model was developed with the dataset comprises of 5000 images of 960 different plants belong to 12 species and it achieves the prediction with 99.48% [94].

Convolutional neural network (CNN) is a subdivision of artificial neural network (ANN) that takes the advantages of the spatial data for the given inputs. Recently, CNN-based neural network model has attracted much research for object or pattern recognition and classification, face identification, offering vision to robots, and in self-driven vehicles. It has a standard structure that comprises of convolutional and pooling layers alternatively. The fully connected layer is the last layer of the model and a SoftMax classifier is also in the final layer as depicted in Figure 6.6. The input data is transformed into the output data volume for neuron activation toward the last layer that maps the given input image data to a one-dimensional feature vector [95].

The CNN-based deep learning model is widely used for accurate and precise disease detection in smart agriculture, so that the disease spread can be controlled at an early stage. Several researches have been carried out to find the diseases in the crops using CNN approach and few works are listed in Table 6.3.

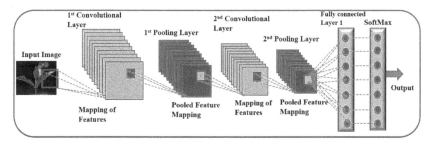

Figure 6.6 Architecture of convolutional neural network (CNN)-based deep learning model.

Table 6.3 List of research works carried out on plant disease detection using deep learning methods(continued).

Research work	Year	Deep learning method	Diseases identified	Crops studied
[96]	2019	A global pooling dilated CNN (GPDCNN) based deep learning model is proposed to identify 6 common cucumber diseases with totally 600 diseased images.	Black spot, downy mildew, gray mold, anthracnose, angular leaf spot, and powdery mildew.	Cucumber crop
[97]	2020	A performance improved algorithm YoLo V3 CNN is developed for identification of tomato diseases and pest with 93.39% accuracy and 20.39 ms detection time.	Early blight, late blight, gray mold, leaf mold, leaf miner, whitefly.	Tomato
[98]	2021	An improved faster R-CNN-based deep learning system is modeled to differentiate weeds and crops in strawberry and pea fields. The accuracy of weed identification is 94.37%.	Weeds detection	Pea and Strawberry fields
[99]	2021	A hybrid model based on convolutional auto encoder (CAE) and convolutional neural network (CNN) for bacterial spot disease in the peach plants. Totally, 9914 parameters are used and the training and testing accuracy of 99.35% and 98.38% has been achieved in disease identification respectively.	Bacterial spot disease	Peach plants
[100]	2021	A novel CNN-based model for rice leaf disease detection is developed with the dataset of 4199 rice leaf disease images. The model achieved training accuracy of 99.78% training accuracy and 97.35 testing accuracy.	Blast, brownspot, sheath blight, bacterial leaf blight, and tungro	Rice crop
[101]	2021	A deep learning model with submodels of DenseNet-121, SE-ResNet-50, ResNeSt-50 with the dataset of 33,026 images to identify the rice leaf diseases with 91% accuracy.	Rice leaf blast, Rice bacterial stripe, Rice false smut, Rice brown spot Rice neck blast, and Rice sheath blight.	Rice crop

Table 6.3 List of research works carried out on plant disease detection using deep learning methods

[102]	2021	A novel CNN-based deep learning model is developed to identify wheat disease with the training accuracy of 97.88% and testing accuracy of 98.62%.	Wheat loose smut, fusarium head blight, tan spot, powdery mildew, leaf rust, karnal bunt crown and root rot, black chaff, and wheat streak mosaic.	Wheat crop
[103]	2021	An MAF module for optimization of mainstream CNN-based deep learning method to identify maize leaf disease with an accuracy of 97.41%.	Maculopathy, rust, and blight	Maize crop
[104]	2021	An improved attention-based CNN deep learning model to identify apple leaf disease with 81,700 diseases images and recognition accuracy of 98.92% is achieved.	Mosaic, black rot, scab, rust, glomerella leaf spot, apple leaf mites, apple litura moth.	Apple
[105]	2021	An automated system using mask region-based CNN deep learning model (Mask R-CNN) for detection of potato leaf diseases with 1423 images and an accuracy of 98%.	Blight disease	Potato

6.4 Autonomous Vehicles in Smart Agriculture

6.4.1 IoT-based tractors

As the labor resources for cropping and harvesting is substantially reducing, it is essential to include heavy machineries and vehicles in the agriculture sector which can work almost 40 times faster than the labor with less cost [106]. With the emerging technological advancements, the manufacturers offer tractors equipped with auto-driven and cloud computing facilities. While comparing with labor personnel, it helps to avoid overlaps in the field area with better precision and accuracy particularly during the spray process of agrochemicals to maintain the nutrients level or to control the pests. Thus, it attracts and motivates many researchers to develop fully automatic tractors for future demands. It is also estimated that there would be a demand of around 700,000 tractors with autosteer or tractor assistance in 2028 and around 40,000 fully automated tractors in 2038 [107]. A low-cost monitoring device is developed by the manufacturer, Hello Tractor, which can be placed on the tractor with associated software and tools. This helps the farmers to implement automation in the tractors for crop monitoring with affordable cost. It also forms a communication between the farmers and the manufacturer for services [108]. A tractor equipped with video cameras and LIDAR sensor systems to detect the object for avoiding collisions. These autonomous tractors were tested during the plantation of soybeans [109]. A tractor connected to a car was developed in France by standards group ETSI, for avoiding collisions and accidents due to farming vehicles [110]. A local analytics engine was proposed to work on the farmer's tractor to adjust the local input data instead of working in the cloud. All the analytics and suggestions are considered to modify the existing data in real time based on the field conditions. It enables the system for tractors to connect it with the internet and create a model for displaying the information to farmers [111].

6.4.2 Harvesting robots

Harvesting is one of the crucial parts of the agricultural production which defines the crop yields and efficiency. The harvesting process can be a single time or on daily basis depending upon the type of the crop. To decide the right time for harvesting is very critical, else it can greatly affect the production. Due to shortage of labor, there is a huge decline in the production of crops and, the major cost is invested in labor-intensive farms. To overcome these issues, the agricultural experts anticipate the contribution of agriculture robots. It not only reduces the labor pressure but also offers flexibility for harvesting. As the inclusion of robotics makes the harvesting process more precise, it attracts the farm experts in the recent decades. Considering the demand, many researchers have done their research on robotics to detect the fruit maturation with its shape, size, and color [112], [113]. Highly performing advanced sensors collect the information about a specific fruit of crop which helps for automatic harvesting. As the fruits are overlapped partially or fully under the leaves, branches, and other fruits, it is not easy to detect the right target for harvesting [114]. Most of the robotics system monitors and involves in the agricultural practices based on the various sophisticated tools such as computer vision, image processing, and the deep or machine learning techniques.

This technique needs specific tools and advanced sensor systems to identify the fruit conditions for harvesting as there are different shapes, sizes, and colors for a single-fruit variety. For these complicated tasks, several robots are developed in smart agricultural research. Some of the developed robots are SW6010 for smart harvesting, Octinion for strawberries, SWEEPER for pepper harvesting, and FFRobot for fruits to pick around 10,000 fruits in an hour [115]–[118]. As the strawberries are consumed fruits almost throughout the year, investment on labor is high during harvesting and packaging procedures and grown mostly in the controlled environment such as green houses, the robots are developed to move in a defined path within the green house for automatic harvesting. Tektu T-100 robot is developed for strawberry harvesting with electric rechargeable batteries and also with zero noise and emission inside the tunnels. The pickers mounted on the robot can be positioned over the plant rows and collect the fruits efficiently in short time [119].

6.4.3 UAVs in agriculture

The IoT has shown the remarkable success in many fields, but the implementation in agriculture requires better communication facilities such as base stations or Wi-Fi, which are not sufficiently available and limits the growth of IoT in agriculture sector. These facilities are still lesser in developing countries and the remote villages. The parameters measured by wireless sensors cannot be transmitted with insufficient and unreliable communication facilities. In such cases, an unmanned aerial vehicle (UAV) provides an alternative solution, and it communicates with the sensors deployed over the wide farming areas and acquires data for further analysis. The UAVs are also called as drones equipped with high resolution cameras and sensor systems, which can fly over large area of fields [120]. Thus, the UAVs provide a real-time, cheaper, faster surveillance for large scale with a sophisticated data acquisition and communication system for crop production. Satellites and airplanes options are used to obtain the macroscopic view of aerial landscape images, but it has limited quality and resolution that cannot provide the effective information for decision- making. With the use of these techniques, it is also difficult to collect the image data frequently based on the requirement and affected by bad weather [121]. The above-mentioned issues can be eliminated with the use of UAVs that takes better-quality images with the camera attached. It helps to acquire microscopic view of images to assess crop conditions, field mapping, and disease detection, which provides timely suggestions to the farmers for necessary steps toward better quality yields. The data collection by UAVs are least affected by weather conditions except rain and collects images multiple times. Due to the better performance, the UAVs play a vital role in the implementation of precision agriculture and create highest revenue compared to the other agricultural robots. The drones represent a unique way of collecting the field level data whenever needed. It provides a real-time procedure for a farmer to obtain the knowledge of field conditions at that moment with appropriate hardware and software technologies.

The agricultural drones are categorized as fixed-wing and multi-rotor drones, available in different ranges based on cost, payload capacity, and hardware facilities [122]. As the fixed-wing drones are crash-tolerant and have long range flight capacity, it has wide field area coverage, for example, senseFly's eBee SQ and DATAhawk [123], [124]. Multi-rotor drones are easy to operate, faster setup, vertical take-off and landing, and wind planning is not required. It is considered as a better choice where the detailed images are required at low altitude, for example, DJI Matrice 200 [125]. The thermal- and hyperspectral imaging sensors are mostly used in the drones to collect information about the soil or plants. Water quantity is measured by the thermal sensors in which the leaves with more water appear cooler. Near infrared (NIR) sensors are also working with same principle, to differentiate NIR reflectance and visible reflectance, that is, NDVI [126]. Hyperspectral imaging sensors differentiate weeds and crops by measuring the wavelength of reflected lights [127]. Due to their wide range of applications, and high performance, UAVs are used in smart agriculture for monitoring of crop health status, planting, and dispensing agrochemicals, etc. With the use of automation and GPS facilities, it has the ability to transform the agricultural practices to the modernized level. Some of the important factors to be considered before the implementation of drones are payload and battery utilization. Though the UAVs have wide range of applications and flexibility in agriculture, it must carry heavy loads while spraying the pesticides and fertilizers. In such situations, the battery utilization must be optimal to extend the fly time and to cover the large area. Thus the optimization of payload and choosing suitable weather conditions or wind directions are important for uninterruptible flying of drones. To overcome the limitations, tethering connection can be used to supply power through longer cables to fly for longer distance so that it does not need to carry heavy batteries.

6.5 Challenges in Smart Agriculture

As listed in Table 6.1, the sensors systems mounted around the agricultural fields provide updated and timely information to the farmers that help them adapt required practices for proper crop growth. The wireless sensor system can also send the information regarding diseased plants to agricultural experts for analysis. With advanced image identification and digital image processing techniques, the plant health status is monitored from remote locations, but the resolution of images and time-to-time update of plant status with an advanced sensor system with affordable cost needs to be addressed still. The communication technologies, as listed in Table 6.2, create a broad scope for optimized production in agriculture. It can adapt to recent technologies such as cloud computing, fog/edge computing, Big Data analytics, and artificial intelligence. These technologies greatly influence smart agriculture. On the other hand, it has several difficulties in real-time implementation. The hardware devices of the technologies implemented in agriculture are directly affected by exposure to sunlight, wind, humidity, and other natural calamities. The devices should be robust and reliable with low battery power consumption over a long time. The IoT system implementation in the large open fields requires many sensors to monitor climate, crop conditions, yield conditions, etc. Advanced sensor systems and drones can collect different data from the fields. The interoperability for data communication between structures such as gateways, service-oriented architecture, and open application programming interfaces is to be improved with a standard framework for IoT-based agriculture. Wireless networks are inexpensive and flexible compared to wired networks. It is still challenging in smart agriculture. Due to the timely changes in the fields, the background noises and interferences affect the reliability of data communication. If a particular sensor node is disconnected, which is responsible for routing the transmission, it can cause the entire network shutdown.

Power drain of IoT is another important issue that limits the lifetime. Renewable energy resources such as solar power and wind power can also be utilized to power IoT devices. Wide area coverage is another issue, which can be solved by LPWAN technology, but the installation cost of LPWAN in agriculture is high. Another most concerning factor in smart agriculture is security and privacy issues in data communication. The other related issues such as privacy, authentication and control of access, storage of information, and management are also critical to be considered. Improving the performance of IoT systems with a decreased cost is the main objective of research in agricultural technology. Though the prices of IoT systems have reduced, the sensor's cost is still high in the market. UAVs are most widely used in smart agricultural practices for monitoring plant growth, spraying agrochemicals, and irrigation efficiently and precisely, compared to traditional machinery. With the latest technologies, the UAVs and other autonomous vehicles are equipped with various sensors including 3D imaging cameras, provide farmers with precise information about the crops for efficient management to produce better yields, as listed in Table 6.4. There are several difficulties involved in the implementation of these technologies in agriculture. The latest lightweight sensors can be mounted on the UAVs, but they cannot carry heavy loads, especially while dispensing agrochemicals over a large field area. Along with optimizing payload, it is necessary to concentrate on the power supply given to the autonomous vehicles used in agriculture. The technological advancements have to address the difficulties involved in the existing technologies used for smart agriculture. Most farmers are in remote villages or rural areas in developing countries. They are least capable of utilizing the technological innovation in agriculture due to their less education. Thus, more focus should be given developing farmers' education and ensuring frequent training for technological updates.

6.6 Conclusions and Future Directions

Smart agriculture is a modern procedure to improve crop quality and quantity using the latest technologies such as wireless sensor networks (WSN), internet of things (IoT), cloud computing, big data analytics, artificial intelligence, and deep learning. These technologies help manage and control various agricultural practices for better crop production. The main objective of smart agriculture is to offer support for efficient decision-making in various farming procedures. The emerging technologies are used to monitor the processes such as soil conditions, crop conditions, yield monitoring, and disease management. The various sensor systems are deployed over the fields to measure parameters such as soil properties, crop type, water level, crop moisture

Table 6.4 List of autonomous vehicles and its functions for smart agriculture.

Work	Method	Application	Main objective
[106], [107]	IoT-based tractors	Without labor interventions, it helps to avoid overlaps in the field area with better precision and accuracy particularly during the spray process of agrochemicals and to maintain the nutrients level or to control the pests.	Tractors equipped with auto-driven and cloud computing facilities help the farmers to implement automation in the tractors for crop monitoring with affordable cost.
[112], [113]	Harvesting robots	Reduces the labor pressure and also offers flexibility for harvesting. As the inclusion of robotics makes the harvesting process more precise, it attracts the farm experts in the recent decades.	Detection of fruit maturation with its shape, size, and color for automatic harvesting
[120], [121]	UAVs in agriculture	UAVs provide a real-time, cheaper, faster surveillance for large scale with a sophisticated data acquisition and communication system for crop production.	It communicates with the sensors deployed over the wide farming areas and acquires data for further analysis. The UAVs are also called as drones equipped with high-resolution cameras and sensor systems, which can fly over the large area of fields.

level, environmental temperature, humidity, and plant health status. The measurement data can be stored, displayed, or transmitted through wireless technology such as IoT, supported by cellular communication, cloud computing, fog computing, etc. Artificial intelligence techniques are mainly used to improve the accuracy and precision of the information retrieved from the sensor data so that effective decisions can be taken in agricultural practices. The chapter discussed various technologies involved in smart agriculture, which also presents the challenges in implementing these technologies in real-time. It also discussed the future direction to be focused on reducing the cost of hardware and software technologies, the performance of software algorithms, robustness of the sensors, educating the farmers, etc. It helps the farmers to implement smart agriculture with user-friendly applications of affordable cost. Hence, the highest productivity with better quality can be achieved to address the future food production demands.

References

[1] M. Ayaz, M. Ammad-Uddin, Z. Sharif, A. Mansour, and E.-H. M. Aggoune, "Internet-of-Things (IoT)-Based Smart Agriculture: Toward Making the Fields Talk," IEEE Access, vol. 7, pp. 129551–129583, 2019, doi: 10.1109/ACCESS.2019.2932609.

[2] D. Bhattacharyay, "Future of Precision Agriculture in India," in Protected Cultivation and Smart Agriculture, New Delhi Publishers, 2020.

[3] A. McBratney, B. Whelan, T. Ancev, and J. Bouma, "Future Directions of Precision Agriculture," Precis. Agric., vol. 6, no. 1, pp. 7–23, Feb. 2005, doi: 10.1007/s11119-005-0681-8.

[4] J. C. Doornkamp, "Book reviewsǎ: Howard, J. A. and Mitchell, C. W. 1985; Phytogeomorphology. New York: John Wiley. xiii + 222 pp. £46.00," Prog. Phys. Geogr. Earth Environ., vol. 10, no. 4, pp. 609–610, Dec. 1986, doi: 10.1177/030913338601000417.

[5] H. Kaspar, T.; Colvin, T.; Jaynes, D.; Karlen, D.; James, D.; Meek, D.; Pulido, D. B. tler, "Relationship between six years of corn yields and terrain attributes," Precis. Agric., vol. 4, no. 1, pp. 87–101, 2003, doi: 10.1023/a:1021867123125.

[6] B. M. Margaret Oliver, Thomas Bishop, Ed., Precision Agriculture for Sustainability and Environmental Protection. Routledge, 2018.

[7] C. Anderson, "Agricultural Drones Relatively cheap drones with advanced sensors and imaging capabilities are giving farmers new ways to increase yields and reduce crop damage," 2016.

[8] C. D. and C. Jones, "Interpretation of soil test reports for agriculture."

[9] 2019. Accessed: Apr. 15, "Available: https://www.agrocares.com/en/products/lab-in-the-box/."

[10] J. Martínez-Fernández, A. González-Zamora, N. Sánchez, A. Gumuzzio, and C. M. Herrero-Jiménez, "Satellite soil moisture for agricultural drought monitoring: Assessment of the SMOS derived Soil Water Deficit Index," Remote Sens. Environ., vol. 177, pp. 277–286, May 2016, doi: 10.1016/j.rse.2016.02.064.

[11] and T. G. T.-G. Vågen, L. A. Winowiecki, J. E. Tondoh, L. T. Desta, "Mapping of soil properties and land degradation risk in Africa using MODIS reflectance," Geoderma, vol. 263, pp. 216–225, 2016.

[12] P. V. Santhi, N. Kapileswar, V. K. R. Chenchela, and C. H. V. S. Prasad, "Sensor and vision based autonomous AGRIBOT for sowing seeds," in 2017 International Conference on Energy, Communication, Data Analytics and Soft Computing (ICECDS), Aug. 2017, pp. 242–245, doi: 10.1109/ICECDS.2017.8389873.

[13] H. Karimi, H. Navid, B. Besharati, H. Behfar, and I. Eskandari, "A practical approach to comparative design of non-contact sensing techniques for seed flow rate detection," Comput. Electron. Agric., vol. 142, pp. 165–172, Nov. 2017, doi: 10.1016/j.compag.2017.08.027.

[14] S. d'Oleire-Oltmanns, I. Marzolff, K. Peter, and J. Ries, "Unmanned Aerial Vehicle (UAV) for Monitoring Soil Erosion in Morocco," Remote Sens., vol. 4, no. 11, pp. 3390–3416, Nov. 2012, doi: 10.3390/rs4113390.

[15] "Ice, Snow, and Glaciers and the Water Cycle." [Online]. Available: available: https://water.usgs.gov/edu/watercycleice.html.

[16] "Water for Sustainable Food and Agriculture by FAO.," p. Available: https://www.fao.org/3/a-i7959e.pdf.

[17] M. Motoshita et al., "Consistent characterisation factors at midpoint and endpoint relevant to agricultural water scarcity arising from freshwater consumption," Int. J. Life Cycle Assess., vol. 23, no. 12, pp. 2276–2287, Dec. 2018, doi: 10.1007/s11367-014-0811-5.

[18] T. G. P. A. K. V. de Oliveira, H. M. E. Castelli, S. J. Montebeller, "Wireless sensor network for smart agriculture using ZigBee protocol," in Proc. IEEE 1st Summer School Smart Cities (S3C), pp. 61–66.

[19] "The UN Decade for Deserts and the Fight Against Desertification: Impact and Role of Dry Lands."

[20] "Irrigation & Water Use.," p. Accessed: Sep. 3, 2019. [Online]. Available: https.

[21] J. L. and C. Fredrick, "Decision process for the application of variable rate irrigation," Dallas, TX, USA, 2012.

[22] C. Romero-Trigueros et al., "Effects of saline reclaimed waters and deficit irrigation on Citrus physiology assessed by UAV remote sensing," Agric. Water Manag., vol. 183, pp. 60–69, Mar. 2017, doi: 10.1016/j.agwat.2016.09.014.

[23] G. S. H. Kiiski, H. Dittmar, M. Drach, R. Vosskamp, M. E. Trenkel, R. Gutser, "Fertilizers, 2. types in Ullmann's Encyclopedia of Industrial Chemistry," 2009.

[24] "FAO. Forests and Agriculture: Land-Use Challenges and Opportunities," 2019.

[25] "Why IOT is Reinventing Plant Fertilization," p. https://www.iof2020.eu/latest/news/2017/09/whythe-, 2019.

[26] L. G, R. C, and G. P, "An automated low cost IoT based Fertilizer Intimation System for smart agriculture," Sustain. Comput. Informatics Syst., vol. 28, p. 100300, Dec. 2020, doi: 10.1016/j.suscom.2019.01.002.

[27] V. V. M. S. P. Benincasa, S. Antognelli, L. Brunetti, C. Fabbri, A. Natale, "Reliability of NDVI derived by high resolution satellite and UAV compared To in-field methods for the evaluation of early crop N status and grain yield in wheat," Exp. Agric., vol. 54, no. 4, pp. 604–622, 2018.

[28] J. B.-K. H. Liu, X. Wang, "Study on NDVI optimization of corn variable fertilizer applicator," Agric. Eng, vol. 56, no. 3, pp. 193–202, 2018.

[29] S. A. N. Khan, G. Medlock, S. Graves, "GPS guided autonomous navigation of a small agricultural robot with automated fertilizing system," 2018.

[30] A. J. Steidle Neto, S. Zolnier, and D. de Carvalho Lopes, "Development and evaluation of an automated system for fertigation control in soilless tomato production," Comput. Electron. Agric., vol. 103, pp. 17–25, Apr. 2014, doi: 10.1016/j.compag.2014.02.001.

[31] "Great Famine, FAMINE, IRELAND [1845–1849].," 2019.

[32] H. A. Bruns, "Southern Corn Leaf Blight: A Story Worth Retelling," Agron. J., vol. 109, no. 4, pp. 1218–1224, Jul. 2017, doi: 10.2134/agronj2017.01.0006.

[33] "Keeping Plant Pests and Diseases at Bay: Experts Focus on Global Measures," 2019.

[34] R. Pohanish, Sittig's Handbook of Pesticides and Agricultural Chemicals. Elsevier Wordmark.

[35] R. P. R. Waskom, T. Bauder, "Best management practices for agricultural pesticide use," 2017.

[36] S. Kim, M. Lee, and C. Shin, "IoT-Based Strawberry Disease Prediction System for Smart Farming," Sensors, vol. 18, no. 11, p. 4051, Nov. 2018, doi: 10.3390/s18114051.

[37] "Semios Integrated Pest Management.," 2019.

[38] A. Wietzke et al., "Insect pollination as a key factor for strawberry physiology and marketable fruit quality," Agric. Ecosyst. Environ., vol. 258, pp. 197–204, Apr. 2018, doi: 10.1016/j.agee.2018.01.036.

[39] S.-O. Chung, M.-C. Choi, K.-H. Lee, Y.-J. Kim, S.-J. Hong, and M. Li, "Sensing Technologies for Grain Crop Yield Monitoring Systems: A Review," J. Biosyst. Eng., vol. 41, no. 4, pp. 408–417, Dec. 2016, doi: 10.5307/JBE.2016.41.4.408.

[40] N. Torbick, D. Chowdhury, W. Salas, and J. Qi, "Monitoring Rice Agriculture across Myanmar Using Time Series Sentinel-1 Assisted by Landsat-8 and PALSAR-2," Remote Sens., vol. 9, no. 2, p. 119, Feb. 2017, doi: 10.3390/rs9020119.

[41] Z. Wang, K. Walsh, and B. Verma, "On-Tree Mango Fruit Size Estimation Using RGB-D Images," Sensors, vol. 17, no. 12, p. 2738, Nov. 2017, doi: 10.3390/s17122738.

[42] P. Udomkun, M. Nagle, D. Argyropoulos, B. Mahayothee, and J. Müller, "Multi-sensor approach to improve optical monitoring of papaya shrinkage during drying," J. Food Eng., vol. 189, pp. 82–89, Nov. 2016, doi: 10.1016/j.jfoodeng.2016.05.014.

[43] "Global Smart Farming Market to Reach $23.14 Billion by 2022."

[44] A. Hemmat, A. R. Binandeh, J. Ghaisari, and A. Khorsandi, "Development and field testing of an integrated sensor for on-the-go measurement of soil mechanical resistance," Sensors Actuators A Phys., vol. 198, pp. 61–68, Aug. 2013, doi: 10.1016/j.sna.2013.04.027.

[45] D. J. Cocovi-Solberg, M. Rosende, and M. Miró, "Automatic Kinetic Bioaccessibility Assay of Lead in Soil Environments Using Flow-through Microdialysis as a Front End to Electrothermal Atomic Absorption Spectrometry," Environ. Sci. Technol., vol. 48, no. 11, pp. 6282–6290, Jun. 2014, doi: 10.1021/es405669b.

[46] S. C. Murray, "Optical sensors advancing precision in agricultural production," Photon. Spectra, vol. 51, p. 48, 2018.

[47] F. J. García-Ramos, M. Vidal, A. Boné, H. Malón, and J. Aguirre, "Analysis of the Air Flow Generated by an Air-Assisted Sprayer Equipped with Two Axial Fans Using a 3D Sonic Anemometer," Sensors, vol. 12, no. 6, pp. 7598–7613, Jun. 2012, doi: 10.3390/s120607598.

[48] M. A. M. Yunus and S. C. Mukhopadhyay, "Novel Planar Electromagnetic Sensors for Detection of Nitrates and Contamination in Natural Water Sources," IEEE Sens. J., vol. 11, no. 6, pp. 1440–1447, Jun. 2011, doi: 10.1109/JSEN.2010.2091953.

[49] Q. Kong, H. Chen, Y. Mo, and G. Song, "Real-Time Monitoring of Water Content in Sandy Soil Using Shear Mode Piezoceramic Transducers and Active Sensing—A Feasibility Study," Sensors, vol. 17, no. 10, p. 2395, Oct. 2017, doi: 10.3390/s17102395.

[50] A. de la Piedra, A. Braeken, and A. Touhafi, "Sensor Systems Based on FPGAs and Their Applications: A Survey," Sensors, vol. 12, no. 9, pp. 12235–12264, Sep. 2012, doi: 10.3390/s120912235.

[51] I. Molina, C. Morillo, E. García-Meléndez, R. Guadalupe, and M. I. Roman, "Characterizing Olive Grove Canopies by Means of Ground-Based Hemispherical Photography and Spaceborne RADAR Data," Sensors, vol. 11, no. 8, pp. 7476–7501, Jul. 2011, doi: 10.3390/s110807476.

[52] "Managing Calibration Curves under John Deere Tractors in Yield Monitor Systems," IOWA State University, Ames, IA, USA.

[53] G. Pajares, A. Peruzzi, and P. Gonzalez-de-Santos, "Sensors in Agriculture and Forestry," Sensors, vol. 13, no. 9, pp. 12132–12139, Sep. 2013, doi: 10.3390/s130912132.

[54] J. D. D. AndÃžjar, A. Ribeiro, C. F. Quintanilla, J. Dorado, "Assessment of a ground-based weed mapping system in maize," Precis. Agric., 2009.

[55] A.-K. Mahlein, "Plant Disease Detection by Imaging Sensors – Parallels and Specific Demands for Precision Agriculture and Plant Phenotyping," Plant Dis., vol. 100, no. 2, pp. 241–251, Feb. 2016, doi: 10.1094/PDIS-03-15-0340-FE.

[56] I. del-Moral-Martínez et al., "Mapping Vineyard Leaf Area Using Mobile Terrestrial Laser Scanners: Should Rows be Scanned On-the-Go or Discontinuously Sampled?," Sensors, vol. 16, no. 1, p. 119, Jan. 2016, doi: 10.3390/s16010119.

[57] E. Moureaux, C., Ceschia, E., Arriga, N., Béziat, P., Eugster, W., Kutsch, W. L., and Pattey, "Eddy covariance measurements over crops," in Eddy Covariance: A Practical Guide to Measurement and Data Analysis, 1st ed., D. Aubinet, M., Vesala T., and Papale, Ed. Springer Atmospheric Sciences Serie, 2011.

[58] A. Crabit, F. Colin, J. S. Bailly, H. Ayroles, and F. Garnier, "Soft Water Level Sensors for Characterizing the Hydrological Behaviour of Agricultural Catchments," Sensors, vol. 11, no. 5, pp. 4656–4673, Apr. 2011, doi: 10.3390/s110504656.

[59] A. K. E. Mohamed, "Analysis of telematics systems in agriculture," Dept. Mach., Utilization, CULS, Prague, Czech Republic, 2013.

[60] R. Patmasari, I. Wijayanto, R. S. Deanto, Y. P. Gautama, and H. Vidyaningtyas, "Design and Realization of Automatic Packet Reporting System (APRS) for Sending Telemetry Data in Nano Satellite Communication System," J. Meas. Electron. Commun. Syst., vol. 4, no. 1, p. 1, Jun. 2018, doi: 10.25124/jmecs.v4i1.1692.

[61] K. P. S. J S Dvorak, M L Stone, "Object Detection for Agricultural and Construction Environments Using an Ultrasonic Sensor," J. Agric. Saf. Health, vol. 22, no. 2, pp. 107–119, Apr. 2016, doi: 10.13031/jash.22.11260.

[62] I. R. Hegazy and M. R. Kaloop, "Monitoring urban growth and land use change detection with GIS and remote sensing techniques in Daqahlia governorate Egypt," Int. J. Sustain. Built Environ., vol. 4, no. 1, pp. 117–124, Jun. 2015, doi: 10.1016/j.ijsbe.2015.02.005.

[63] B. Research., "An Introduction to LPWA Public Service Categories: Matching Services to IoT Applications," 2016.

[64] G.-Z. H. and C.-L. Hsieh, "Application of integrated control strategy and Bluetooth for irrigating romaine lettuce in greenhouse," IFACPapersOnLine, vol. 49, no. 16, pp. 381–386, 2016.

[65] B. Jonathan, "Evaluation of Bluetooth low energy in agriculture environments: An empirical analysis of BLE in precision agriculture," Malmö Högskola Univ., Malmö, Sweden, 2016.

[66] S. Y. Deniz Taşkın, Cem Taşkın, "Developing a Bluetooth Low Energy Sensor Node for Greenhouse in Precision Agriculture as Internet of Things Application," Adv. Sci. Technol. Res. J., vol. 12, no. 4, pp. 88 – 96, 2018, doi: 0.12913/22998624/100342.

[67] S. C. M. G. R. Mendez, M. A. M. Yunus, "A WiFi based smart wireless sensor network for an agricultural environment," in Proc. 5th Int. Conf. Sens. Technol., 2011, pp. 405-410.

[68] J. I. J. Petäjäjärvi, K. Mikhaylov, M. Hämäläinen, "Evaluation of LoRa LPWAN technology for remote health and wellbeing monitoring," in Proc. 10th Int. Symp. Med. Inf. Commun. Technol. (ISMICT), 2016, pp. 1–5.

[69] Z. M. N. Zhu, Y. Xia, Y. Liu, C. Zang, H. Deng, "Temperature and humidity monitoring system for bulk grain container based on Lora wireless technology," in Proc. ICCCS, 2018, pp. 102–110.

[70] A. Lavric, A. I. Petrariu, and V. Popa, "Long Range SigFox Communication Protocol Scalability Analysis Under Large-Scale, High-Density Conditions," IEEE Access, vol. 7, pp. 35816–35825, 2019, doi: 10.1109/ACCESS.2019.2903157.

[71] G. Alfian, M. Syafrudin, and J. Rhee, "Real-Time Monitoring System Using Smartphone-Based Sensors and NoSQL Database for Perishable Supply Chain," Sustainability, vol. 9, no. 11, p. 2073, Nov. 2017, doi: 10.3390/su9112073.

[72] A. Botta, W. de Donato, V. Persico, and A. Pescapé, "Integration of Cloud computing and Internet of Things: A survey," Futur. Gener. Comput. Syst., vol. 56, pp. 684–700, Mar. 2016, doi: 10.1016/j.future.2015.09.021.

[73] M. S. Mekala and P. Viswanathan, "CLAY-MIST: IoT-cloud enabled CMM index for smart agriculture monitoring system," Measurement, vol. 134, pp. 236–244, Feb. 2019, doi: 10.1016/j.measurement.2018.10.072.

[74] R. B. S. Singh, I. Chana, "Agri-info: Cloud based autonomic system for delivering agriculture as a service," Int. Things, vol. 9, no. 100131, 2020.

[75] A. Monteiro, S. Santos, and P. Gonçalves, "Precision Agriculture for Crop and Livestock Farming—Brief Review," Animals, vol. 11, no. 8, p. 2345, Aug. 2021, doi: 10.3390/ani11082345.

[76] S. A. F. Bonomi, R. Milito, J. Zhu, "Fog computing and its role in the internet of things," in Proc. 1st Edition of the MCC Workshop on Mobile Cloud Computing," in Helsinki, Finland, pp. 13–16.

[77] A. F. S. M. A. Zamora-Izquierdo, J. Santa, J. A. Martínez, V. Martínez, "Smart farming IoT platform based on edge and cloud computing," Biosyst. Eng, vol. 177, pp. 4–17, 2019.

[78] J. Liu et al., "A microfluidic based biosensor for rapid detection of Salmonella in food products," PLoS One, vol. 14, no. 5, p. e0216873, May 2019, doi: 10.1371/journal.pone.0216873.

[79] R.-Y. Chen, "An intelligent value stream-based approach to collaboration of food traceability cyber physical system by fog computing," Food Control, vol. 71, pp. 124–136, Jan. 2017, doi: 10.1016/j.foodcont.2016.06.042.

[80] B. T. X. W. Chen, H. H. Wang, "Multidimensional agroeconomic model with soft-IoT framework," Soft Comp, vol. 24, pp. 12187–12196, 2020.

[81] J. Muangprathub, N. Boonnam, S. Kajornkasirat, N. Lekbangpong, A. Wanichsombat, and P. Nillaor, "IoT and agriculture data analysis for smart farm," Comput. Electron. Agric., vol. 156, pp. 467–474, Jan. 2019, doi: 10.1016/j.compag.2018.12.011.

[82] F.-H. Tseng, H.-H. Cho, and H.-T. Wu, "Applying Big Data for Intelligent Agriculture-Based Crop Selection Analysis," IEEE Access, vol. 7, pp. 116965–116974, 2019, doi: 10.1109/ACCESS.2019.2935564.

[83] L. Lambrinos, "Internet of things in agriculture: A decision support system for precision farming," in Proc. IEEE Int. Conf. Dependable, Autonomic and Secure Computing, Int. Conf. Pervasive Intelligence and Computing, Int. Conf. Cloud and Big Data Computing, Int. Conf. Cyber Science and Technology Congress, 2019, pp. 889–892.

[84] R. da R. R. U. J. L. dos Santos, G. Pessin, C. A. da Costa, "Agriprediction: A proactive internet of things model to anticipate problems and improve production in agricultural crops," Comp. Electron. Agric, vol. 161, pp. 202–213, 2019.

[85] J. Ruan, H. Jiang, X. Li, Y. Shi, F. T. S. Chan, and W. Rao, "A Granular GA-SVM Predictor for Big Data in Agricultural Cyber-Physical Systems," IEEE Trans. Ind. Informatics, vol. 15, no. 12, pp. 6510–6521, Dec. 2019, doi: 10.1109/TII.2019.2914158.

[86] A. L. Diedrichs, F. Bromberg, D. Dujovne, K. Brun-Laguna, and T. Watteyne, "Prediction of Frost Events Using Machine Learning and IoT Sensing Devices," IEEE Internet Things J., vol. 5, no. 6, pp. 4589–4597, Dec. 2018, doi: 10.1109/JIOT.2018.2867333.

[87] K. Jha, A. Doshi, P. Patel, and M. Shah, "A comprehensive review on automation in agriculture using artificial intelligence," Artif. Intell. Agric., vol. 2, pp. 1–12, Jun. 2019, doi: 10.1016/j.aiia.2019.05.004.

[88] J. S. M. Ghahramani, Y. Qiao, M. C. Zhou, A. O'Hagan, "AI-based modeling and data-driven evaluation for smart manufacturing processes," IEEE/CAA J. Autom. Sin., vol. 7, no. 4, pp. 1026–1037, 2020.

[89] A. Rajput and V. B. Kumaravelu, "Fuzzy logic-based distributed clustering protocol to improve energy efficiency and stability of wireless smart sensor networks for farmland monitoring systems," Int. J. Commun. Syst., vol. 33, no. 4, p. e4239, Mar. 2020, doi: 10.1002/dac.4239.

[90] L. Liu et al., "PestNet: An End-to-End Deep Learning Approach for Large-Scale Multi-Class Pest Detection and Classification," IEEE Access, vol. 7, pp. 45301–45312, 2019, doi: 10.1109/ACCESS.2019.2909522.

[91] F. Bu and X. Wang, "A smart agriculture IoT system based on deep reinforcement learning," Futur. Gener. Comput. Syst., vol. 99, pp. 500–507, Oct. 2019, doi: 10.1016/j.future.2019.04.041.

[92] D. R. Vincent, N. Deepa, D. Elavarasan, K. Srinivasan, S. H. Chauhdary, and C. Iwendi, "Sensors Driven AI-Based Agriculture Recommendation Model for Assessing Land Suitability," Sensors, vol. 19, no. 17, p. 3667, Aug. 2019, doi: 10.3390/s19173667.

[93] P. Jiang, Y. Chen, B. Liu, D. He, and C. Liang, "Real-Time Detection of Apple Leaf Diseases Using Deep Learning Approach Based on Improved Convolutional Neural Networks," IEEE Access, vol. 7, pp. 59069–59080, 2019, doi: 10.1109/ACCESS.2019.2914929.

[94] S. S. A.-N. B. A. Ashqar, B. S. Abu-Nasser, "Plant seedlings classification using deep learning," Int. J. Acad. Inf. Syst. Res., vol. 3, no. 1, pp. 7–14, 2019.

[95] J. Gua et al., "Recent Advances in Convolutional Neural Networks," Pattern Recognit., vol. 77, pp. 354–377, 2018.

[96] S. Zhang, S. Zhang, C. Zhang, X. Wang, and Y. Shi, "Cucumber leaf disease identification with global pooling dilated convolutional neural network," Comput. Electron. Agric., vol. 162, pp. 422–430, Jul. 2019, doi: 10.1016/j.compag.2019.03.012.

[97] J. Liu and X. Wang, "Tomato Diseases and Pests Detection Based on Improved Yolo V3 Convolutional Neural Network," Front. Plant Sci., vol. 11, Jun. 2020, doi: 10.3389/fpls.2020.00898.

[98] S. Khan, M. Tufail, M. T. Khan, Z. A. Khan, and S. Anwar, "Deep learning-based identification system of weeds and crops in strawberry and pea fields for a precision agriculture sprayer," Precis. Agric., vol. 22, no. 6, pp. 1711–1727, Dec. 2021, doi: 10.1007/s11119-021-09808-9.

[99] P. Bedi and P. Gole, "Plant disease detection using hybrid model based on convolutional autoencoder and convolutional neural network," Artif. Intell. Agric., vol. 5, pp. 90–101, 2021, doi: 10.1016/j.aiia.2021.05.002.

[100] S. M. M. Hossain et al., "Rice Leaf Diseases Recognition Using Convolutional Neural Networks," in International Conference on Advanced Data Mining and Applications ADMA 2020: Advanced Data Mining and Applications, 2021, pp. 299–314, doi: 10.1007/978-3-030-65390-3_23.

[101] R. Deng et al., "Automatic Diagnosis of Rice Diseases Using Deep Learning," Front. Plant Sci., vol. 12, Aug. 2021, doi: 10.3389/fpls.2021.701038.

[102] L. Goyal, C. M. Sharma, A. Singh, and P. K. Singh, "Leaf and spike wheat disease detection & classification using an improved deep convolutional architecture," Informatics Med. Unlocked, vol. 25, p. 100642, 2021, doi: 10.1016/j.imu.2021.100642.

[103] M. Zhang, X., Qiao, Y., Meng, F., Fan, C., & Zhang, "Identification of maize leaf diseases using improved deep convolutional neural networks.," IEEE Access, vol. 6, pp. 30370–30377, 2018.

[104] P. Wang, T. Niu, Y. Mao, Z. Zhang, B. Liu, and D. He, "Identification of Apple Leaf Diseases by Improved Deep Convolutional Neural Networks With an Attention Mechanism," Front. Plant Sci., vol. 12, Sep. 2021, doi: 10.3389/fpls.2021.723294.

[105] J. Johnson et al., "Enhanced Field-Based Detection of Potato Blight in Complex Backgrounds Using Deep Learning," Plant Phenomics, vol. 2021, pp. 1–13, May 2021, doi: 10.34133/2021/9835724.

[106] D. Popescu, F. Stoican, G. Stamatescu, L. Ichim, and C. Dragana, "Advanced UAV–WSN System for Intelligent Monitoring in Precision Agriculture," Sensors, vol. 20, no. 3, p. 817, Feb. 2020, doi: 10.3390/s20030817.

[107] C. Cambra Baseca, S. Sendra, J. Lloret, and J. Tomas, "A Smart Decision System for Digital Farming," Agronomy, vol. 9, no. 5, p. 216, Apr. 2019, doi: 10.3390/agronomy9050216.

[108] N. Ahmed, D. De, and I. Hussain, "Internet of Things (IoT) for Smart Precision Agriculture and Farming in Rural Areas," IEEE Internet Things J., vol. 5, no. 6, pp. 4890–4899, Dec. 2018, doi: 10.1109/JIOT.2018.2879579.

[109] A. Muminov, D. Na, C. Lee, H. Kang, and H. Jeon, "Modern Virtual Fencing Application: Monitoring and Controlling Behavior of Goats Using GPS Collars and Warning Signals," Sensors, vol. 19, no. 7, p. 1598, Apr. 2019, doi: 10.3390/s19071598.

[110] I. Potamitis, I. Rigakis, N.-A. Tatlas, and S. Potirakis, "In-Vivo Vibroacoustic Surveillance of Trees in the Context of the IoT," Sensors, vol. 19, no. 6, p. 1366, Mar. 2019, doi: 10.3390/s19061366.

[111] S. Liu, L. Guo, H. Webb, X. Ya, and X. Chang, "Internet of Things Monitoring System of Modern Eco-Agriculture Based on Cloud Computing," IEEE Access, vol. 7, pp. 37050–37058, 2019, doi: 10.1109/ACCESS.2019.2903720.

[112] T. S. U. Yaqub, A. Al-Nasser, "Implementation of a hybrid wind-solar desalination plant from an internet of things (IoT) perspective on a network simulation tool," Appl. Comput. Inf., vol. 15, no. 1, pp. 7–11, 2019.

[113] C. M. Angelopoulos, G. Filios, S. Nikoletseas, and T. P. Raptis, "Keeping data at the edge of smart irrigation networks: A case study in

strawberry greenhouses," Comput. Networks, vol. 167, p. 107039, Feb. 2020, doi: 10.1016/j.comnet.2019.107039.

[114] A. Goap, D. Sharma, A. K. Shukla, and C. Rama Krishna, "An IoT based smart irrigation management system using Machine learning and open source technologies," Comput. Electron. Agric., vol. 155, pp. 41–49, Dec. 2018, doi: 10.1016/j.compag.2018.09.040.

[115] R. N. R. and B. Sridhar, "IoT based smart crop-field monitoring and automation irrigation system," in Proc. 2nd Int. Conf. Inventive Systems and Control, Coimbatore, India, 2018, pp. 478–483.

[116] B. Keswani et al., "Adapting weather conditions based IoT enabled smart irrigation technique in precision agriculture mechanisms," Neural Comput. Appl., vol. 31, no. S1, pp. 277–292, Jan. 2019, doi: 10.1007/s00521-018-3737-1.

[117] N. G. S. Campos, A. R. Rocha, R. Gondim, T. L. Coelho da Silva, and D. G. Gomes, "Smart & Green: An Internet-of-Things Framework for Smart Irrigation," Sensors, vol. 20, no. 1, p. 190, Dec. 2019, doi: 10.3390/s20010190.

[118] G. Severino, G. D'Urso, M. Scarfato, and G. Toraldo, "The IoT as a tool to combine the scheduling of the irrigation with the geostatistics of the soils," Futur. Gener. Comput. Syst., vol. 82, pp. 268–273, May 2018, doi: 10.1016/j.future.2017.12.058.

[119] X. Y. K. Huang, K. L. Li, L. Shu, "Demo abstract: High voltage discharge exhibits severe effect on ZigBee-based device in solar insecticidal lamps internet of things," in IEEE INFOCOM 2020- IEEE Conf. Computer Communications Workshops, 2020, pp. 1266–1267.

[120] M. A. Ferrag and L. Maglaras, "DeliveryCoin: An IDS and Blockchain-Based Delivery Framework for Drone-Delivered Services," Computers, vol. 8, no. 3, p. 58, Aug. 2019, doi: 10.3390/computers8030058.

[121] C. Dupraz, H. Marrou, G. Talbot, L. Dufour, A. Nogier, and Y. Ferard, "Combining solar photovoltaic panels and food crops for optimising land use: Towards new agrivoltaic schemes," Renew. Energy, vol. 36, no. 10, pp. 2725–2732, Oct. 2011, doi: 10.1016/j.renene.2011.03.005.

[122] A. Weselek, A. Ehmann, S. Zikeli, I. Lewandowski, S. Schindele, and P. Högy, "Agrophotovoltaic systems: applications, challenges, and opportunities. A review," Agron. Sustain. Dev., vol. 39, no. 4, p. 35, Aug. 2019, doi: 10.1007/s13593-019-0581-3.

[123] J. L. Xue, "Photovoltaic agriculture-new opportunity for photovoltaic applications in China," Renew. Sustain. Energy Rev., vol. 73, pp. 1–9, 2017.

[124] Y. H. S. F. Yang, L. Shu, Y. Liu, K. L. Li, K. Huang, Y. Zhang, "Poster: Photovoltaic agricultural internet of things the next generation of smart farming," in Proc. Int. Conf. Embedded Wireless Systems and Networks, United States, 2019, pp. 236–237.

[125] Z. A. J. H. Sharma, A. Haque, "Maximization of wireless sensor network lifetime using solar energy harvesting for smart agriculture monitoring," Ad Hoc Netw., vol. 94, no. 101966, 2019.

[126] R. da Rosa Righi, G. Goldschmidt, R. Kunst, C. Deon, and C. André da Costa, "Towards combining data prediction and internet of things to manage milk production on dairy cows," Comput. Electron. Agric., vol. 169, p. 105156, Feb. 2020, doi: 10.1016/j.compag.2019.105156.

[127] Y. Wang et al., "High-voltage output triboelectric nanogenerator with DC/AC optimal combination method," Nano Res., vol. 15, no. 4, pp. 3239–3245, Apr. 2022, doi: 10.1007/s12274-021-3956-0.

7

Plant Feature Extraction for Disease Classification

Amar Kumar Dey[1] and Abhishek Sharma[2]

[1]Department of Electronics & Telecommunication Engineering, Bhilai Institute of Technology, India
[2]Department of Electrical & Electronics, Ariel University, Israel
E-mail: amardeyhope@gmail.com; abhishek15491@gmail.com

Abstract

Agricultural production and forestry rely on plant health status. It is the major source of livelihood and revenue generation in India. Thus, maintaining good plant health requires continuous and expert monitoring, including early disease diagnosis in plants. With industrial and technological development, human experts are replaced by machine experts to perform these challenging repetitive tasks and early-stage plant disease diagnoses with greater efficiency. It became even more challenging for human experts when the disease symptoms are almost similar but require different treatments. Plant diseases damage 20–40% of crops each year.

Recent machine vision approaches provide an effective plant disease diagnosis, fetcher which are analogous to human expert approach. Thus, proper fetcher extraction is the key to efficient disease early-stage diagnosis.

The study focuses on a thorough examination of machine learning-based extraction methods, as well as their benefits and drawbacks. It covers a wide range of features, based on shape, texture, and color, for different diseases in diverse cultivations.

Keywords: Machine vision, texture feature, color features, leaf disease, imageprocessing.

7.1 Introduction

Plants and fruit harvest are the prime sources of liveliness for both humans and animals. Farm production offers food security and occupation also generates export revenue, in agriculture-based developing nations, where agriculture output employs over 50% of the population, generates export earnings, and ensures food security [1]. The leaves of many plants and shrubs are beneficial to humankind due to their salutary characteristics.

Although agriculture employs nearly half of India's population, it only accounts for 17.5% of the country's GDP [2]. This sector not only provides employment but also ensures global food security. Infected crops result in a global loss of 15–25% of total production per year which is a serious issue [3].

Effective plant disease controlling requires automated disease identification. Disease identification is automated using a hybrid of man, a visual system, and an intelligent machine. Thus, feature extraction plays a key role in automating disease detection with accuracy and precision [4].

It entails the creation of a database of pathogen or infectious agents, as well as the automatic identification of diseases using digital imaging and machine vision for plant ailment detection. Researchers have tried many alternatives in both image processing procedures and machine learning models due to the possibility of various alternatives at different stages of plant disease identification, as illustrated in Figure 7.1. An image-based crop disease detection system typically comprises of the above four steps in its architecture. Image analyzer, features extraction, classifier, and disease detection [5].

An image analyzer was used to process a set of input images (both damaged and healthy) and extract unique features [6]. The classifier was then given these features, as well as the information about whether the image was of a diseased or normal leaf. The classifier next learns the relationship between the extracted features and the likely disease presence conclusion [7].

Each step is extremely significant and has distinct characteristics in diagnosing diseases, based on signs patent on plant leaves. The feature abstraction step, on the other hand, is critical in the identification of different diseases, as it distinguishes one type of infection, from other infection types having similar symptoms or evident spots/lesions [8].

The dimensionality cutback technique uses feature abstraction to split and reduce a big set of original data into smaller categories. Generally, the original data consists of many variables; to process such variability a high computational facility is utilized. So, by choosing and merging variables into features, feature extraction removes the redundancy from the data and makes

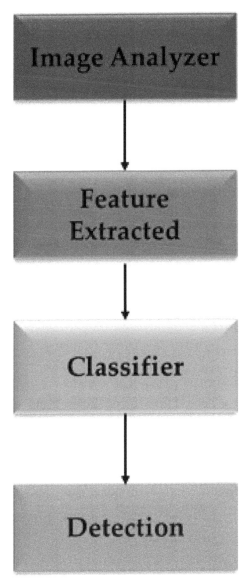

Figure 7.1 Basic blocks used for plant disease detection.

it compact by reducing complexity. These properties are simple to utilize while characterizing the actual dataset precisely and uniquely [9].

It is conceivable to have almost identical detectible spots on leaf images at times, which adds to the complexity of such systems. However, extracting significant features using the most suitable and applicable scheme can completely resolve the problem of lesions with equal visibility [10].

This research aims at several feature abstraction techniques that are used to identify plant diseases. A literature review was conducted and discussed in Section 7.2. Section 7.3 of the paper presents the general features representation and characteristics of a conventional plant disease detection system. In Section 7.4, various texture feature extraction techniques are explored in plant disease detection. It also provides an overview of the statistical method, gray-level run-length matrix, and histogram of gradient magnitudes for texture classification. The advantage and limitations of each approach are also summarized. Color feature extraction techniques based on the color histogram approach and color co-occurrence matrix approach are discussed in Section 7.5. In Section 7.6, various shape-based features were explored. Section 7.7 concludes this study with a research scope for the future.

7.2 Literature Review

Features play an important role in identifying plant disease. Archana *et al.* [11] make use of gray-level co-occurrence matrix (GLCM) a most commonly used texture feature to identify the brown spot, rice blast, and bacterial disease in rice crops. The intensity value at the pixel of interest identifies the texture's qualities. The extracted texture was given to the SVM classifier. Four different classes (healthy along with three diseases mentioned above) were identified with classification accuracy above 99.2%.

Shrivastava *et al.* [12] developed an SVM classifier to identify rice plant disease. Fourteen different color features were extracted to classify healthy, bacterial blight, rice blast, and sheath blight disease in a rubber plant. The classifier accuracy was above 95.4%, which shows color features are promising in detecting plant disease.

Lilhore *et al.* [13] implemented an enhanced CNN network to classify cassava leaf disease. Shape-based featured input was given to classify brown streak, green mottle, mosaic, and healthy leaf. In the proposed method, a shape-based technique first extracts the curve receipt using leaf stem and afterward determines the inconsistencies using a Jeffrey divergence estimate method. The accuracy of the classifier was above 99.3%.

Thus, the choice of the feature selection depends and varies from species to species and the classifier used.

7.3 Features Representation and Characteristics

The most basic properties or characteristics utilized to recognize a picture are called image features. One image's attributes can be utilized to identify it from others [14]. The goal of feature extraction is to extract visual attributes that can be used to differentiate this image from others. Four categories of image characteristics are commonly utilized in feature extraction methods: color (gray) image, shape and texture features along with spatial relations. Features extracted from images should be able to describe things both abstractly and concretely, providing a solid foundation for later classification [15]. The following traits should be present in good features:

1. Uniqueness: Each item must have its distinct representation that allows it to be distinguished from other items.
2. Integrity: The features may explain the entire object's properties.
3. Being invariant under the geometric manipulation: To avoid results being altered, obtained features may not vary with deviations in size or orientation.
4. Agility: The contrasts between similar things can be easily reflected using descriptive methods.
5. Abstractness: Many image details can be used to derive relevant feature information. The object description should be straightforward and effective.

The representation and extraction of picture features are the first steps in image analysis. Good characteristics allow you to fully identify the object you are looking for. Effective feature extraction methods can accurately extract features, simplifying classifier construction and resulting in more accurate classification results [16].

7.4 Texture Features

The texture has been considered a measure of roughness, distinction, polarity, line-likeness, uniformity, and smoothness by several authors [17]–[19], or as natural sceneries with semi-repetitive pixel patterns. A similarity clustering may also be detected in the texture in a test image. Further, we classify texture as statistical texture, dynamic texture, and temporal texture. The texture is the most widely used feature for detecting leaf disease based on the machine vision approach. Researchers now have access to a huge variety of texture feature extraction methods [20]. Furthermore, new ways are continually being developed, many of which are based on current scientific advancements.

7.4.1 Statistical methods for texture classification

Gray-level co-occurrence matrix (GLCM) is the commonly used statistical texture feature. It was proposed by [21]. It consists of a set of 14 different features namely contrast, sum of squares, correlation, angular second moment, difference variance, sum variance, variance, difference moment, sum average, sum entropy, entropy, difference entropy, measure of correlation, and correlation coefficients [22]. But only five (shown in Table 7.1 of these 14 measures are sufficient, according to Sharma et al. [23]. The number of occurrences of picture pixels followed by the same or some other gray level is used to calculate GLCM. The output of GLCM obtained from an image will show the frequency of the gray level pair of an input image.

Gray-level co-occurrence Matrix (GLCM), also known as gray-level spatial dependence [24] is an arithmetical method for assessing texture properties

Table 7.1 Significant five Haralick texture features

S. No.	Haralick features	Definition	Equation
1	Contrast	Contrast is a statistic that quantifies the difference in intensity between pixel and neighbor throughout the image; it is 0 for a uniform image.	$= \sum_{i=1}^{M} \sum_{j=1}^{N} (i-j)^2 P(i,j)$
2	Correlation	A pixel's correlation is a measure of how strongly it is connected to its neighbors across the image.	$= \sum_{i=1}^{M} \sum_{j=1}^{N} \frac{(i-\mu)(j-\mu)P(i,j)}{\sigma^2}$
3	Entropy	Entropy measures the image's intricacy, this complicated texture has a higher entropy.	$= \sum_{i=1}^{M} \sum_{j=1}^{N} P(i,j)$
4	Energy	The total of squared components is energy, for a uniform image its value is one.	$= \sum_{i=1}^{M} \sum_{j=1}^{N} P^2(i,j)$
5	Homogeneity	The spatial homogeneity of a pixel pair is calculated. The homogeneity should be high if the gray values of each pixel combination are similar.	$= \sum_{i=1}^{M} \sum_{j=1}^{N} \frac{P(i,j)}{1+|i-j|}$

The count of pixels in x direction in the given image is M, while the count of pixels in y direction is N. $N_x = 1,2,3,..... M$ along x-axis, and $N_y = 1,2,3....... N$ along the y-axis. Assumer gray pixel i and j.

characterized by the spatial membership of pixels. By estimating how frequently a pixel with the intensity value correlates to a pixel with the value j, a GLCM matrix is formed. The frequency of separating two pixels in a specific vector that appears in an image is called GLCM. The arrangement in the matrix is determined by the distance and orientation between the pixels, such as lateral, perpendicular, orthogonal, and anti-orthogonal relationships [25].

7.4.1.1 Advantages and limitations

For simple circumstances when the textures are distinguishable, the GLCM approach produces better results. Also, the GLCM method is simple to apply and produces excellent results in a wide range of areas [26]. The GLCMs, on the other hand, are particularly subjective to the size of an image texture that is produced due to their large dimensionality. As a result, gray levels are frequently reduced. Furthermore, it was found that the GLCM-based technique offers faster processing and reduces complexity, when processing small images, but consumes a lot of memory [27]. Furthermore, the GLCM characteristics are not ideal for photos with a lot of noise.

7.4.2 Gray-level run-length matrix

The number of pairs of the gray-level score and their length of runs in a given region of interest is the statistic in gray-level run-length matrix (GLRLM) [28] (ROI). A gray-level run is a compilation of pixels with an equal gray level that are scattered in the ROI in a specific order and collinearly [29]. The length of the gray level run is the count of pixels in that particular set. As a result, such a set is defined by the gray level score and the length of the gray-level run. A GLRLM is a matrix-based 2D plot that captures the occurrence of all different combinations of gray-level scores and gray-level runs in an ROI.

Gray-level runs in a picture are measured using run-length measurements. The primitive (run) is a group of pixels in the same direction that have the same gray level. A primitive is identified by its gray level G, length L, and orientation. The count of runs with pixels intensity of gray level G and length L is defined as a component of the GLRLM $E(G,L)$ for a picture. A feature vector of GLRLM indices is produced for every histogram $H(G,L)$. Several picture texture properties can be defined by specifying $P(G,L)$, the occurrence of a run-length. Different types of primitive (run) [22] are tabulated in Table 7.2.

Table 7.2 Description of different types of primitive along with mathematical relations

S. No.	Primitive type	Description	Equation
1	Short primitive emphasis	It is a characteristic of minute-grained textures.	$= \sum_{G=0}^{n_G - 1} \sum_{\iota=1}^{L} \frac{P(G,L)}{\iota^2}$
2	Long primitive emphasis	It relates to rough surfaces.	$= \sum_{G=0}^{n_G - 1} \sum_{\iota=1}^{L} P(G,L) . \iota^2$
3	Primitive length uniformity	It indicates that a few spin outliers are dominating the histogram.	$= \sum_{G=0}^{n_G - 1} \sum_{\iota=1}^{L} [P(G,L)]^2$

n_G is the count of gray level and ι is the maximum spin.

7.4.3 Advantages and limitations

The primitive length method, on the other hand, has the advantage of ignoring background information. Because it is less impacted by the background, primitive length data are more important for local structure detection [30]. This is especially beneficial when the texture analysis windows are somewhat small.

When the number of gray levels is increased, the GLRLM will tend to have numerous runs. As a result, the structure information is lost [31].

7.4.4 Histogram of gradient magnitudes

The variation of gradient directions is utilized as a feature in this feature descriptor. Because the amplitude of gradients at edges and corners is significant, picture gradients are useful. Since the edges and corners of an object have a lot more information about its shape than flat areas [32].

The variation of the local slope of pixel intensity was neglected while a histogram of gradient magnitudes is produced.

The magnitude gradient indicates the strength of the edge by the amount of difference between pixels in the surrounding area. The plot is produced across the gradient levels of pixels, but it ignores the gradient orientation. The approach is rotation-invariant as a result of this. A complete workflow of the above approach is discussed in [15], [33].

7.4.4.1 Advantages and limitations

The rotation invariance and low computation time of the histogram of gradient magnitudes are its two key advantages: The amount of time it takes to compute an image is proportional to the number of pixels in the image under consideration [15]. Furthermore, the approach has been demonstrated to outperform alternative texture descriptors in texture classification and picture segmentation, such as LBP and rotation invariant LBP [32].

7.5 Color Features Extraction

Translation and rotation of the color feature provide a significant image representation. The standard approach of extracting color features divides the complete color space into a fixed number of bins. Each bin represents a distinct color, and each pixel is assigned to the closest colored bin, with the size of the bin being expressed as a percentage of the total color in the image [34]. Two approaches to extracting color features are color histogram and color moment [35].

The image is represented by a color histogram from unique perspectives. It depicts the color bin frequency distribution in a picture. It keeps track of identical pixels and stores them. Color bins are classified based on color intensity histogram and global bin. The color bin is a global color descriptor that examines each constant color in a picture. It is utilized to tackle issues like movement, rotation, and angle of view changes. The local color bin concentrates on specific areas of the image. In contrast to the global color bin, the local color bin takes into account the spatial pattern of pixels. Red, green, and blue (RGB), the color feature obtained from the disease sample image is shown in Figure 7.2. The intensity values for the background, healthy part, and the rotten part corresponding to each color channel are shown in Figure 7.2 (b)–7(d). Thus, leaf disease has its unique color features, which help in identifying the disease using an leaf image.

7.5.1 Advantage and disadvantages

Color histograms are useful for searching and retrieving image databases because they are simple to generate and insensitive to slight fluctuations in the image [36]. Along with these advantages, it has two major limitations. No consideration is given to overall spatial data. The histogram is not unique, because two separate photos with identical color distribution produce comparable histograms, whereas the same view images with varying light exposure produce different histograms [37].

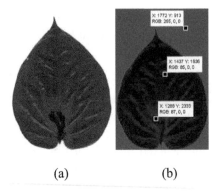

(a)   (b)

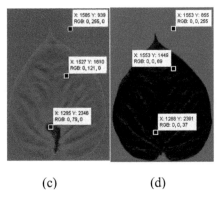

(c)   (d)

Figure 7.2   Color features extraction of infected leaf sample: (a) Original RGB image, (b) Red color channel, (c) Green color channel, (d) Blue color channel.

7.5.2 Color co-occurrence matrix (CCM)

In the color co-occurrence matrix [38] color variation between neighboring pixels is used to compute the likelihood of the same color pixel occurrence among adjacent pixels of the image, and this likelihood is used for the image characteristics evaluation [32].

Pixels make up an image, and each pixel represents four neighboring pixel colors. As shown in Figure 7.3, a three-by-three convolution mask was created for each pixel $G(i,j)$ in an image. It was then subdivided into four blocks of two-by-two grids (pixels), each of which included pixel $G(i,j)$. Taking the top left corner pixel p_1 first, scan sequences of these grids are analyzed, generating seven such patterns known as scan pattern sequence motifs, as illustrated in Figure 7.4.

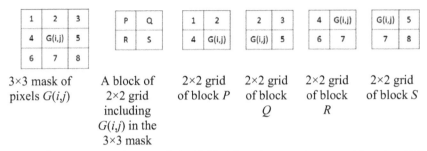

3×3 mask of pixels G(i,j)

A block of 2×2 grid including G(i,j) in the 3×3 mask

2×2 grid of block P

2×2 grid of block Q

2×2 grid of block R

2×2 grid of block S

Figure 7.3   A three-by-three mask is divided into four two-by-two grids using CCM approach

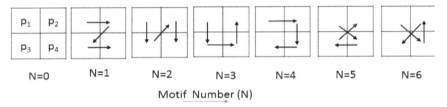

N=0   N=1   N=2   N=3   N=4   N=5   N=6

Motif Number (N)

Figure 7.4   The possible scan pattern motifs.

The image was now displayed as four images of scan pattern motifs, which were then divided into four two-dimensional image size matrices. Motifs of scan sequence produce picture attributes and the color co-occurrence matrix based on these four criteria.

7.5.2.1 Advantage and disadvantages

An image's CCM, which is typically represented as a histogram of intensity values, is a global feature that does not require knowledge of the image's component objects. Furthermore, CCM is unaffected by view or resolution, and color comparison can be done automatically without the need for human interaction [33].

The color histogram information of the complete image is empirically stored in the diagonal elements of CCM, while the shape information is theoretically stored in the non-diagonal parts of CCM. However, because the number of diagonal elements in CCM is far higher than the number of non-diagonal elements, indexing images based on CCM without modification suppresses the shape information (color corners). Thus CCM has poor evaluation over corner elements which may sometime contain important fetchers [34].

7.6 Shape-based Feature Extraction

The existing approaches to shape-based feature extraction are presented in this section. The contour and physical form of an object is referred to as its shape. One of an object's most essential characteristics is its shape. Most real-world objects do not even have regular geometrical shapes, making it difficult to characterize them. To express shape information, we associate unknown shapes with known items, such as how circular/spherical the item is. For example, in Figure 7.5, the perimeter boundary of a subject, the diameter enclosed within the perimeter, circular subject, triangular subject, or other shapes, of the subject all come under the category of shape features [35]. A shape is an object's exterior or visual aspect that is defined by boundaries.

Shape-based features are classified into two: one is a geometrical feature and the other is a moment feature. The geometrical features are perimeter, area, Euler number, hole, roundness, and distance. Moment-based shape features are center of gravity, orientation, boundary rectangle area, and eccentricity [35], [36].

a. Distances (Object length)

It is the shortest distance measured between two collinear pixels (x_1,y_1) and (x_2,y_2). It can be defined using eqn (7.1) as:

$$d=\sqrt{(x_1-x_2)^2+(y_1-y_2)^2} \tag{7.1}$$

b. Center of gravity

The centroid is another name for the center of gravity. Its position concerning the shape should be fixed. The centroid, which is comparable to the

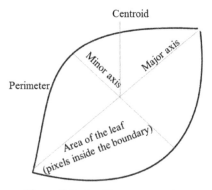

Figure 7.5 Leaf image geometry.

circle's center in complex polygons, indicates the object's center. The leaf centroid is shown in the figure and is calculated using eqn (7.2a) & (7.2b), respectively:

$$x_c = \frac{1}{\sum_{x=0}^{N} \sum_{y=0}^{M} f(x,y)} \sum_{x=0}^{N} \sum_{y=0}^{M} x f(x,y) \tag{7.2a}$$

$$y_c = \frac{1}{\sum_{x=0}^{N} \sum_{y=0}^{M} f(x,y)} \sum_{x=0}^{N} \sum_{y=0}^{M} y f(x,y) \tag{7.2b}$$

c. Area (A):

The area of an item will be a count of pixels enclosed in the perimeter of that item; as shown in the figure, all of the pixels within the item correspond to its area, i.e., area is a count of pixels in a closed perimeter. The area enclosed by the leaf shape in the above figure is given by eqn (7.3):

$$A = \sum_{x=1}^{n} \sum_{y=1}^{n} I(x,y) \tag{7.3}$$

d. Hole:

A gap in an image is referred to as a hole. For example, the English capital letter "B" has two holes and "P" has one hole.

e. Perimeter (p):

The perimeter of a leaf is the number of pixels that make up the leaf boundaries. The perimeter of a black and white image is the amount of foreground pixels that touch the image background. If $b_1, \ldots, b_N$ is a boundary variable, then perimeter can be obtained by using eqn (7.4) as:

$$p = \sum_{i=1}^{N-1} |b_i - b_{i+1}| \tag{7.4}$$

f. Major-axis Angle

The angle obtained between the major axis and x-axis of an image can be used to determine the object's major axis angle, it varies from 0° to 360°. It results in object orientation measurement. It is given by eqn (7.5):

$$Major - axis\ angle = \tan^{-1}\left(\frac{y_2 - y_1}{x_2 - x_1}\right) \tag{7.5}$$

g. Roundness

The ratio of area covered by an item to the area of a circle with an equal convex circumference can be used to estimate roundness or circularity (area-to-perimeter ratio) that ignores local irregularities. It is described in eqn (7.6):

$$Roundness = \frac{4\pi.\, Area}{(\text{perimeter})^2} \qquad (7.6)$$

h. Euler number

The Euler number can be calculated using the amount of holes (H) and connected components (C) in a picture and is given by eqn (7.7).

$$Eul.\, number = C - -H \qquad (7.7)$$

i. Eccentricity

It is the dimension of an object's short axis divided by the dimension of its long axis. Eccentricity is also known as ellipticity and is given by eqn (7.8).

$$Eccentricity = \frac{Minor\ axis\ length}{Major\ axis\ length} \qquad (7.8)$$

j. Boundary rectangle area

It is the area of the smallest rectangle that enclosed the target object and is given by eqn (7.9):

$$Boundary\ rectangle\ area = Major\ axis\ length\ \times Minor\ axis\ length \qquad (7.9)$$

7.6.1 Advantage and disadvantages

The shape is frequently associated with the object, which has a semantic value. As a result, the shape feature is a higher-level characteristic than the color and texture features [37]. The pre-segmentation effect affects the accuracy of the shape feature extraction output. This technique, also, lacks a solid mathematical model. The result is unreliable when the target is of irregular shape. When the target object is described in detail, there is a large demand for calculation and storage. Many of the objectives represent target shape information that is not the same as human perceptions [38].

7.7 Conclusion and Future Scope

A variety of features based on textures, color, and shape were explored in the present study. The feature selection (FS) process is the most important part of ML-based classification. This process aims to select the most valid and significant features in the classification process. The aim of feature selection is to find the smallest number of features from the data source that are relevant to the model's construction. There was redundancy in texture, color, and shape-based features which are taken care as follows:

- In texture, redundancy was removed by using GLCM 14 Haralick texture features were reduced to 5 significant features.
- In color-based features, redundancy was taken care of by CCM.
- Geometrical and moment feature in shape feature extraction further facilitate invariant features detection and hence considered as high-level features extraction than texture and color.

Pests and diseases damage roughly 20–40% of crops each year in the current environmental setting [39]. As a future scope of the present work, hybrid multi-features (texture color and shape) are combined to extract the complex object in practice that needs attention.

References

[1] G. Dhingra, V. Kumar, H. D. Joshi, 'Study of digital image processing techniques for leaf disease detection and classification', Multimedia Tools and Applications, 77(15), pp.19951-20000, 2018.
[2] S. J. J. Jui, A. M. Ahmed, A. Bose, N. Raj, E. Sharma, J. Soar, M. W. I. Chowdhury, 'Spatiotemporal Hybrid Random Forest Model for Tea Yield Prediction Using Satellite-Derived Variables', Remote Sensing, 14(3), 805, 2022.
[3] D. N. Singh, J. S. Bohra, Tyagi, V., T. Singh, T. R Banjara, G. Gupta, 'A review of India's fodder production status and opportunities', Grass and Forage Science, 77(1), 1-10, 2022.
[4] M. Sharma, C. J. Kumar, A. Deka, 'Early diagnosis of rice plant disease using machine learning techniques', Archives of Phytopathology and Plant Protection, 55(3), 259-283, 2022.
[5] S. Ashwinkumar, S. Rajagopal, V. Manimaran, B. Jegajothi, 'Automated plant leaf disease detection and classification using optimal

MobileNet based convolutional neural networks', Materials Today: Proceedings, 51, 480-487, 2022.

[6] D. Shah, N. Vora, C. Vora, B. Sonawane, 'Image-Based Plant Disease Detection', Springer, In Data Intelligence and Cognitive Informatics, pp. 651-666, Singapore, 2022.

[7] A. Rahman, M. Al Foisal, M. Rahman, M. Miah, M. F. Mridha, 'Deep-CNN for Plant Disease Diagnosis Using Low Resolution Leaf Images', Springer, In Machine Learning and Autonomous Systems, pp. 459-469, Singapore, 2022.

[8] M. Ji, Z. Wu, 'Automatic detection and severity analysis of grape black measles disease based on deep learning and fuzzy logic', Computers and Electronics in Agriculture, 193, 106718, 2022.

[9] A. K. Dey, M. Sharma, M. R. Meshram, 'Development of ANN and ANFIS Classifier for Betel Leaf Pathogen Detection', Journal of The Institution of Engineers (India): Series B, pp. 1-8, 2022.

[10] A. Singh, H. J. SV, D. Aishwarya, J. S. Jayasree, 'Plant Disease Detection and Diagnosis using Deep Learning', In IEEE, International Conference for Advancement in Technology (ICONAT), pp. 1-6, Jan., 2022.

[11] K. S. Archana, S. Srinivasan, S. P. Bharathi, R. Balamurugan, T. N. Prabakar, A. Britto, 'A novel method to improve computational and classification performance of rice plant disease identification', The Journal of Supercomputing, pp. 1-21, 2022.

[12] V. K. Shrivastava, M. K. Pradhan, 'Rice plant disease classification using color features: a machine learning paradigm', J Plant Pathol 103, pp. 17–26, 2021.

[13] U. K. Lilhore, A. L. Imoize, C. C. Lee, S. Simaiya, S. K. Pani, N. Goyal, C. T. Li, 'Enhanced convolutional neural network model for cassava leaf disease identification and classification', Mathematics, 10(4), 580, 2022.

[14] W. Yu, L. Deng, Y. H. Wu, 'Fatigue driving behavior recognition method based on image and vehicle feature change', Advances in Transportation Studies, 2022.

[15] V. K. Vishnoi, K. Kumar, B. Kumar, 'A comprehensive study of feature extraction techniques for plant leaf disease detection', Multimedia Tools and Applications, 81(1), pp. 367-419, 2022.

[16] S. Dananjayan, Y. Tang, J. Zhuang, C. Hou, S. Luo, 'Assessment of state-of-the-art deep learning based citrus disease detection techniques using annotated optical leaf images', Computers and Electronics in Agriculture, 193, 106658, 2022.

[17] M. R. Keyvanpour, S. Vahidian, Z. Mirzakhani, 'An analytical review of texture feature extraction approaches', International Journal of Computer Applications in Technology, 65(2), pp. 118-133, 2021.

[18] A. Sharma, A. Sharma, M. Averbukh, S. Rajput, V. Jately, S. S. Choudhury, B. Azzopardi, 'Improved moth flame optimization algorithm based on opposition-based learning and Lévy flight distribution for parameter estimation of solar module', Energy Reports, 8, 6576-6592, 2022.

[19] IN. Venkatesvara Rao, D. Venkatavara Prasad, M. Sugumaran, 'Real-time video object detection and classification using hybrid texture feature extraction', International Journal of Computers and Applications, 43(2), pp-119-126, 2021.

[20] S. Sachar, A. Kumar, 'Survey of feature extraction and classification techniques to identify plant through leaves', Expert Systems with Applications, 167, 114181. 2021.

[21] B. Pathak, D. Barooah, 'Texture analysis based on the gray-level co-occurrence matrix considering possible orientations', International Journal of Advanced Research in Electrical, Electronics and Instrumentation Engineering, 2(9), pp. 4206-4212, 2013.

[22] A. Sharma, A. Sharma, J. K. Pandey, M. Ram, 'Swarm intelligence: foundation, principles, and engineering applications', CRC Press, 2022.

[23] A. Ramola, A. K. Shakya, D. Van Pham, 'Study of statistical methods for texture analysis and their modern evolutions', Engineering Reports, 2(4), e12149, 2020.

[24] V. S. Thakare, N. N. Patil, 'Classification of texture using gray level co-occurrence matrix and self-organizing map'. In IEEE international conference on electronic systems, signal processing and computing technologies, pp. 350-355. Jan. 2014.

[25] N. Iqbal, R. Mumtaz, U. Shafi, S. M. H. Zaidi, 'Gray level co-occurrence matrix (GLCM) texture based crop classification using low altitude remote sensing platforms', Peer J Computer Science, 7, e536, 2021.

[26] A. Singh, A. Sharma, S. Rajput, A. Bose, X. Hu, 'An Investigation on Hybrid Particle Swarm Optimization Algorithms for Parameter Optimization of PV Cells', Electronics, 11(6), 909, 2022.

[27] A. P. Saikia, R. Roy, S. Datta, S. Gope, S. Deb, K. R. Singh, 'Classification of Medicinal Plant Species Using Neural Network Classifier: A Comparative Study', In Springer, Proceedings of the International Conference on Computational Intelligence and Sustainable Technologies, pp. 371-383, Singapore, 2022.

[28] P. Goyal, R. Rani, K. Singh, 'State-of-the-Art Machine Learning Techniques for Diagnosis of Alzheimer's Disease from MR-Images: A Systematic Review', Archives of Computational Methods in Engineering, 1-44, 2021.

[29] H. Sabrol, S. Kumar, 'Plant leaf disease detection using adaptive neuro-fuzzy classification', In Springer, science and information conference, pp. 434-443, Cham, April 2019.

[30] A. K. Rath, J. K. Meher, 'Disease detection in infected plant leaf by computational method', Archives of phytopathology and plant protection, 52(19-20), pp. 1348-1358, 2019.

[31] A. Blessy, D. D. C. J. W. Wise, 'Detection of Affected Part of Plant Leaves and Classification of Diseases Using CNN Technique', International Journal of Engineering and Techniques, 4(2), pp. 823-829, 2018.

[32] U. Shafi, R. Mumtaz, Z. Shafaq, S. M. H. Zaidi, M. O. Kaifi, Z. Mahmood, S. A. R. Zaidi, 'Wheat rust disease detection techniques: a technical perspective', Journal of Plant Diseases and Protection, pp. 1-16, 2022.

[33] E. Acar, O. F. Ertugrul, E. Aldemir, A. Oztekin, 'Automatic identification of cassava leaf diseases utilizing morphological hidden patterns and multi-feature textures with a distributed structure-based classification approach', Journal of Plant Diseases and Protection, pp. 1-17, 2022.

[34] V. Chaudhari, H. H. Dawoodi, M. P. Patil, Banana Leaf Disease Recognition Based on Local Binary Pattern. In Springer, Smart Trends in Computing and Communications, pp. 653-661, Singapore, 2022.

[35] M. S. Al-gaashani, F. Shang, M. S. Muthanna, M. Khayyat, A. El-Latif, 'Tomato leaf disease classification by exploiting transfer learning and feature concatenation', IET Image Processing, 2022.

[36] P. Kaur, S. Harnal, R. Tiwari, S. Upadhyay, S. Bhatia, A. Mashat, A. M. Alabdali, 'Recognition of Leaf Disease Using Hybrid Convolutional Neural Network by Applying Feature Reduction', Sensors, 22(2), 575, 2022.

[37] M. V. Appalanaidu, G. Kumaravelan, 'Plant leaf disease detection and classification using machine learning approaches: a review', Innovations in Computer Science and Engineering, pp. 515-525, 2021.

[38] A. Sharma, S. Choudhary, R. K. Pachauri, A. Shrivastava, D. Kumar, 'A review on artificial bee colony and it's engineering applications', Journal of Critical Reviews, 7(11), pp. 4097-4107, 2021.

[39] F. O. Sunmola, O. A. Agbolade, 'Design of Shallow Neural Network Based Plant Disease Detection System', European Journal of Electrical Engineering and Computer Science, 5(4), pp. 5-9, 2021.

[40] R. Damayanti, D. F. A. Riza, A. W. Putranto, R. J. Nainggolan, 'Vernonia Amygdalina Chlorophyll Content Prediction by Feature Texture Analysis of Leaf Color', In IOP Conference Series: Earth and Environmental Science, IOP Publishing. 757(1), pp. 012026. May, 2021.

[41] D. Sindhu, S. Sindhu, 'Image Processing Technology Application for Early Detection and Classification of Plant Diseases', International Journal of Computer Ences and Engineering, 7(5), 92-97. 2019.

[42] S. Adam, A. Amir, 'Fruit Plant Leaf Identification Feature Extraction Using Zernike Moment Invariant (ZMI) and Methods Backpropagation. In IEEE, International Conference on Informatics, Multimedia', Cyber and Information System (ICIMCIS), pp. 225-230, October, 2019.

[43] S. Mahajan, A. Raina, X. Z. Gao, A. Kant Pandit, 'Plant recognition using morphological feature extraction and transfer learning over SVM and AdaBoost. Symmetry', 13(2), 356, 2021.

[44] H. D. Gadade, D. K. Kirange, 'Machine Learning Based Identification of Tomato Leaf Diseases at Various Stages of Development', In IEEE. 5th International Conference on Computing Methodologies and Communication (ICCMC), pp. 814-819, April, 2021.

[45] P. K. Sujatha, J. Sandhya, J. S. Chaitanya, & R. Subashini, 'Enhancement of segmentation and feature fusion for apple disease classification', In IEEE, Tenth International Conference on Advanced Computing (ICoAC), pp. 175-181, December, 2018.

[46] B. Sahoo, S. K. Rath, S. K. Mahanta, M. Arakha, 'Nanotechnology Mediated Detection and Control of Phytopathogens', In Springer, Bio-Nano Interface, pp. 109-125, Singapore, 2022.

8

Development Methodologies for the Internet of Things: For all Commercial and Industrial Needs

Yazeed Alzahrani

School of Computing and Information Technology,
University of Wollongong, Australia

Abstract

The chapter majorly focuses on the system development methodologies of internet of things (IoT). A detailed literature survey is presented for the discussion of various challenges in the development of software, design, and deployment of hardware. Due to the increase in demand for electronic gadgets and internet connectivity, the security threats are also increasing day by day. Therefore, the demand of advanced development methodologies is also increasing day by day for high-quality and secure software products. The chapter discusses the development methodologies for different IoT applications. The methodologies are analyzed for two different parameters—physical infrastructure and emerging technologies. Security is a vital aspect of software quality, especially in this day and age, as the majority of software are distributed over the internet. Furthermore, as sensors deal with data, the security concerns for IoT are substantially greater than for traditional system techniques. Software development techniques are intended to increase software quality by incorporating actions that encourage quality in the final product. Individual developers are typically overlooked in most studies on producing high-quality and safe software. In this chapter, we analyze the secure-software development methodology for different IoT applications. The conclusion has provided the study an overall analysis and evaluation of the research.

Keywords: Internet of things, emerging technologies, physical infrastructure, system security, research methodology.

8.1 Introduction

Internet of things (IoT) and artificial intelligence (AI) have changed the way we live, work, and learn, and the demand for secure IoT communication with low- energy consumption and high data rate continues to increase rapidly. Currently, cellular networks and wireless local area networks coexist and support a diverse set of mobile services. Meanwhile, numerous new forms of wireless networks, such as cognitive radio networks and wireless sensor networks are still being created to address the needs of developing applications with new communication requirements. When constructing these networks, it is critical to consider not only how to obtain the desired capabilities for new applications but also how to achieve the required security (against various attacks/threats) and optimum bandwidth while conserving energy. This investigation requires an interdisciplinary effort that encompasses areas of security, communication, networking, and information theory. Through information-theoretic research, we wish to find the fundamental limits of protocols and develop new protocol designs to access the medium that ultimately achieve these limits with a high level of security. This can help to solve issues with the current internet of things for monitoring as well as other applications. The development of the protocol is a big challenge while developing software for various applications. The features/properties of the development can be divided into two categories— global characteristics and local characteristics. Majorly, there are two key requirements of the development methodologies for different applications as given below:

- Physical infrastructure
- Emerging technologies

In this chapter, we have discussed the development methodologies for the different IoT applications.

8.2 Research Questions

This study aims to answer the following questions:

- **RQ1:** The physical infrastructure required for the development of different IoT applications?

- **RQ2:** What are the new methods and emerging technologies that can improve the development methodology for IoT?
- **RQ3:** What are the different distributed security systems that can further improve the security of IoT applications?
- **RQ4:** What are the different research methods that can improve the IoT software and hardware systems?

To answer these research questions, the development methodology of the proposed work is discussed for five IoT applications. In all five IoT applications, the methodology is analyzed for two important parameters as given below:

- Physical infrastructure
- Emerging technologies

8.3 Development Methodologies

The development methodologies are discussed for five applications—smart cities, railway monitoring, agriculture field monitoring, automatic driving, forest monitoring.

8.3.1 Development methodologies for IoT-based smart cities

Cities 2.0, often known as smart cities, are digital-age representations of urban life [1]. In the coming years, suburban and rural populations are projected to migrate to urban areas, resulting in a tremendous population concentration in the city center. Cities are projected to benefit from emerging paradigms such as Industry 4.0 [2]. A major component is incorporating the internet of things (IoT) paradigm as the backbone of society [3]. IoT-enabled services will generate vast volumes of data that may be utilized to support and optimize critical infrastructure while also providing new insights and breakthroughs. The great bulk of these data, however, will be sensitive and should be managed discreetly so that individual liberty and privacy are not jeopardized. The challenge today is figuring out how to build and deploy effective and dependable massively networked systems. One area where we might learn anything beneficial is the deployment of monitoring techniques in smart campuses. A college or university campus is a scaled-down version of a city, with a fairly confined community large enough to encounter many of the technological, social, and human challenges that arise at the city scale. However, no comprehensive and systematic survey of monitoring smart campus systems has been undertaken to our knowledge. This essay was inspired by the lack of research that seeks to characterize the state-of-the-art smart campus surveillance. As a result, the essay investigates IoT-based surveillance systems in smart campuses, which, while similar to smart cities, have some unique characteristics that demand additional security and privacy safeguards. We developed a taxonomy and grading technique for each of these systems. The functionality of the researched systems was evaluated using physical infrastructure, supporting technologies, software analytics, system security, and research methods. The weights collected by the taxonomy enable a reliable comparison of state-of-the-art systems. In addition, the strategy makes it easier to draw relevant conclusions and deductions, as well as providing insights and guidance into the critical services supplied by a smart campus monitoring system. Finally, the study reveals a number of research projects aiming at developing future smart campus monitoring systems.

8.3.1.1 Spatial (Geographical) IoT

Smart campus monitoring is improved by the proliferation of sensors and actuators integrated to support geographic IoT awareness. Geographic position prediction and time of arrival estimation are possible thanks to stochastic processes based on IoT networks [4]. In such systems, it is possible to use a publish–subscribe utility to link sensor activity to specific geospatial data sources of observed measures for monitoring purposes [5]. Edge computing models focus on data collection, capturing only the most important information from IoT devices. These types of localized data sources are used to provide geospatial-based services [6]. Disaster management services are provided using IoT data streams that are geospatially annotated to facilitate big data analytics aimed at offering campus recovery [7]. Geospatial modeling is utilized to support safe student transportation within smart campus by exploiting geo-located sensors' infrastructure design to create a secure IoT-enabled architecture. Hassine *et al.* [9] focus on a survey on geospatial IoT, which leverages personalized location-based services that are context-aware to provide viable geospatial analytical approaches and monitoring applications for smart campus physical infrastructure. Narendra *et al.* [10] propose an event-driven architecture for asynchronous transactions across a campus sensor network by utilizing spatiotemporal data sources for online analytical streaming processing. Geospatial analysis can be utilized to examine and monitor the campus area using data from wireless IoT sensors and actuators. This type of technology is used to build and maintain a secure smart campus public space architecture [11]. Location awareness is supported by a system described in [12], which uses geospatial IoT-driven apps to provide an integrated solution for campus surveillance. Kotronis *et al.* [13] present a software architecture for providing smart campus integrated micro-services to students that allows for geospatial data sources and analytical analysis. For successful campus monitoring utilities, several geographic IoT services are essential.

8.3.1.2 Smart campus surveillance

Sensors and actuators may be easily integrated with IoT technology for effective smart campus surveillance. Students are unobtrusively watched in such an atmosphere to protect their privacy and human rights. Students must be aware that they are being observed in order to provide well-being in their workplace, according to ethical and legal criteria. Monitoring public spaces is an effective deterrent to delinquent behavior, resulting in a safer environment for everyone [14], [15]. Surveillance systems in IoT-enabled smart campuses are also designed to catch such behavior and better understand the individual reasons and core causes. Inferences drawn from the collected data can then be used to guide delinquent behavior prevention, prediction, and early warning before it occurs, acting as a security shield for today's smart campus. We examine a large number of systems in the smart campus monitoring sector to determine their strengths and limitations in this article. The goal is to create a foundation for categorizing contemporary technologies offered in research efforts and patents based on their surveillance value. However, before we can report on the results of this survey, we need to compare systems based on the suggested taxonomy's study dimensions. To conduct a comparative analysis, we begin by defining a concrete taxonomy that takes advantage of the available systems. This taxonomy will serve as the foundation for mapping any smart campus surveillance system, allowing for comparisons to comparable systems in the literature. Readers and researchers will be able to discover any flaws in current research and propose efficient strategies for dealing with new frameworks in the subject using the proposed taxonomy and classification.

The development methodology for smart cities can be classified into five different sections. The first module is the physical infrastructure of the smart cities as shown in Figure 8.1. There are three important parts of the physical infrastructure. The first part deals with smart shopping malls. As shopping is one of the important parts of today's life. Due to the pandemic (COVID-19), the scenario of shopping is completely changing. The basic requirement of the shopping mall is to develop a smooth rush-free shopping environment for

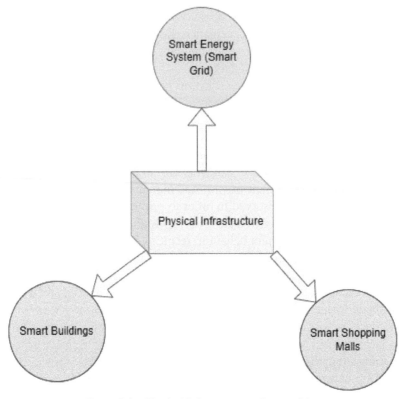

Figure 8.1   Physical infrastructure of smart cities.

customer satisfaction. The secondary requirement of the business organization is to generate high revenue from the system. Customer satisfaction will also increase the revenue in a long term. Therefore, the major focus of smart shopping malls is to provide a high-quality service in a rush-free environment. This door-to-door delivery through online shopping is already popular and is used by customers. However, there are two major challenges with this technology. The first challenge is that in some cases, the online view of the product is comparably far different than the exact view of the product. This affects the trust of the customer in the organization as well as it also increases the losses of the organization due to the return arrangement of the product. The challenge can be addressed by the advancement of virtual reality (VR) and augmented reality (AR). The second challenge is to provide the online details of the customer density in the mall for the efficient management of the customer density in the mall.

The second important part of the physical architecture of smart cities is to develop smart buildings for the home and offices. As the demand for the data rate is increasing day by day, and therefore the researchers are working to a higher frequency range. The 5G already shifted the mm-wave range and in 6G, the researchers are working on 150 GHz. A number of researchers are working in the terahertz range for efficient data transmission. However, the higher frequency shift causes many drawbacks for the infrastructure. The range of the signals reduces for the higher frequency. This will increase the demand for base stations and will increase the density of base stations in a particular area. However, the size of the base station will also be reduced. Thus, in modern smart buildings, we can install the base stations for the buildings. This will improve the communication quality of the system.

The third important part of smart cities is to develop a smart energy system (smart grid). The distribution of energy is one of the important challenges for smart cities. The IoT-based smart grid method can be used for power distribution and also for smart metering. For this, each smart building can be connected to smart grids. The smart grid can be connected to the IoT cloud and users can control all the appliances through the IoT cloud.

The next important part of the smart cities is new emerging technologies as shown in Figure 8.2. The new emerging technologies can be categorized into five different parallel sections. The first module is the adaptability with 5G telecommunication technology. The 5G supports various methods for the integration of other emerging technologies. The high-bandwidth availability at the mm-wave range provides short-range high data rate communication. Cognitive radio is the second strong field for data communication. The VHF, UHF, and other traditional radio channel frequencies are not utilized nowadays due to digital communication methods. Also, many other bands are not optimally utilized all the time. Many of the devices do not transmit the data all the time. The cognitive radio devices can sense the unutilized bandwidth range and the data can be transmitted through the free channel. The cognitive radio can also be utilized in 5G telecommunication range. Apart from that core, IoT technologies can also play an important role in smart cities. Active monitoring of temperature, humidity, wind speed, and other air pollution parameters is very important for air quality index and weather forecasting. Similarly, passive monitoring of device performance is also important for the internal monitoring of the local devices.

The third important structure of the important smart city is system security. The security of the system is one of the most important parts of any

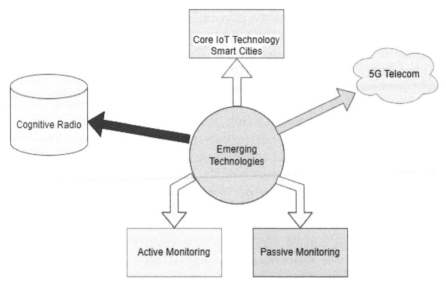

Figure 8.2 Emerging technologies for smart cities.

system. The security requirement depends upon the type of the application. The concept of the smart city is based on the cloud and the internet. Therefore, security standards should be very high. In the smart cities, the complete banking system, as well as local government data, will also utilize cloud services. Thus, the security requirement depends upon some important factors. In today's scenario, blockchain provides distributed security which is safer as compared to the other security methods. To analyze the security of the network, a few trust parameters can be considered based on the security requirement, for example, latency, efficiency, etc. The proposed application-specific model can be analyzed for the different attack models. The different standard attack models help to analyze the security of the protocol.

Software analysis is also one of the important parts of all IoT applications. Software plays an important role in the design and development of any machine. Because the smart city is completely dependent on the cloud and internet, the optimal utilization of the resources is a big challenge. Apart from the application requirement, the other important part of the software design is that it should be able to fulfill the device requirement and field requirement. Device interfacing is also an important challenge for advanced IoT-based smart cities. The interfacing of the devices should be backward compatible. The smart surveillance system can also be used in smart cities. The defaulters can be traced and it will help the police also to trap the attackers. The efficient software can perform the data analysis for the surveillance system. The last important part is that the software should be compatible with the emerging technologies needed.

The last and most important part of all the IoT applications is the scope of future development. Various research methods can be used for the performance improvement and development of software. However, the IoT integration with 5G is still in the development stage. Therefore, the information pre-processing is important for the future development of the tool.

8.3.2 Development methodologies for IoT-based automatic driving

The IoT-based automatic driving is now the future of the transportation industry. As we know that automatic driving is only possible through severe control from the cloud and efficient monitoring of the roads. All the vehicles are controlled through the server with the help of control signals.

The control signal is very important for the control of vehicles, thus, the transmission of control signals should be transmitted through a highly reliable channel. 5G telecommunication technology provides very highly reliable communication through the dedicated channel. The automatic driving development methodology can also be categorized into two parts. However, the system security, software analytics, and research methodology are the same as already discussed in the smart cities. In this section, the physical infrastructure requirement and emerging technologies for automatic driving are discussed.

The physical architecture of the automatic driving requires modification in the roads as well as in the vehicles as shown in Figure 8.3. For the smart roads, various sensors are required on the roads, for example, pressure sensor, FBG sensor, alignment sensor, infrared sensors, ultrasonic sensor, etc. Similarly, to make efficient communication between vehicle-to-vehicle and vehicle-to-server requires roadside devices as well as smart vehicles. The roadside devices continuously communicate with the server and transmit the control signals to the vehicles.

The emerging parallel technologies are also very important to discuss as the integration with the other technologies makes the system more powerful as compared to the independent system. The integration is shown in Figure 8.4. The 5G telecommunication integration is one of the primary

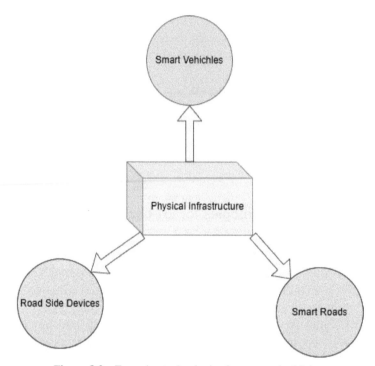

Figure 8.3   Emerging technologies for automatic driving.

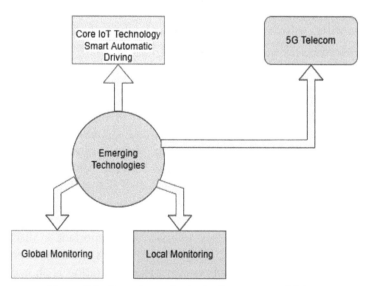

Figure 8.4   Emerging technologies for automatic driving.

requirements of the current VANET. The 5G high-frequency range, that is, mm-wave can communicate directly roadside with devices. Also, IoT can be integrated with VANET. IoT with ML and AI can make the VANET system very powerful. IoT integration will also make the system more powerful for real-time applications. Apart from this, regional monitoring is also very important for the transportation system. Therefore, the integration of VANET with the existing monitoring system is important. The integration of VANET can be done with weather forecasting monitoring, traffic monitoring, road health monitoring, and a smart city spatial map.

8.3.3 Development methodologies for IoT-based agriculture field monitoring

The IoT-based agriculture field monitoring will be the future of efficient fulfillment of future needs. The agriculture field monitoring improves the efficiency of crop development and also fulfills the optimal need for pesticides and also other agricultural needs. The agriculture sensors can be classified into event-driven and continuous monitoring nodes. The agriculture field development methodology can also be categorized into five parts. However, the system security, software analytics, and research methodology are the same as already discussed in the smart cities. In this section, the physical infrastructure requirement and emerging technologies for agriculture field monitoring are discussed.

The physical devices are the primary requirement for the agriculture field monitoring as shown in Figure 8.5. Smart sensors and actuators play an important role in the smart agricultural field. The drone or smart actuators can be placed on the monitoring field and these devices operate as per the instruction given by the headquarters.

Emerging technologies also play a big role in the improvement of the devices in the future as shown in Figure 8.6. The integration of the devices with the newly available technology devices is the challenge. The 5G telecommunication integration is one of the primary requirements of the agriculture field monitoring. The 5G high-frequency range, that is, mm-wave can communicate with the intermediate devices. Also, IoT can be developed for field monitoring. IoT with ML and AI can make field monitoring systems very powerful. IoT integration will also make the system more powerful for real-time applications. Apart from this continuous monitoring and event monitoring, nanofabrication can further improve the performance in terms of sensor accuracy as well as energy consumption of the nodes.

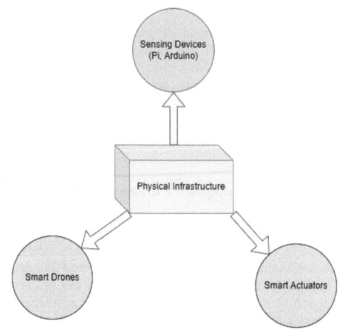

Figure 8.5 Physical infrastructure of agriculture field monitoring.

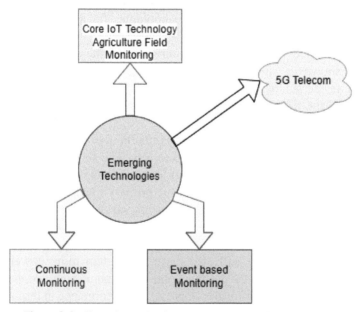

Figure 8.6 Emerging technologies for agriculture field monitoring.

8.3.4 Development methodologies for IoT-based railway monitoring

Railway monitoring is one of the most important fields where the integration of different emerging fields can play a big role in IoT-based monitoring. The development methodology of railway monitoring focuses on heterogeneous data traffic application, for example, railway monitoring, and telemedicine applications. A large number of different types of sensors are required for railway condition monitoring applications. The traffic generation rate of the sensors is different and it is based on their requirements for railway bridge monitoring, landslide monitoring, track condition monitoring, etc. The sensors are placed on different track locations based on their requirements. Analysis of the data traffic generation rate shows that for some places traffic generation rate is low and for other places, it is very high (20—120 Kbps). So, the networks should be analyzed for low, medium, and high traffic generation rates in the range of 20–120 Kbps. The development methodology of railway monitoring can also be categorized into five parts. However, the system security, software analytics, and research methodology are the same as already discussed in the smart cities. In this section, the physical infrastructure requirement and emerging technologies for railway monitoring are discussed.

The physical architecture of railway monitoring is dependent on smart IoT devices as shown in Figure 8.7. The most important development parts are smart stations, smart train, and smart track monitoring systems. Bogie monitoring and engine monitoring are two different sections of train monitoring. Railway track condition monitoring can also be categorized into different subsections.

Technology is changing day by day. In today's perspective, the five major fields are growing very rapidly as shown in Figure 8.8. IoT and artificial intelligence (AI) are growing very rapidly. The IoT-based cloud system with a strong machine learning algorithm can be used for the advanced IoT platform. 5G is also advancing day by day. The advancement of 5G is rapidly increasing to meet the user requirement. The bandwidth requirement can also be fulfilled by utilizing the unutilized bands. For this, one can use cognitive radio. The advancement in sensor technology is also changing day by day. Therefore, the energy requirement of both event-based monitoring as well as continuous monitoring sensors will decrease and the lifetime of the overall network will increase.

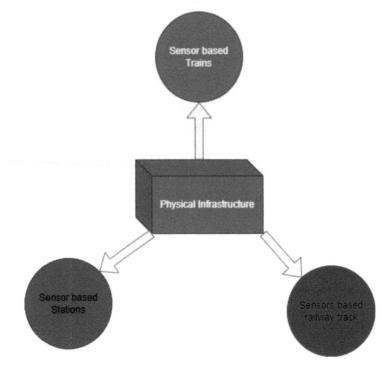

Figure 8.7 Physical infrastructure of railway monitoring.

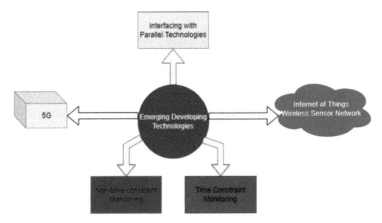

Figure 8.8 Emerging technologies for railway monitoring.

8.3.5 Development methodologies for IoT-based forest monitoring

The development methodology of forest monitoring is majorly focused on event detection in the monitoring field. The sensors sense the data and transmit the data to the sensor network. The bus system is used to transmit that data to the micro-controller. The sensor analyst is used to analyze that data to further process that data. At the server end, the monitoring station performs this task. After the analysis of the data, the monitoring station transmits the command to the actuators to perform the necessary action. The sensors distributed over the forest monitoring field can detect event monitoring as well as continuous monitoring data. The data traffic generated by event monitoring sensors can be used for the detection of an event.

The task of the physical devices for the forest monitoring application can be categorized into two parts—wildfire monitoring as well as an early prediction as shown in Figure 8.9. The second task is to save the forest from

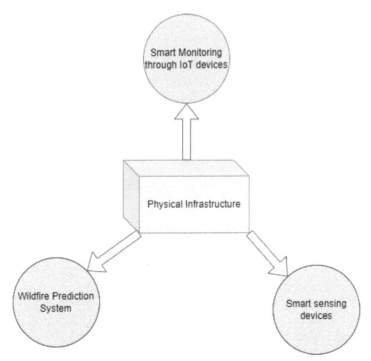

Figure 8.9 Physical infrastructure of forest monitoring.

unmeaningful, undesired, and criminal activities. For this, three different types of IoT networks and devices are required as shown in Figure 8.9.

Emerging technologies also play a big role in the improvement of the devices in the future as shown in Figure 8.10. The integration of the devices with the newly available technology devices is the challenge. The 5G telecommunication integration is one of the primary requirements of forest monitoring. The 5G high-frequency range i.e., mm-wave can communicate with the intermediate devices. Also, IoT can be developed for field monitoring. IoT with ML and AI can make field monitoring systems very powerful. IoT integration will also make the system more powerful for real-time applications.

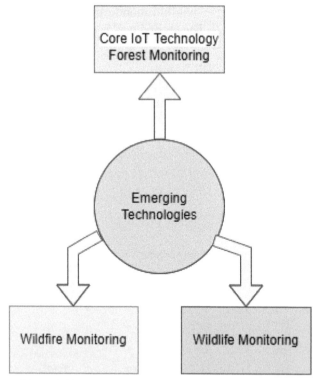

Figure 8.10 Emerging technologies for forest monitoring.

8.4 Conclusion

In this chapter, the development steps are discussed for five different IoT applications. By using the above steps, advanced hardware and software can be developed for the IoT application. All the five IoT applications are discussed for physical infrastructure, emerging technologies, system security, software analytics, and research methodology. As IoT systems provide a plethora of benefits, efficiently implementing such systems becomes increasingly vital. Because IoT systems include many software, hardware, and communication components, they are more difficult to design than standard software systems. Therefore, efficient development of IoT systems requires systematic approaches that come into existence mainly in the form of system development methodologies.

References

[1] Liu, Xing and Baiocchi, Orlando, A comparison of the definitions for smart sensors, smart objects and Things in IoT, 2016 IEEE 7th Annual Information Technology, Electronics and Mobile Communication Conference (IEMCON), 1–4, 2016, IEEE.

[2] Windekilde, I., others (2017). From Reactive to Predictive Services: the Internet of Things (IoT) Enabled Product Service Systems (PSS) of Innovative and Sustainable Business Model. Marketing i Zarzkadzanie, 50(4), 117–127.

[3] Li, N., Sun, M., Bi, Z., Su, Z. Wang, C. (2014). A new methodology to support group decision-making for IoT-based emergency response systems. Information systems frontiers, 16(5), 953–977. Gartner Identifies the Top 10 Strategic Technology Trends for 2019.

[4] Noronha, A., Moriarty, R., O'Connell, K., Villa, N. (2014). Attaining IoT Value: How to move from connecting things to capturing insights. Cisco, Republica Tcheca.

[5] Gluhak, A., Krco, S., Nati, M., Pfisterer, D., Mitton, N., Razafindralambo, T. (2011). A survey on facilities for experimental internet of things research. Journal is required.

[6] Foteinos, V., Kelaidonis, D., Poulios, G., Vlacheas, P., Stavroulaki, V., Demestichas, P. (2013). Cognitive management for the internet of things: A framework for enabling autonomous applications. IEEE vehicular technology magazine, 8(4), 90–99.

[7] Larrucea, X., Combelles, A., Favaro, J., Taneja, K. (2017). Software engineering for the internet of things. IEEE Software, 34(1), 24–28.

[8] Reggio, G. (2018). A UML-based proposal for IoT system requirements specification. In Proceedings of the 10th International Workshop on Modelling in Software Engineering (pp. 9–16).

[9] Hassine, T., Khayati, O., Ghezala, H. (2017). An IoT domain meta-model and an approach to software development of IoT solutions. In 2017 International Conference on Internet of Things, Embedded Systems and Communications (IINTEC) (pp. 32–37).

[10] Narendra, N., Ponnalagu, K., Ghose, A., Tamilselvam, S. (2015). Goal-driven context-aware data filtering in IoT-based systems. In Intelligent Transportation Systems (ITSC), 2015 IEEE 18th International Conference on (pp. 2172–2179).

[11] Mezghani, E., Exposito, E., Drira, K. (2017). A model-driven methodology for the design of autonomic and cognitive IoT-based systems: application to healthcare. IEEE Transactions on Emerging Topics in Computational Intelligence, 1(3), 224–234.

[12] Sosa-Reyna, C., Tello-Leal, E., Lara-Alabazares, D. (2018). Methodology for the model-driven development of service oriented IoT applications. Journal of Systems Architecture, 90, 15–22.

[13] Kotronis, C., Nikolaidou, M., Dimitrakopoulos, G., Anagnostopoulos, D., Amira, A., Bensaali, F. (2018). A Model-based Approach for Managing Criticality Requirements in e-Health IoT Systems. In 2018 13th Annual Conference on System of Systems Engineering (SoSE) (pp. 60–67).

[14] Mazzini, S., Favaro, J., Baracchi, L. (2015). A Model-Based Approach Across the IoT Lifecycle for Scalable and Distributed Smart Applications. In Intelligent Transportation Systems (ITSC), 2015 IEEE 18th International Conference on (pp. 149–154).

[15] Duan, R., Chen, X., Xing, T. (2011). A QoS architecture for IOT. In Internet of Things (iThings/CPSCom), 2011 International Conference on and 4th International Conference on Cyber, Physical and Social Computing (pp. 717–720).

[16] Sharma, V., Das, S., Kewaley, S. (2015). Design Thing'ing: methodology for understanding and discovering Use cases in IoT scenarios. In Proceedings of the 7th International Conference on HCI, IndiaHCI 2015 (pp. 113–115).

[17] Ide, M., Amagai, Y., Aoyama, M., Kikushima, Y. (2015). A lean design methodology for business models and its application to IoT business model development. In 2015 Agile Conference (pp. 107–111).

[18] Fortino, G., Russo, W., Savaglio, C. (2016). Agent-oriented modeling and simulation of IoT networks. In 2016 federated conference on computer science and information systems (FedCSIS) (pp. 1449–1452).

[19] Manate, B., Fortis, F., Moore, P. (2014). Applying the Prometheus methodology for an Internet of Things architecture. In 2014 IEEE/ACM 7th International Conference on Utility and Cloud Computing (pp. 435–442)

[20] Zambonelli, F. (2016). Towards a Discipline of IoT-Oriented Software Engineering.. In WOA (pp. 1–7).

[21] Fortino, G., Guerrieri, A., Russo, W., Savaglio, C. (2015). Towards a development methodology for smart object-oriented IoT systems: A metamodel approach. In 2015 IEEE international conference on systems, man, and cybernetics (pp. 1297–1302).

[22] A Model-Based Approach Across the IoT Lifecycle for Scalable and Distributed Smart Applications - IEEE Conference Publication.

[23] Giang, N., Blackstock, M., Lea, R., Leung, V. (2015). Developing IoT applications in the Fog: A Distributed Dataflow approach. In 2015 5th International Conference on the Internet of Things (IOT) (pp. 155–162).

[24] Venticinque, S., Amato, A. (2018). A methodology for deployment of IoT application in fog. Journal of Ambient Intelligence and Humanized Computing, 1–22.

[25] Puhlmann, F., Slama, D. (Year is required!). An IoT Solution Methodology. Journal is required!.

[26] Udoh, I., Kotonya, G. (2018). Developing IoT applications: challenges and frameworks. IET Cyber-Physical Systems: Theory Applications, 3(2), 65–72.

[27] Seydoux, N., Drira, K., Hernandez, N., Monteil, T. (2016). IoT-O, a core-domain IoT ontology to represent connected devices networks. In European Knowledge Acquisition Workshop (pp. 561–576).

[28] Kotis, K., Katasonov, A. (2012). An iot-ontology for the representation of interconnected, clustered and aligned smart entities. Technical report, VTT Technical Research Center, Finland VTT Technical Research Center, Finland.

[29] Ouchani, S. (2018). Ensuring the Functional Correctness of IoT through Formal Modeling and Verification. In International Conference on Model and Data Engineering (pp. 401–417).

[30] Kravari, K., Bassiliades, N. (2018). A rule-based eCommerce methodology for the IoT using trustworthy Intelligent Agents and Microservices. In International Joint Conference on Rules and Reasoning (pp. 302–309).

[31] Elyasaf, A., Marron, A., Sturm, A., Weiss, G. (2018). A Context-Based Behavioral Language for IoT.. In MODELS Workshops (pp. 485–494).

[32] Kravari, K., Bassiliades, N. (2018). A rule-based eCommerce methodology for the IoT using trustworthy Intelligent Agents and Microservices. In International Joint Conference on Rules and Reasoning (pp. 302–309).

[33] Kammüller, F., Augusto, J., Jones, S. (2017). Security and privacy requirements engineering for human centric IoT systems using eFRIEND and Isabelle. In Software Engineering Research, Management and Applications (SERA), 2017 IEEE 15th International Conference on (pp. 401–406).

[34] Stetsuyk, E., Maevsky, D., Maevskaya, E. (2018). METHODOLOGY OF GREEN SOFTWARE DEVELOPMENT FOR THE IOT DEVICES1. Tablet, 20(1), 46.

[35] Krco, S., Pokric, B., Carrez, F. (2014). Designing IoT architecture (s): A European perspective. In Internet of Things (WF-IoT), 2014 IEEE World Forum on (pp. 79–84).

[36] Hossain, M., Fotouhi, M., Hasan, R. (2015). Towards an analysis of security issues, challenges, and open problems in the internet of things. In Services (SERVICES), 2015 IEEE World Congress on (pp. 21–28).

[37] Bugeja, J., Jacobsson, A., Davidsson, P. (2016). On privacy and security challenges in smart connected homes. In Intelligence and Security Informatics Conference (EISIC), 2016 European (pp. 172–175).

[38] Mavropoulos, O., Mouratidis, H., Fish, A., Panaousis, E., Kalloniatis, C. (2016). Apparatus: Reasoning about security requirements in the internet of things. In International Conference on Advanced Information Systems Engineering (pp. 219–230).

[39] Udoh, I., Kotonya, G. (2017). Developing IoT applications: challenges and frameworks. IET Cyber-Physical Systems: Theory Applications.

[40] Ali, S., Jarwar, M., Chong, I. (2018). Design Methodology of Microservices to Support Predictive Analytics for IoT Applications. Sensors, 18(12), 4226.

[41] Fortino, G., Russo, W., Savaglio, C., Shen, W., Zhou, M. (2017). Agent-oriented cooperative smart objects: From IoT system design to implementation. IEEE Transactions on Systems, Man, and Cybernetics: Systems(99), 1–18.

[42] Giang, N., Blackstock, M., Lea, R., Leung, V. (2015). Developing IoT applications in the fog: a distributed dataflow approach. In Internet of Things (IOT), 2015 5th International Conference on the (pp. 155–162).

[43] Carvalho, R., Andrade, R., Oliveira, K., Kolski, C. (2018). Catalog of Invisibility Requirements for UbiComp and IoT Applications. In 2018 IEEE 26th International Requirements Engineering Conference (RE) (pp. 88–99).

[44] Dhouib, S., Cuccuru, A., Le F\'evre, F., Li, S., Maggi, B., Paez, I., Rademarcher, A., Rapin, N., Tatibouet, J., Tessier, P., others (2016). Papyrus for IoT–A Modeling Solution for IoT. CEA, LIST, Laboratory of Model Driven Engineering for Embedded Systems, 91191.

9

Bio-inspired Multilevel ICHB-HEED Clustering Protocol for Heterogeneous WSNs

Prateek Gupta[1], Amrita[2], Himansu Sekhar Pattanayak[3], Gunjan[4], Lalit Kumar Awasthi[5], and Vachik S. Dave[6]

[1]University of Petroleum & Energy Studies, India
[2]Banasthali Vidyapith, India
[3]Bennett University Greater Noida, India
[4]SRM University Delhi-NCR, India
[5]National Institute of Technology, India
[6]Walmart Global Tech, USA
E-mail: cseprateek@gmail.com1; saiamrita27@gmail.com;
himansusekharpattanayak@gmail.com; gunjan.rehani@gmail.com;
lalitdec@gmail.com; vachik.dave25@gmail.com

Abstract

A good clustering approach in the context of a wireless sensor network (WSN) is to organize the sensor nodes (SNs) in the network to optimize the consumption of energy as well as ensure a longer operational lifespan. The HEED protocol (Hybrid Energy-Efficient Distributed) is a clustering technology that is frequently used in WSNs. ICHB-HEED is a smart cluster head election protocol using BFOA (bacterial foraging optimization algorithm). In this paper, we propose a multi-level ICHB-HEED (MLICHBHEED) protocol consisting of varying energy levels in a heterogeneous model based on the ICHB-HEED protocol, i.e., MLICHBHEED-1, MLICHBHEED-2, MLICHBHEED-3, MLICHBHEED-4, and MLICHBHEED-5 up to five levels of heterogeneity, which can further be extended to any level of heterogeneity as per the WSN's requirement. The experiments show that when the level of heterogeneity increases, the energy dissipation rate lowers down, allowing nodes to stay active for longer periods, resulting in a considerable increase in network lifespan.

Keywords: Wireless sensor network, HEED, clustering, optimization, ICHB algorithm, ICHB-HEED, bio-inspired.

9.1 Introduction

Nowadays, a broad array of wireless topologies supports several applications because fixed infrastructure is no longer required and becomes obsolete in today's era of wireless connection-based topologies. Wireless solutions are able to work around the wired system's shortcomings. Wireless communication is widely utilized for numerous applications, which include agriculture monitoring, micro-climate monitoring, volcano surveillance, habitat monitoring, military actions, and disaster relief management, among many others. Microsensors were developed as a result of recent developments in MEMS (micro-electromechanical systems) and VLSI (very large scale integration). WSNs are created when wireless SNs are joined together in a loosely coupled fashion in any type of network. The vast majority of these SNs are responsible for data processing and transmission functions. Event generation, query generation, and periodic data collection serve as the methods that SNs use to obtain data about their environment, which is then sent to BS (base station) for decision-making and processing [1]–[3]. The tiny size of SNs means that they have limited battery life, storage, and processing capabilities and that they are more vulnerable if they are left unattended [4, 5]. The major goal is to build a low-energy protocol that can keep the sensors in the network working longer [6, 7]. This network of SNs, commonly referred to as an ad-hoc network, is an essential component for the creation of energy-efficient protocols that use an ad-hoc network of SNs. Selecting the correct cluster head (CH), an optimal cluster size, cluster management, and data routing without redundancy are all vital. To assist in resolving these challenges, developing a clustering-based protocol that deals with these issues will be useful [5, 8].

Several routing protocols have been created based on clustering methods, LEACH (Low-energy adaptive clustering hierarch [9], Adaptive periodic TEEN (APTEEN), and Modified TEEN (MTEEN) [10] are the most popular among them. On the other hand, it has certain disadvantages, such as the development of more CHs as a result of untraversed nodes that are converted to CHs, subsequently after preliminary CHs are finalized [11]. As the number of packets increases, it results in a surge of broadcast overhead. This results in apparent energy depletion in SNs, which results in a reduction in network life. The additional stress placed on CHs, particularly those located close to the BS, increases the likelihood of hot spots forming. An essential thing to consider when developing a method is minimizing energy use. Providing quality of service (QoS) and managing competing concerns such as coverage, life duration, and performance are critical factors in providing great customer service [12, 13, 15]. To rectify these challenges, many optimization methodologies are highly acclaimed, which incorporate biological inspiration, artificial intelligence, expert knowledge, and meta-heuristic processes, such as PSO (Particle swarm optimization), ACO (ant colonial-based optimization), BFOA (bacterial foraging optimization algorithm), GA (genetic algorithms), multi-objective optimization [14] and a fuzzy logic FLS. In [16], the ICHB method, which is based on the BFOA (bacterial foraging optimization algorithm) for selecting suitable network CHs focusing on residual energy as the key attribute in a few variations of OHEED, has been developed and tested. In [17], multi-level network model for heterogeneous WSN was proposed. To have an energy-efficient ICHB-HEED protocol, we present the multi-level ICHB-HEED (MLICHBHEED) protocol for non-homogenous networks constituted of SNs.

The remainder of this article follows the following order: Discussion of relevant literature is presented in Section 9.2; discussion of the proposed work is presented in Section 9.3. Section 9.4 discusses the results of the proposed approach, and finally, the work is concluded in Section 9.5.

9.2 Review of Literature

To distribute load and manage energy in WSNs, several clustering techniques have been devised in the past. The well-known LEACH method, invented by Heinzelman *et al.*, is considered the first clustering technique based on a distributed model. There are two stages to this method. SNs elect CH using a probabilistic approach during the setup phase, and in the steady-state phase, the transmission of the own observed data of the SNs happens to its associated CH. In a single hop, each CH, also known as a collection node, collects and sends data to the BS. If each SN has a probability value connected with it, an SN with lesser energy can also become a CH. Despite this, the network distributes CHs in an unbalanced manner. This method is a widely used benchmark, which also serves as a source of inspiration for many algorithm refinement methods, including TL-LEACH [20], LEACH-B

[44], LEACH-M [18], and LEACH-C [45]. Loscri *et al.* [20] concluded that several writers have used multi-hop routing to spread traffic evenly. In PEGA-SIS (power-efficient gathering for information systems), efficient SNs with minimal power usage [21] were addressed. In this instance, only one leader node (i.e., CH) communicates with the BS in each round. The various SNs transmitted signals. Data packets are sent to proximity nodes. An SN is a data-gathering device. The closer it gets to the node's leader, the more it combines sensed data with the received data with its sensed data, and fused data packets are transmitted toward the next node until it reaches the node's leader. HEED, a multi-hop clustering method developed by Younis and Fahmy [22], used an independent-observation design to evaluate the impact of CHs election on outcomes. The choice was based on two factors: communication cost inside the cluster and remaining energy, which depends on the number of nodes around each node. It was impossible to calculate intra-cluster communication costs unless all of the network information was known. Integrated HEED (iHEED) was suggested by Younis and Fahmy [23] as an improvement to HEED, which utilized data aggregation procedures. The operations, such as AVG and MAX, were employed. This procedure utilizes data aggregators. There is a cluster-based routing algorithm called MiCRA [24]. Another variation on HEED was discovered by Subramanian. The network has two layers of CH formation, the first being within the Internet, the second being outside the Internet. When considering CH selection for the second level, MiCRA and HEED placed the same level of CH selection; however, the first-level CHs are only considered. MiCRA used this technique to ensure that each node uses the energy in a balanced way which results in the nodes of the network running for an extended period. For better routing of data and clustering in the WSNs, the suggested approach [25], employed gradient values to divide nodes into uneven cluster sizes. To assist in minimizing control packet energy consumption in WSNs, DHCRA (distributed hierarchical clustering routing algorithm) was developed [26]. A new partially static clustering technique called EESSC (energy-efficiency semi-static cluster) employed a specialized packet head to communicate data to sensor nodes and updated each SN's energy information during the transmission process, according to a study conducted by Du *et al.* [27]. Gherbi *et al.* [28] created a unique technique known as HEBM (hierarchical energy balancing multipath) routing system that enhanced WSN load balancing and lowered energy use, prolonging the network's lifespan. Passino [29] was the first to present BFOA, based on the social behavior of E. Coli bacteria. Bacterial foraging has provided a plethora of answers to these and other technical and computational issues, such as optimum control,

harmonic estimation, and channel equalization, but there has been minimal WSN research. Kulkarni *et al.* [30] developed a span-based node localization approach that is distributed, iterative, and employs a localization of node algorithm, a bio-inspired technique to search for nodes in a random and irregular manner, as well as a probabilistic search optimization algorithm known as PSO. Using BFOA, researchers Li *et al.* [31] proposed a method known as the low-energy intelligent clustering protocol (LEICP), which they claimed can be used to identify the locations of methane hydrate chambers in each cycle of the LEACH process. The fuzzy logic system (FLS) is an optimization approach that works well in real-world situations, as described by Negnevitsky [32], if no complete and dependable information is available. Using linguistically based rules in a natural way helps FLS make important real-time decisions that affect quality [32]. Gupta *et al.* [33] presented three variables utilized in CH selection: battery level, node concentration, and base station distance. By considering local distance and energy variables, Kim *et al.* [34] proposed FLS that decreases overhead output while CH selection takes place in LEACH. EAUCF [35] considered CH diameter and distance from BS employing FLS to address cluster diameter estimate issues. A new method proposed by UCFIA [36] for handling the hot spots problem relies on clustering models that are uneven (for utilization of FLS) and an ACO-based inter-cluster routing method. Clustering-based routing protocols (such as type-2 FLS) for inter-cluster communication, as well as improved ACO algorithms, were proposed, which was known as CRT2FLACO [37]. The multi-objective fuzzy-based clustering method developed by Sert *et al.* [38] provided a solution to the energy hole and hot spot issues encountered in both stationary and dynamic networks. A novel distributed unequal clustering (DUCF) protocol was introduced by Baranidharan and Santhi [39] for improved EAUCF. Another layered protocol for unequal clustering— LEBUCR was proposed recently [40]. WSNs and clustering-based energy-efficient bio-inspired approaches were later discussed in clustering-based heterogeneous optimized-HEED protocols [41].

9.3 Proposed Work

9.3.1 Network model for multilevel energy heterogeneity

There are several assumptions for the WSNs, which we have listed below [16, 41].

- SNs remain static and stationary after the deployment.
- As there is no GPS module, the SNs are location unaware.

- Each SN has a unique identification number (UID) associated with it.
- All the SNs are capable of generating, transmitting, and processing the data.
- The initial energy levels may be different for each SN.
- There is no scope to recharge the battery after the deployment.
- The BS has unlimited energy with high processing power, and it is preferably placed at the center of the network.
- The network contains a BS, preferably at the center of the network, and is assumed to have the high processing power and unlimited energy.
- The wireless links between the nodes are assumed to be symmetric and bidirectional; the amount of energy spent for transmitting a packet from SN SN_i to SN_j is the same as transmitting a packet from SN SN_j to SN_i.
- SNs usually combine the data received from other SNs and re-transmit.
- SNs broadcast the message to their neighbors B_{msg}; upon receiving the broadcast, the recipient SNs reply with R_{msg}. The node density is calculated from the total R_{msg} is received by the sender SN.
- The beacon message BCN_{msg} is sent by the BS to all the SNs at the beginning, and each SN can identify the distance to BS with the RSSI (received signal indicator).

We have designed the model for the *n-level* of heterogeneity. Here we discuss *1-level* to *5-levels* of heterogeneity of energy levels in the network at the starting of the process for the SNs (SNs). Let's assume there are N number of SNs in the WSN, and there are l initial energy levels. The value of l can take any positive integer value [42], however we have used $l=1$ to $l=5$ for our work. We assume that there are N_1 nodes with energy $level_1$, N_2 nodes with energy $level_2$, N_3 nodes with energy $level_4$, N_4 nodes with energy $level_4$, N_5 nodes with energy $level_5$. The node numbers in each level depend on the n node numbers in the previous level, and a parameter α Number of nodes at any of the $level_i$ is defined by the equation given below:

$$N_i = N_{i-1} \times \alpha. \tag{9.1}$$

The value of α is set to 0.6 for this work. The number of SNs at different levels is given as:

$$N_2 = N_1 \times \alpha$$
$$N_3 = N_2 \times \alpha = N_1 \times \alpha^2$$
$$N_4 = N3 \times \alpha = N_1 \times \alpha^3$$
$$N_5 = N_4 \times \alpha = N_1 \times \alpha^4. \tag{9.2}$$

The generalized equation can be written as:

$$N_i = N_1 \times \alpha^{i-1}. \tag{9.3}$$

Therefore, the total number of SNs at all the levels is:

$$N = N_1 + N_2 + N_3...N_l = N_1(1 + \alpha + \alpha^2 + \alpha^3...\alpha^{l-1}). \tag{9.4}$$

Since $\alpha < 1.0$, replacing $\alpha = 0.6$ in the above equation we get:

$$N_1 = N(1 + 0.6 + 0.6^2 + 0.6^3...0.6^{l-1}). \tag{9.5}$$

Hence, the value of N_1 is calculated depending on the value of l, given by the equation below:

$$N_1 = \frac{N}{(1 + 0.6 + 0.6^2 + 0.6^3...0.6^{l-1})}. \tag{9.6}$$

As the value of N_1 is calculated, the values of $N_2, N_3,...$etc., can also be calculated using eqn (9.2).

At the beginning of the process, the energy level of the SNs *level*$_1$ is E_1. The energy levels of each SN are initially given by the following equation:

$$E_i = E_1 \times (1 + (1 - i) \times \nabla E). \tag{9.7}$$

The total energy E of WSN is:

$$
\begin{aligned}
&N_1 \times E_1 + N_2 \times E_2... + Nl \times E_l \\
=&N_1 \times E1 + N_2 \times E(1 + \nabla E) + ... + N_2 \times E((1 + (l - 1) \times \nabla E) \\
=&N_1 \times E_1 + N_1 \times \alpha \times E_1(1 + \nabla E) + ... + N_1 \times \alpha^{l-1} \\
&\times E_1(1 + (l - 1) \times \nabla E).
\end{aligned} \tag{9.8}
$$

9.3.2 MLICHBHEED protocol

The multi-level ICHBHEED (MLICHBHEED) protocol is a modification by employing the ICHB algorithm [16] on HEED [22] and employing diverse *l-level* of heterogeneous energy nodes in WSN. Its clustering procedures, data transmission procedures, and data reception are discussed below.

9.3.2.1 Clustering and CH election process

The residual energy controls the CH selection procedure in HEED by MLICHBHEED protocol using the ICHB algorithm [16]. ICHB is a bio-inspired algorithm and is a derivative of the BFOA-based [29] algorithm used for optimization. In this process, the behavior of E. Coli bacterium is searching for nodes with residual energy in the network and is employed as an active cluster head (CH). The E. Coli bacterium starts with any of the SNs and migrates to other SNs (in the chemotaxis mechanism) in pursuit of greater nutrition value. These newer positions of the bacterium (on SN) are used as a CH for the current round in network simulation.

9.3.2.2 ICHB algorithm

The initial position of each bacterium which is generated randomly on a few SNs in the network, is denoted as $Pos(i)=\{\beta_k(i)|k=1,2,3\ldots K\}$. There are K bacteria population. $Pos(i)$ denotes the position of all the bacteria at the ith chemotactic step and $\beta_k(i)$ denotes the position of an individual bacterium. The cost of the bacterium $\beta_k(i)$ situated at any SN has an associated cost denoted as $E(k, i)$. Bacteria population is 5% of the SNs of the network [16].

After the completion of the initialization process, each bacterium creates its own random vectors v_k, which consist of the unique ids of SNs under the communication range R. Each bacterium migrates SN to SN in search of superior nutrition gradient based on its own random vector in the chemotactic process. The shifting of the kth bacterium for the $(i+1)$th step of the chemotaxis process is defined as:

$$Pos_k(i+1) = [S_k^{j+1}(Unique - id)]^{v_k}, \qquad (9.9)$$

where S_k^{j+1} movement of kth bacterium from the current position $(j+1)$th position in random vector v_k The bacterium traverses all the SNs in the random vector during this process while comparing and storing the unique-id and residual energy value of that SN whose residual energy is highest among all visited SNs. This SN acts as the probable CH for this region in the network.

Furthermore, deploying different energy-level SNs in the network, higher energy-level SNs have much higher chances of being elected as CHs, which is a key factor of using multi-level heterogeneity SNs in the network. Due to this, it delays the death of SNs in the network and extends the network lifetime to a greater extent and provides longevity to the network for sending data packets to the BS for decision-making purpose.

These selected CH nodes in the network broadcast an advertising message BCH_{msg} meant for other SNs in their communication range. The SNs, after receiving the message, attach themselves to these CHs. The density parameter is used to decide the CH, if a particular SN receives more than one message from different CHs. The rest of the clustering method is identical to that used in [16]. This brings the cluster creation process to a close for this round. The process iterates until any active node is still left in the network.

9.3.2.3 Process of data transmission and collection

The SNs transmit the data packages with data fetched from the environment to the corresponding CHs. CHs accumulate received data packages from multiple SNs and propagate them in a single-hop way to the BS for decision-making.

The SNs collect data from the environment and transmit it to their corresponding CHs through data packets. CHs aggregate data packets received from several SNs and transmit them to the BS in a single-hop way for decision-making purposes.

9.3.2.4 Energy depletion model for MLICHBHEED protocol

The depleted energy during the entire process is for the data transmission, reception, and fusion process. Both CHs and SNs are part of the process and deplete the energy of the participant entities. We start the discussion on energy dissipation with the radio-energy model [43, 21]. Let E_e be the energy in joules for the radio transmission and reception circuit. Energy consumed by amplification of radio signal for short-distance transmission $(d<=d_0)$ is ϵ_S and for a longer distance $(d>d_0)$ is ϵ_L. Energy consumed for transmission of the data packet of size L bit for short-distance E_{ST} and long-distance E_{LT} are expressed as:

$$E_{ST} = E_e \times L + \epsilon_S \times L \times d^2 \, for \, d \leq d_0$$
$$E_{LT} = E_e \times L + \epsilon_L \times d^4 \, for \, d > d_0. \tag{9.10}$$

Energy utilization for L bit packet reception is:

$$E_R = E_e \times L. \tag{9.11}$$

Threshold value d_0 is expressed as $d_0\sqrt{\frac{\epsilon_S}{\epsilon_L}}$. For fusing the data packets, the required energy E_f is set to the value 5nj/bit/msg.

The WSN of N and SNs is partitioned into C clusters containing an equal number of SNs. There are N/C SNs in each cluster, including a CH. Each cluster member of MLICHBHEED spends its energy sensing data from the environment and sends its assigned CH. After reception of data packets from the assigned SNs, the CH combines the packets and sends them to the BS. A round is now concluded. A round in this process is comprised of sensing, processing, transmitting, receiving, combining, and re-transmitting the data packets till it reaches BS. In a single round, energy spent by an SN in the network is defined as:

$$E_{SN_j} = E_{ST}, \tag{9.12}$$

where E_{SN_j} is the energy consumed by jth member of the cluster and E_{ST} is the energy consumed for short transmission between an SN to its cluster head.

In a single round, the cluster heads require the following amount of energy:

$$E_{CG_C} = \left(\frac{N}{C} - 1\right) \times E_R + \left(\frac{N}{C} - 1\right) \times E_f \times L + E_{LT}, \tag{9.13}$$

where E_{CH_C} is the dissipated energy by CH of the cth cluster, E_R is the energy consumed for receiving, $E_f \times$ is the energy consumed for fusing the data arrived from different SNs with each packet is of L bit, and E_{LT} is the energy required for long-distance transmission from cluster head to BS. Total energy consumed in the round is:

$$E_C = \left(\frac{N}{C} - 1\right) \times E_{SN_j} + E_{CH_C}, \tag{9.14}$$

where E_C represents total energy consumed in a round by cth cluster. Therefore, the total consumption of energy by the network consisting of C clusters is given as:

$$E_{round} = E_c \times C. \tag{9.15}$$

9.4 Results and Discussions

In this section, we provide the simulation results for the implementation of the ICHB-HEED protocol for a heterogeneous network model, proposed as multi-level ICHBHEED (MLICHBHEED), and compare it with the HEED protocol. A heterogeneous WSN can have type-1, type-2, type-3, type-4, and type-5 nodes as shown in Figure 9.1. We have thus named the implementation of ICHB-HEED MLICHBHEED-1, MLICHBHEED-2, MLICHBHEED-3, MLICHBHEED-4, and MLICHBHEED-5 to reflect these five different levels of heterogeneity. This probability is calculated by combining residual energy and node density in the ICHB-HEED protocol. MLICHBHEED uses the same variables for cluster head selection as ICHB-HEED because it is based on the ICHB-HEED protocol. Simulations were performed in MATLAB using 100 SNs randomly distributed across a square area with dimensions of 100×100 meters. Table 9.1 lists the input parameters that we used in our simulations for the simulation configuration.

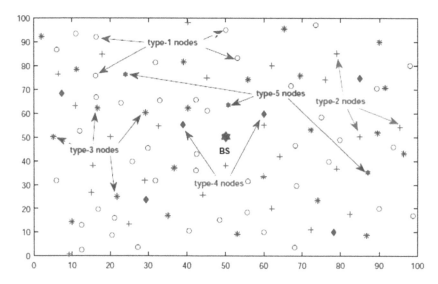

Figure 9.1 Network model showing different types of SNs in the network for MLICHB-HEED protocol at the 5-level of heterogeneity.

Simulation findings for several levels of energy heterogeneity in the network model are discussed. These levels are 1, 2, 3, 4, and 5. There are inequalities in the energy of nodes of types 5, 4, 3, 2, and 1. These inequalities must be satisfied for nodes of types 5, 4, 3, 2, and 1, i.e., $E5 > E4 > E3 > E2 > E1$, and so on. These models, MLICHBHEED-1, MLICHBHEED-2, MLHEED-3, MLICHBHEED-4, and MLICHBHEED-5, all employ the same simulating network environment as the MLHEED-1, MLHEED-2, MLHEED-3, and MLHEED-4 and MLHEED-5. According to this diagram, the energy of various SN types with differing degrees of heterogeneity varies.

Table 9.1   Parameters for simulation

Simulation parameters	Value
Network field	$100 \times 100m^2$
Number of sensors (N)	100
BS location	$(50, 50)$
Initial energy	$0.2\ J$
Threshold distance (d_0)	$70m$
Total number of bacteria (K)	5% of total sensors
Total chemotactic steps	1
Length of swim	Number of sensors in random vector
Energy dissipation required to run transmitter or receiver circuitry (E_e)	$50nJ/bit$
Energy dissipation required by amplifier in transmission of signal at shorter distance (ϵ_S)	$10pJ/bit/m^2$
Energy dissipation by amplifier in transmission of signal at longer distance (ϵ_L)	$0.0013pJ/bit/m^4$
Cluster range (R)	$25m$
Message Size (L)	$4000\ bits$
Data fusion cost (E_f)	$5nJ/bit/message$

In 1-level heterogeneity, the type-1 node has as 0.2 J; in 2-level heterogeneity, the type-1 and type-2 nodes have 0.2 and 0.4 J, respectively; in 3-level heterogeneity, the type-1, type-2, and type-3 nodes have 0.2, 0.4, and 0.5 J, respectively; in 4-level heterogeneity, the type-1, type-2, type-3 and type-4 nodes have as 0.2, 0.4, 0.5, and 0.6 J, respectively; in 5-level heterogeneity the type-1, type-2, type-3, type-4, and type- 5 nodes have as 0.2, 0.4, 0.5, 0.6, and 0.7 J respectively. The number of SNs for different levels of heterogeneity, i.e., type-1 to type-5, for MLICHBHEED and MLHEED implementation is given in Table 9.2, and the energies of various types of nodes for all levels used in our simulations are given in Table 9.3.

9.4.1 Network lifetime

The number of alive nodes in proportion to the number of rounds was used to calculate the network longevity, as demonstrated in Figure 9.2. The nodes that

Table 9.2 Number of sensors in five-level heterogeneity for MLHEED and MLICHBHEED protocols

Protocols	MLHEED-1	MLHEED-2	MLHEED-3	MLHEED-4	MLHEED-5
Type-1 nodes	100	60	52	49	47
Type-2 nodes	NA	40	30	26	24
Type-3 nodes	NA	NA	18	15	14
Type-4 nodes	NA	NA	NA	10	9
Type-5 nodes	NA	NA	NA	NA	6
Protocols	MLICHBHEED-1	MLICHBHEED-2	MLICHBHEED-3	MLICHBHEED-4	MLICHBHEED-5
Type-1 nodes	100	60	52	49	47
Type-2 nodes	NA	40	30	26	24
Type-3 nodes	NA	NA	18	15	14
Type-4 nodes	NA	NA	NA	10	9
Type-5 nodes	NA	NA	NA	NA	6

are alive are those whose energies are greater than zero. As seen in Figure 9.2 and Table 9.4 all nodes in the MLICHBHEED protocol are alive for a longer period span. Network lifespan is extended by using the MLICHBHEED. For example, in MLHEED-1 and MLICHBHEED-1, the first node dies in the first round, whereas the last node dies in the 668th and 794th rounds. Because the ICHB method is implemented, there is no increase in network energy. For example, in MLHEED-5 and MLICHBHEED-5, the first node dies in round 459 and the last node dies in round 2653. The network lifespan increases by 7.9% when the ICHB algorithm is implemented, while the network's energy remains constant.

Table 9.5 shows the total network energy, network lifetime, and the percent increase in lifetime corresponding to the percentage increase in energy, for all variation implementations. The MLHEED-5 and MLICHBHEED-5 provide the highest network lifetime, i.e., boosting network energy by 78% improves network lifetime by 267.96% in the MLHEED-5 and by 234.13% in the MLICHBHEED-5.

Table 9.3   Categorization of energies in five-level heterogeneity for MLHEED and MLICHBHEED protocols

Protocols	MLHEED-1	MLHEED-2	MLHEED-3	MLHEED-4	MLHEED-5
Energy of Type-1 nodes (J)	0.2	0.2	0.2	0.2	0.2
Energy of Type-2 nodes (J)	NA	0.4	0.4	0.4	0.4
Energy of Type-3 nodes (J)	NA	NA	0.5	0.5	0.5
Energy of Type-4 nodes (J)	NA	NA	NA	0.6	0.6
Energy of Type-5 nodes (J)	NA	NA	NA	NA	0.7

Protocols	MLICHBHEED-1	MLICHBHEED-2	MLICHBHEED-3	MLICHBHEED-4	MLICHBHEED-5
Energy of Type-1 nodes (J)	0.2	0.2	0.2	0.2	0.2
Energy of Type-2 nodes (J)	NA	0.4	0.4	0.4	0.4
Energy of Type-3 nodes (J)	NA	NA	0.5	0.5	0.5
Energy of Type-4 nodes (J)	NA	NA	NA	0.6	0.6
Energy of Type-5 nodes (J)	NA	NA	NA	NA	0.7

9.4.2 Total energy consumption

In addition, we calculated the total energy utilized by the network in every round, as shown in Figure 9.3. Instantaneous energy expenditure per round is measured by this metric, i.e., the energy difference between the start and finish of each loop in a network. Nodes of types 1, 2, 3, 4, and 5 have total initial energies of 20, 28, 31, 4, 35, 6, and 38.5 J correspondingly. MLICHBHEED-5 outperforms all other MLICHBHEED levels. This demonstrates that the MLICHBHEED-5 dissipates energy at a considerably slower pace than the other versions.

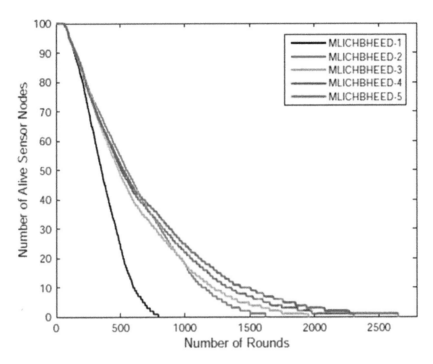

Figure 9.2 Number of remaining alive SNs after each round for MLICHBHEED protocols at varying heterogeneity levels.

Table 9.4 Round number when first and last nodes are dead

Protocols	First node dead	Last node dead
MLHEED-1	167	668
MLHEED-2	192	1156
MLHEED-3	254	1626
MLHEED-4	365	2092
MLHEED-5	459	2458
MLICHBHEED-1	67	794
MLICHBHEED-2	67	1621
MLICHBHEED-3	67	1953
MLICHBHEED-4	67	2306
MLICHBHEED-5	67	2653

Table 9.5 Percentage increase in network energy and the corresponding increase in network lifetime for MLICHBHEED protocol

Protocols	MLHEED-1	MLHEED-2	MLHEED-3	MLHEED-4	MLHEED-5
Total network energy (J)	20	28	31.4	33.7	35.6
Lifetime (J)	668	1156	1626	2092	2458
percentage increase in energy	NA	40	57	68.5	78
percentage increase in lifetime	NA	73.05	143.41	213.17	267.96
Protocols	MLICHBHEED-1	MLICHBHEED-2	MLICHBHEED-3	MLICHBHEED-4	MLICHBHEED-5
Total network energy (J)	20	28	31.4	33.7	35.6
Lifetime (J)	794	1621	1953	2306	2653
percentage increase in energy	NA	40	57	68.5	78
percentage increase in lifetime	NA	104.16	145.97	190.43	234.13

Table 9.6 Number of packets sent to BS after each round

Protocols	Number of packets sent to BS
MLHEED-1	0.62×10^4
MLHEED-2	0.90×10^4
MLHEED-3	1.16×10^4
MLHEED-4	1.71×10^4
MLHEED-5	2.20×10^4
MLICHBHEED-1	0.64×10^4
MLICHBHEED-2	1.3×10^4
MLICHBHEED-3	1.5×10^4
MLICHBHEED-4	1.7×10^4
MLICHBHEED-5	1.9×10^4

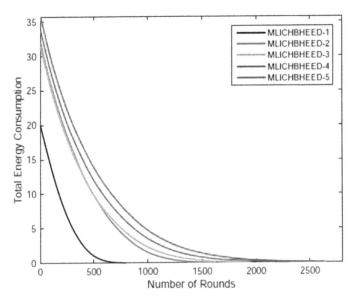

Figure 9.3 Residual energy of the network after each round for MLICHBHEED protocols at varying heterogeneity levels.

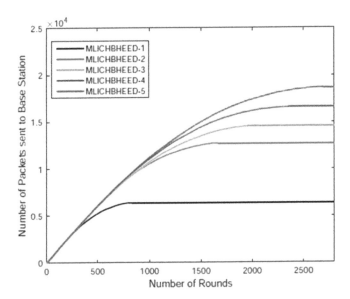

Figure 9.4 Number of packets sent to the BS by different MLICHBHEED protocols in each round.

9.4.3 Number of packets sent to the base station

In Figure 9.4, we have calculated the number of packets transferred to the base station in a round. Figure 9.4 and Table 9.6 show that the MLICHBHEED-5 transmits the most packets to the base station, of all versions. The number of packets transferred to the base station using the MLHEED-1, MLHEED-2, MLHEED-3, MLHEED-4, MLHEED-5, MLICHBHEED-1, MLICHBHEED-2, MLICHBHEED-3, MLICHBHEED-4, MLICHBHEED-5 are respectively, 0.62×10^4, 0.90×10^4, 1.16×10^4, 1.71×10^4, 2.20×10^4, 0.64×10^4, 1.3×10^4, 1.5×10^4, 1.7×10^4 and 1.9×10^4 corresponding to the number of rounds. MLICHBHEED sends more data packets to the base station than MLHEED while consuming the same amount of energy at the same degree of heterogeneity, as can be shown in Table 9.6.

9.5 Conclusion

A multi-level ICHB-HEED (MLICHBHEED) protocol for heterogeneous network models has been discussed in this chapter, which is capable of handling any finite amount of heterogeneity in terms of energy levels provided to the different SNs in the WSNs. Experimental validation of MLICHBHEED involved evaluating up to five levels of heterogeneity in terms of energy one, two, three, four, and five levels of heterogeneity. Increasing network heterogeneity increases longevity by a much larger amount than increasing network energy does. The number of living SNs, energy usage, and packets transmitted to BS were all measured for each diversified version of the MLICHBHEED protocol in terms of heterogeneity. In addition to reducing energy usage, the variants of MLICHBHEED that have been proposed also improve network performance and lifespan of the WSN which shows that employing heterogeneity in terms of energy levels of SNs provides better network performance, network lifespan, and a higher number of data packets to BS for decision-making purpose.

References

[1] I. F. Akyildiz, W. Su, Y. Sankarasubramaniam, and E. Cayirci, "Wireless sensor networks: a survey," Computer networks, vol. 38, no. 4, pp. 393–422, 2002.

[2] A. Verma, R. Mondal, P. Gupta, and A. Kumar, "Neural based energy-efficient stable clustering for multi-level heterogeneous WSNs," 1st International Conference on Secure Cyber Computing and Communication, 2002.

[3] B. Ram, N. Chand, P. Gupta, and S. Chauhan, "A new approach layered architecture based clustering for prolong life of wireless sensor network (WSN)," International Journal of Computer Applications, vol. 15, no. 1, pp. 40–45, 2011.

[4] J. Hill, R. Szewczyk, A. Woo, S. Hollar, D. Culler, and K. Pister, "System architecture directions for networked sensors," ACM Sigplan notices, vol. 35, no. 11, pp. 93–104, 2000.

[5] P. Gupta and A. K. Sharma, "Designing of energy efficient stable clustering protocols based on BFOA for WSNs," Journal of Ambient Intelligence and Humanized Computing, vol. 10, pp. 681–700, 2019.

[6] S. A. El-Said, A. Osamaa, and A. E. Hassanien, "Optimized hierarchical routing technique for wireless sensors networks," Soft Computing, vol. 20, no. 11, pp. 4549–4564, 2016.

[7] P. Gupta and A. Sharma, "M-ICHB based extended stable clustering protocols for three-level heterogeneous WSNs," 1st International Conference on Secure Cyber Computing and Communication, IEEE, 2018.

[8] P. Kumarawadu, D. J. Dechene, M. Luccini, and A. Sauer, "Algorithms for node clustering in wireless sensor networks: A survey," 4th International Conference on Information and Automation for Sustainability, IEEE, pp. 295–300, 2018.

[9] W. B. Heinzelman, A. P. Chandrakasan, and H. Balakrishnan, "An application-specific protocol architecture for wireless microsensor networks," IEEE Transactions on wireless communications, vol. 1, no. 4, pp. 660–670, 2002.

[10] G. Rehani, P. Maheshwari, and A. K. Sharma, "Modified teen for handling inconsistent cluster size problem in WSN," International Conference on Wireless Com- munications, Signal Processing and Networking (WiSPNET). IEEE, pp. 1–6, 2018.

[11] N. Aslam, W. Phillips, W. Robertson, and S. Sivakumar, "A multi-criterion optimization technique for energy efficient cluster formation in wireless sensor networks," Information Fusion, vol. 12, no. 3, pp. 202–212, 2011.

[12] M. Adnan, M. A. Razzaque, I. Ahmed, I. F. Isnin., "Bio-mimic optimization strategies in wireless sensor networks: A survey," Sensors, vol.

14, no. 1, pp. 299–345, 2014.

[13] P. Gupta and A. K. Sharma, "Energy efficient clustering protocol for WSNs based on bio-inspired ICHB algorithm and fuzzy logic system," Evolving Systems, vol. 10, no. 4, pp. 659–677, 2019.

[14] A. K. Sharma, K. Verma., "NSGA-II with ENLU inspired clustering for wireless sensor networks," Wireless Networks, vol. 26, no. 5, pp. 3637–3655, 2020.

[15] A. U. Haque, P. Mandal, J. Meng, and M. Negnevitsky, "Wind speed forecast model for wind farm based on a hybrid machine learning algorithm," International Journal of Sustainable Energy, vol. 34, no. 1, pp. 38–51, 2015.

[16] P. Gupta and A. K. Sharma, "Clustering-based optimized heed protocols for wsns using bacterial foraging optimization and fuzzy logic system," Soft Computing, vol. 23, no. 2, pp. 507–526, 2019.

[17] S. Singh, "Energy efficient multi-level network model for heterogeneous wsns," Engineering Science and Technology, vol. 20, no. 1, pp. 105–115, 2017.

[18] M. Tong and M. Tang, "Leach-b: an improved leach protocol for wireless sensor network," 6th international conference on wireless communications networking and mobile computing (WiCOM). IEEE, pp. 1–4, 2010.

[19] V. Mhatre and C. Rosenberg, "Design guidelines for wireless sensor networks: communication, clustering and aggregation," Ad hoc networks, vol. 2, no. 1, pp. 45–63, 2004.

[20] V. Loscri, G. Morabito, and S. Marano, "A two-levels hierarchy for low- energy adaptive clustering hierarchy (TL-Leach)," IEEE vehicular technology conference, vol. 62, no. 3. IEEE; 1999, p. 1809, 2005.

[21] S. Lindsey and C. S. Raghavendra, "Pegasis: Power-efficient gathering in sensor information systems," in Proceedings, IEEE aerospace conference, vol. 3. IEEE, pp. 3–3, 2002.

[22] O. Younis and S. Fahmy, "Heed: a hybrid, energy-efficient, distributed clustering approach for ad hoc sensor networks," IEEE Transactions on mobile computing, vol. 3, no. 4, pp. 366–379, 2004.

[23] O. Younis and S. Fahmy, "Energy-efficient routing and data aggregation in sensor networks: An experimental study," 2004.

[24] K. K. Khedo and R. Subramanian, "Misense hierarchical cluster-based routing algorithm (micra) for wireless sensor networks," World Academy of Science: Engineering and Technology, vol. 52, pp. 190–195, 2009.

[25] T. Liu, Q. Li, and P. Liang, "An energy-balancing clustering approach for gradient-based routing in wireless sensor networks," Computer Communica- tions, vol. 35, no. 17, pp. 2150–2161, 2012.

[26] M. Sabet and H. R. Naji, "A decentralized energy efficient hierarchical cluster- based routing algorithm for wireless sensor networks," AEU-International Journal of Electronics and Communications, vol. 69, no. 5, pp. 790–799, 2015.

[27] T. Du, S. Qu, F. Liu, and Q. Wang, "An energy efficiency semi-static routing algorithm for wsns based on hac clustering method," Information fusion, vol. 21, pp. 18–29, 2015.

[28] C. Gherbi, Z. Aliouat, and M. Benmohammed, "An adaptive clustering approach to dynamic load balancing and energy efficiency in wireless sensor networks," energy, vol. 114, pp. 647–662, 2016.

[29] K. M. Passino, "Biomimicry of bacterial foraging for distributed optimization and control," IEEE control systems magazine, vol. 22, no. 3, pp. 52–67, 2002.

[30] R. V. Kulkarni, G. K. Venayagamoorthy, and M. X. Cheng, "Bio-inspired node localization in wireless sensor networks," IEEE International Conference on Systems, Man and Cybernetics. IEEE, pp. 205–210, 2009.

[31] Q. Li, L. Cui, B. Zhang, and Z. Fan, "A low energy intelligent clustering protocol for wireless sensor networks," International Conference on Industrial Technology. IEEE, pp. 1675–1682, 2010.

[32] M. Negnevitsky, Artificial intelligence: a guide to intelligent systems, 1st edn. Addison-Wesley Longman Publishing Co., Inc., Boston, 2001.

[33] I. Gupta, D. Riordan, and S. Sampalli, "Cluster-head election using fuzzy logic for wireless sensor networks," 3rd Annual communication networks and services research conference (CNSR'05). IEEE, pp. 255–260, 2005.

[34] J.-M. Kim, S.-H. Park, Y.-J. Han, and T.-M. Chung, "Chef: cluster head election mechanism using fuzzy logic in wireless sensor networks," 10th International Conference on Advanced Communication Technology, vol. 1. IEEE, pp. 654–659, 2008.

[35] H. Bagci and A. Yazici, "An energy aware fuzzy approach to unequal clustering in wireless sensor networks," Applied Soft Computing, vol. 13, no. 4, pp. 1741–1749, 2013.

[36] M. Song and C.-l. Zhao, "Unequal clustering algorithm for wsn based on fuzzy logic and improved aco," The Journal of China Universities of Posts and Telecommunications, vol. 18, no. 6, pp. 89–97, 2011.

[37] W.-X. Xie, Q.-Y. Zhang, Z.-M. Sun, and F. Zhang, "A clustering routing protocol for wsn based on type-2 fuzzy logic and ant colony optimization," Wireless Personal Communications, vol. 84, no. 2, pp. 1165–1196, 2015.

[38] S. A. Sert, H. Bagci, and A. Yazici, "Mofca: Multi-objective fuzzy clustering algorithm for wireless sensor networks," Applied Soft Computing, vol. 30, pp. 151–165, 2015.

[39] B. Baranidharan and B. Santhi, "Ducf: Distributed load balancing unequal clustering in wireless sensor networks using fuzzy approach," Applied Soft Computing, vol. 40, pp. 495–506, 2016.

[40] S. A. K. Gunjan and V. KARAN, "Layered energy balanced unequal clus- tering and routing (lebucr) protocol for wireless sensor networks." Adhoc & Sensor Wireless Networks, vol. 46, 2020.

[41] P. Gupta and A. K. Sharma, "Clustering-based heterogeneous optimized- heed protocols for wsns," Soft Computing, vol. 24, no. 3, pp. 1737–1761, 2020.

[42] S. Singh, S. Chand, and B. Kumar, "Multi-level heterogeneous network model for wireless sensor networks," Telecommunication Systems, vol. 64, no. 2, pp. 259–277, 2017.

[43] W. R. Heinzelman, A. Chandrakasan, and H. Balakrishnan, "Energy-efficient communication protocol for wireless microsensor networks," Proceedings of the 33rd annual Hawaii international conference on system sciences. IEEE, pp. 10, 2000.

[44] M. Tong, and M. Tang, "LEACH-B: An improved LEACH protocol for wireless sensor network," Proceedings of 6th international conference on wireless communications networking and mobile computing (WiCOM), pp 1–4, 2010.

[45] W. B. Heinzelman, A. P. Chandrakasan, H. Balakrishnan, "An application specific protocol architecture for wireless microsensor networks," IEEE Transaction on Wireless Communication 1(4):660–670, 2002.

10

IoT Enabled by Edge Computing for Telecomms and Industry

Manoj Kumar Sharma, Ruchika Mehta, and Rajveer Singh Shekhawat

Manipal University Jaipur, India
E-mail: m0918.sharma@gmail.com; ruchika.mehta@jaipur.manipal.edu;
rajveer.shekhawat.in@gmail.com

Abstract

We are living in the age of fast-evolving technology and it is significantly influencing our society, work culture, etc. Now we are witnessing 5G technology for fast and reliable internet. At the same time, we are increasing the use of IoT in our daily use. As we know, IoT enhances the communication among different digital assets (i.e., sensors, devices) and helps to capture required data and transmit it to fog or cloud. However, IoT devices generate a volume of data at the end-user level and it needs to be processed in a short span of time on the cloud. However, collection of data and transmission to the cloud-based processing is not as efficient as needed. Sending the data to the cloud for processing has a lot of overhead which degrades the quality of service and leaves a negative impact on IoT applications and their network performance. To eliminate such negative influence different cloud computing techniques (i.e., edge computing, fog computing, etc.) are in practice. Edge computing is playing a crucial role with IoT paradigms, especially in local networks, where we have to take decisions instantly instead of sending information to centralized systems for decision-making. Edge computing is a perfect example of a processing system. Edge computing is very useful in such applications where instant or local decision-making can enhance the productivity and quality of service (i.e., supply chain management, agriculture, resource utilization decisions, energy consumption, etc.). In the chapter, we are presenting a comprehensive exploration of IoT applications with edge computing and their architectures.

Keywords: Decentralized, edge computing, IoT, machine learning, sensors, wireless, 5G.

10.1 Introduction to IoT and Edge Computing

In the growing age of human society and industrialization, technologies like IoT play an important role. IoT frameworks are playing important roles in smart city transformation (smart traffic control, lighting on the streets, waste collection and management, etc.). IoT construct a centralized and distributed connected components (devices) environment which assess data from sensors and other devices transmit that data information to processing units; transmit instructions to devices or machines, monitor operations and devices, control the quality of process and product and finally we can say that it creates an intelligent environment which has less human intervention and more automation to achieve all goals like high production at low cost, monitoring, storage, transportation, quality, safety and security, distribution, etc. In an IoT network, millions of nodes are connected to each other and share data. In IoT-enabled systems, time sensitivity is very important, delay in operation is not allowed, all the operations are to be completed in a given time frame and this behavior is very useful in some applications like medical, transportation, electricity, etc. However, such time sensitivity is very hard to achieve with traditional cloud computing services where all the data has to be collected at the central system and processed there. A colossal amount of raw information is produced by millions of devices which become very hard to handle centrally. So, we have to involve new-age computing techniques like edge computing framework to decentralize the processing work and locally the data can be processed and used. In IoT technology, device-to-device data communication is very important, where devices share data locally and work with different types of networks. Another type of communication is device to cloud, where IoT gadgets collect data and transmit it to the cloud-based storage or processing unit where it centrally processes and different remote applications can access the row data as well as process outcomes. The next level of communication is machine to gateway which helps to perform operations between cloud services and IoT gadgets like a middleware bridge. However, IoT has played an important role in device interconnection, information gathering and exchange but in the cloud the data should be sent to the central server for processing and this is a major issue where a large amount of data is captured by the devices and has to be processed in a time-specific manner. However, not all data is always required to share with other pears and need not be processed at central processing server only, it can be handled at local level also and the advantage of such local processing is that decision can be taken faster as and when needed, and the processing burden on the central processing server also can be reduced. However, IoT-based cloud computing is undoubtebly enhancing the modern industries but it has various practical limitations also (i.e., connectivity, data latency, bandwidth, security, etc.). Data security is one of the challenging problems especially when data is in transmission to long-range distance and most of the time cyberattacks take place during data transmission; the protection of the IoT data during transmission is a big challenge. Network connectivity is also a major issue, as the IoT framework is more about the automation, which demands the in-time operations, and operations depends on the data analysis and instructions. Interruption in the network data transmission due to network problem itself, or due to power failure, may cause data loss or delay in decisions, which may not be wearable to the efficient business processes. Data latency is one more critical issue in IoT-based centralized cloud computing frameworks. Network bandwidth is fully responsible for the data latency and we cannot keep network bandwidth constant, if a number of nodes have started transmitting data at the same time, the bandwidth will be divided among all nodes and data transmission rate will suffer. Edge computing is important to IoT networks to resolve the above-discussed issues and to do some level of decentralized decision power distribution. Nowadays, researchers and industries are drawing their focus toward the efficient implementation of edge computing to enhance the security of the data while transmission and reduce decision-taking time. In terms of security issues in IoT architecture, the network latency and service quality can be improved by edge computing. However, local data computing nodes have less computation power as compared to central server, even though various small decisions can be taken at that level only. Edge computing can be used in peer-to-peer network fashion, ad-hoc network fashion, etc., and it reduces the data traveling distance and time, which directly increases the data security while transmission [1], [2]. There are some advantages of IoT based on edge computing, one of them is data security: while data transmission threat to the data is a biggest challenge, edge computing helps to secure the network and also improve overall data confidentiality. Because, at a time it is very hard for attackers to down all the processing nodes and corrupt or all the data from all processing nodes in a single commend. Another advantage of edge computing is minimized operational costs: just by storing and processing the data on local nodes the transmission, storage, and processing costs at the central processing unit are reduced.

Another advantage of IoT-based edge computing is unlimited scalability of the IoT networks without consideration of existing storage capabilities and incurred cost. We experience one more advantage using edge computing that is improved performance because in centralized processing cloud, data traveling time increases and performance accuracy decreases but in edge computing most of the processing happens at end nodes or at local nodes which save the data traveling time and at the same time storage problem also resolves. However, with the decreasing time requirement, centralized storage and processing helps the business rapidly at low cost with great reliability and efficiency. However, along with advantages of edge computing, resource management is a critical issue and user privacy, optimization, offloading tasks, accessibility of edge nodes, etc., are also important issues [1]–[3]. Simple edge computing issues are explained in Figure 10.1. It has already discussed in previous sections the advantages of edge computing (i.e., security, cost saving, etc.) over cloud computing. However, fog computing is another type of IoT-based computing framework which is faster but smaller. In fog computing, a series of IoT nodes transmit data packets in a decentralized manner, and data can be accessed, stored, computed, processed, and transmitted over the network. Such an approach helps in time efficiency of the operations. The data security is also ensured by implementing protocol gateways. It exhibits various advantages over centralized cloud, that is, privacy, mobility, integration, and data latency, like edge computing.

10.1.1 IoT and edge-computing scenarios

IoT devices are intelligent enough to capture and transmit data to the cloud storage either in wired or wireless manner and an IoT framework is shown in Figure 10.2.

In an IoT framework there are four important components:

(a) **Things**: A device has the internet connecting capability along with data capturing, transmitting, and retrieval.
(b) **Data acquisition**: Acquiring the data from the environment is another important aspect of IoT-based framework (i.e., temperature, humidity, distance, pressure, movement, etc.)

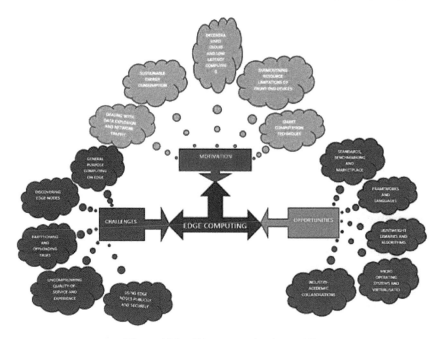

Figure 10.1 Edge computing issues [4].

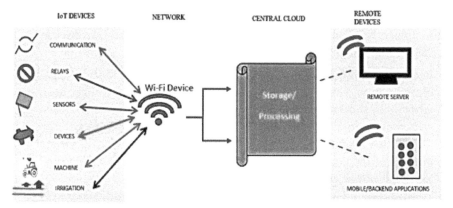

Figure 10.2 IoT Framework [4], [5].

(c) **Data processing**: Acquired data has to be processed and analyzed, this is one of the important aspects of the IoT framework.

(d) **Communication**: Communication among different IoT nodes in device to device, device to user, device to server manner is a very important aspect of IoT framework.

Different types of hardware devices are needed to complete the IoT framework to carry out above-explained operations [4], [5].

  (i) **Sensors**: Sensors are IoT devices to sense the physical changes in the environments and transmit collected data to the cloud server for processing.
 (ii) **Microcontrollers**: Microcontrollers are electronic circuits to control the operations through different devices.
(iii) **Basic devices**: Data storage and processing units like cell phones, laptops, and tablets are very important components of the IoT framework.
 (iv) **Datasheets**: Datasheets keep the record of hardware component functionality.
  (v) **Integrated circuits**: ICs are microcontrollers themselves which have pre-fused instructions.
 (vi) **Wireless network**—IoT nodes can communicate remotely.
(vii) **Internet**—It provides a means of communication to the IoT components.
(viii) **Cloud server**—It is a centralized facility for data storage and processing, which can be accessed by any of the IoT nodes remotely.

Figure 10.3 is representing lifecycle of IoT devices. Edge computing hardware is to alleviate data centers and cloud burdens. So, edge computing hardware must have sufficient storage, compact in size, rugged in nature, rich network connectivity, and power connectivity. An specific hardware devices are needed to implement the edge computing [6].

  (i) **Rugged and fan-less computers**: Computing hardware must be capable enough to withstand volatile environments (i.e., temperature, dust, vibrations, shocks, debris, etc.), so, its design must be totally closed and compact to overcome the problems of debris, dirt, and dust.
 (ii) **Meet the performance requirements**: Edge computing hardware must meet the performance criteria for which they are deployed (i.e., low–high power efficiency, etc.).
(iii) **Mountable**: Edge computing components or devices work under low-space conditions so they must be compact in size and mountable as and where required.
 (iv) **Rugged storage**: In edge computing, data collection, processing, and analyzing happen on local nodes so the processing device must have sufficient storage capacity.
  (v) **Rich I//O**: Edge computing devices usually connect with different devices or equipment so they must have multiple I/O ports.

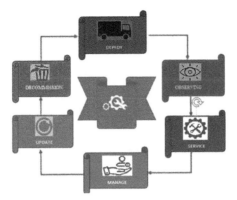

Figure 10.3   Lifecycle of IoT devices [4], [5].

(vi) **Wired/wireless connectivity**: Edge computing devices usually pace in remote areas and their connectivity is the major issue so they must have both wireless and wired connectivity capabilities.

(vii) **Wide power range**: Edge computing components work in varying power environments so the power-wearing capacity of the devices must be in a range of 9–50 VDC, etc.

(viii) **Security**: Edge computing devices work in remote areas where data security is a big challenge, so they must be connected with trusted platform modules only which have crypto-processors to make it tamper-resistant.

(ix) **Real-time process**: In the complex industrial operations, medical operations, etc., time bound data analysis and processing is a prime requirement. So, edge computing devices must have real-time data processing capability.

10.1.2 IoT versus edge computing

IoT is very useful and beneficial when we have existing data sources and devices and can be easily analyzed and shared from the central processing cloud. However, it is a superficial situation, usually IoT devices spread out remotely and they capture a high amount of data, which has to travel long to the central processing cloud. This data traveling is affected by the network speed, cyberattacks, bandwidth, availability of network, electricity, and other connectivity issues. Such hurdles are the cause of transmission delay and decision delay. Here, edge computing plays a crucial role, in which various decisions which are not mandatorily required to be taken by a central cloud-based system can be taken at local nodes, which helps to take in time decisions, save traveling time, and very less chances to have cyberattacks. Such local nodes can process the collected information and can easily spread out to nearby systems to take appropriate actions in designated time. It is less affected by the network bandwidth and availability of electricity which is a major issue in remote locations. Information disruption and information loss issues are also overcome with distributed processing through local nodes. Finally, increasing response time of the local nodes is directly proportional to the system operation performance [7], [8].

10.1.3 Industry 4.0

A rapid increment in the world population and modernization of the society has increased the need of quality of services and enhanced quality production of the daily usages and industrial usages products. Industry 4.0 is a digital revolution in the industrial practices which helps to develop new-age organizational control strategies, production strategies, preservation, transportation, distribution, and more over the quality control. Industry 4.0 is mainly focusing on the new-age automation, information exchange, and cyber-physical systems development and IoT, cloud computing techniques like age computing are helping to create modern smart production units so they can be more competitive in the global market. The ultimate goal of Industry 4.0 concept is to enhance and automate the decision- making process in decentralized way, do real-time monitoring of the assets, processing, quality control, etc., enhance the real-time information exchange from raw material to the market demand and final distribution of the products. A full value supply chain is an essential ingredient in Industry 4.0 which includes the origins and supply of raw material and components which are required in different forms and at different stage of manufacturing and along with manufacturing this supply change enhance to the final dispersal of the products to the market and end user. So, end-to-end supply chain creation, monitoring, and efficient implementation is the ultimate way to sustain the business in this competitive global market and it is very useful to the end users also as they have the facility to pick and choose the best-quality product. A simple technological representation of Industry 4.0 is given in Figure 10.4. However, Industry 4.0 has information- sensitive and information-intensive manufacturing in a closely connected global environment of systems, services peoples, big data analysis tools, processes, IoT, cloud computing, etc. IoT and cloud computing

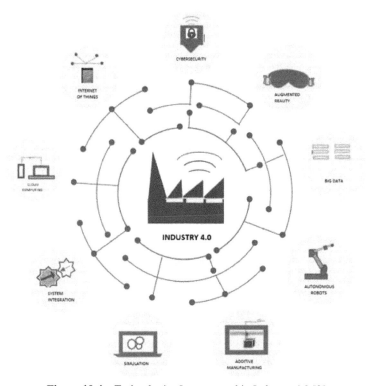

Figure 10.4 Technologies Incorporated in Industry 4.0 [9].

techniques are the key of industry 4.0 revolution. However, the concept of Industry 4.0 cannot be developed overnight. We have a long way to go to achieve the desired goals which need a lot of stages to climb up and various strategic decisions to follow.

Cyber-physical system is the basis to develop new-era capabilities in design and development, prototyping of products, remote decision-making and control, quality of services and diagnosis, environment and process monitoring, maintenance, trace and track the process and supply chain, health and pressure monitoring of the systems and structure, plan new innovative practices in real time. A real time end-to-end data modeling and mapping is an integral part of Industry 4.0 life cycle and value addition in the integration of operational technologies and information technology is the backbone of Industry 4.0, because real-time decision-making can easily enhance the quality production and transformation of the industry. The vertical and horizontal integration is next level modernization of the industry to achieve the desired goals. With vertical integration the traditional automation view will completely disappear and different systems and applications will dramatically change, and some of them will be replaced with the emerging techniques/technologies with the help of the IoT platform. Another horizontal integration which is not a hierarchical view like vertical integration, it is about end-to-end value change, where a real-time information flow is very important from the initial stage, logistics, transportation, distribution, and handover of the product to the end user. However, security of the systems, products, people, and quality control are the great concerns of any product and its production process, which are also clearly addressed in Industry 4.0. Security in Industry 4.0 encompasses data security, network security, cybersecurity, data integrity, control system security, employee's security, assets security, and infrastructure security. The future scope of Industry 4.0 is to adopt new-age business strategies and models which can enhance the product quality and quantity of the products and enhance the security of data and systems with enhanced quality and profitability [9].

10.1.4 Cloud computing and edge computing

However, cloud computing v/s edge computing is not a debatable point and there is no direct competition between these two technologies. Such hybrid solutions provide more computing flexibility to the operational unit as a tandem. Cloud computing provides various services like computing processors (i.e., GPU, TPU), cloud-based data storage, software development, resource access through internet, etc. Cloud computing services are scalable, services can be rented by the user and back-end processing is managed by the service provider. Some basic cloud service models are as:

 (i) **Platform as a service (PaaS)**: In PaaS, we can deploy our applications on the cloud and need not manage the platform, which is automatically managed by cloud service provider (i.e., AWS, Azure, etc.)

 (ii) **Software as a service (SaaS)**: In this service we can access applications which are already available on cloud platforms.

(iii) **Infrastructure as a service (IaaS)**: It is a very important service provided by the cloud platform where we have the flexibility to manage OS, applications, bandwidth, etc., as per our requirement.

Some of the popular cloud deployment models are explained here:

 1. **Community cloud**: In this similar interest and requirement institutions share the cloud services.

2. **Private cloud**: A specific institution can create its own cloud for its own purposes.
3. **Public cloud:** A cloud service provider provides the rented basis service access to both public and private customers.
4. **Hybrid cloud**: It consists of different clouds and data and applications can be shared among the clouds.

Despite various challenges, cloud computing allows one to start a small business at low cost and gradually enhance it. So, scalability is an important aspect which is provided by cloud computing. Reliability and availability of the services is another benefit which is provided by cloud computing. Moreover, cloud maintenance is low, accessibility is high and it helps to reduce operational and infrastructure cost.

Here, edge computing plays a crucial role, in which various decisions which are not mandatorily required to be taken by a central cloud-based system can be taken at local nodes, which helps to take in time decisions, save traveling time, and very less chances to have cyberattacks. Such local nodes can process the collected information and can easily spread out to nearby systems to take appropriate actions in designated time. It is less affected by the network bandwidth and availability of electricity which is a major issue in remote locations. Information disruption and information loss issues are also overcome with distributed processing through local nodes. Finally, increasing response time of the local nodes is directly proportional to the system operation performance [8], [9]. However, it is important to use both IoT and edge computing together, because the 5G network is providing advanced internet experience with upgraded bandwidth, data latency, etc., it enhances the cloud computing capabilities [10].

10.1.5 Cloud computing and edge computing: use cases

In edge computing, storage and processing capabilities are provided wherever they are needed and industry can be benefited with such scenarios. There are certain use case examples as given below:

(i) **Autonomous vehicles**: Autonomous vehicles are a very important use case of cloud and edge computing. With the help of edge computing, cloud service driver-less vehicles convoy can travel together one after another by avoiding road congestion, save fuel, etc., by just following the front vehicle, which is controlled by the driver, rest will be communicating to each other with low-data latency. A simple edge computing-based vehicle convoy is given in Figure 10.5.

(ii) **Smart grid**: A sophisticated application of edge computing is a smart grid system which helps to efficiently manage energy composition. IoT devices are placed with different machines and systems in the processing units, which continuously monitor the energy consumption and requirements of the machines. Such sensors can monitor real-time health of the device/machine, and along with this the peak hour requirement of the electricity is also pre-estimated, which helps the grid management to take proactive measures to fulfill the requirement.

(iii) **Healthcare and patient monitoring**: Healthcare infrastructure and decisions are very time sensitive and have good opportunities for edge computing. With the help of edge computing services, various decisions like glucose monitoring, breathing monitoring, pulse monitoring, heart rate monitoring, blood pressure monitoring, etc., are very simple requirements in healthcare to be monitored for in time-preventive safety measures for patient health. However, this can be handled with third-party cloud, but it may cause decision delay and security of patient health data but with the help of edge computing, on-site storage and process of data is possible.

(iv) **Virtualized radio and cloud gaming**: Virtualization of the mobile network reduces the cost and enhances the flexibility. Such networks can perform complex operations at low-data latency. Cloud gaming is one of

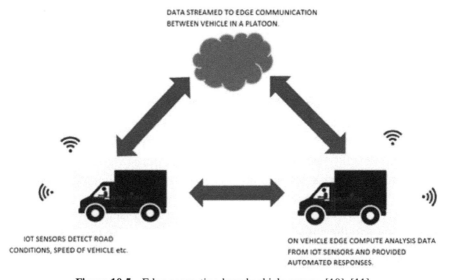

Figure 10.5   Edge computing-based vehicle convoy [10], [11].

the edge computing examples, which streams the live feed of a game directly on the device. This edge computing framework helps to reduce data latency and provides immersive gaming. A simple cloud-based gaming is shown in Figure 10.6.

(v) **Smart city [11] IoT** has been very popular because it integrates with various sensing devices, cameras, processing units, etc., which helps us to make real-time decisions. However, in smart cities, cloud-based IoT is not much beneficial due to decision delay at centralized cloud processing. Instead of it, edge computing with IoT is very good to develop smart city infrastructure in which real-time information processing and decision- making can happen with the help of local edge processing devices, which can further share such locally stored and processed data among peer nodes. In smart city scenario all smart devices like, traffic control, vehicle movement, congestion on roads, peak hours monitoring of traffic and alternate routes identification, pollution monitoring and preventing measures, synchronization of water supply for domestic usages and industry usage, water pressure monitoring, identification of more or less water demanding areas etc., can be easily handled by IoT technologies. At another side, criminal activities identification through surveillance cameras and real-time alarming systems etc., are very important operations which can be easily handled and synchronized their operations with great operational efficiency. The main functioning of an edge computing is to store and process most of the data which is collected by various IoT devices locally and disperse the information and decision to peers locally so the burden on centralized cloud processing

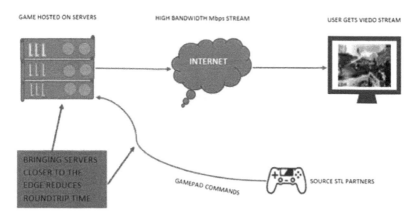

Figure 10.6 Cloud gaming using edge computing [10], [11].

and storage can be minimized and operational decisions can be taken at great time efficiency.

As per the above discussion, edge computing can contribute in smarty city development as:

- The data latency issue is completely resolved with the help of edge computing because data can be stored and processed locally, so delay in transmission and decision can be minimized.
- Data security is a big question in network transmission. As edge computing locally stores and processes the data, the chances of cyberattacks are also minimized.
- However, edge computing works in a decentralized fashion so the storage and processing burden on the centralized processing cloud and storage can be minimized.
- Sustainability of the services is a big challenge in cloud-based services. Edge computing processes data locally and in case any node fails to process or store data it can be easily handled by the peer nodes locally.

(vi) Smart transportation system [12]: Improving traffic efficiency and fulfilling the traveler's needs are the basic requirements of smart transportation. An edge computing-based IoT system addresses the traveler's

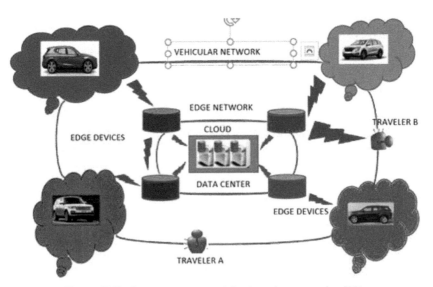

Figure 10.7 Smart transport model using edge computing [12].

preferences (i.e., fair, travel timing, travel distance, delay in departure and arrival, etc.) [12]. The utilities and benefits of edge computing-based public transport can be conveyed to the travelers. Reduced time delay and real-time decision-making are other attractions to the travelers. However, in the public transport system the distribution of the vehicles at different stations is another challenge; due to lack of real-time decisions, an uneven vehicle distribution happens which can be easily overcome with edge-based cloud service. A simple smart transport model using edge computing is given in Figure 10.7.

10.2 IoT and Edge Computing Framework

Edge computing is distributed in nature, there data stored and processed locally with minimized time response and transmission bandwidth of IoT-enabled network. Edge computing overcomes the saturation issues in network and data latency. It helps to minimize storage and processing burdens on centralized cloud storage and processing units. Fog computing is an extension of edge computing where a considerable amount of storage, computation, and local communication accomplishes. Edge computing is determined to process data locally or send to the centralized cloud processing unit. A typical edge computing architecture is given in Figure 10.8. Communication of edge devices is given in Figure 10.8. The edge devices are connected first with edge nodes and then with centralized cloud units. Edge nodes are the establishing links between edge devices and clouds. Thus, edge components are interfaced among local devices and clouds.

An edge computing architecture is given in Figure 10.9. The basic essential features of work are input, processing, and finally outcome of the processes. There are terms which we use in IoT-based edge computing.

(i) **Data source**: In IoT-enabled edge computing framework, a local node collects the data from devices and stores it there, known as the data source.

(ii) **AI**: After data collection, evaluation of the uncovered patterns from the data is a next-level operation. However, it is a layered process to get final acceptable outcomes. Identification of the useful information from the previous processing stage is an actionable stage. It consists of visualization, panel controls, alerts, etc. However, instead of centralized cloud server, edge computing servers are near to the end user and edge architecture is distributed as far as end storage and processing, where

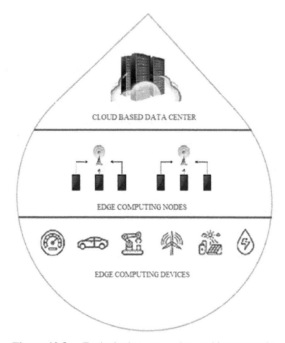

Figure 10.8   Typical edge computing architecture [13].

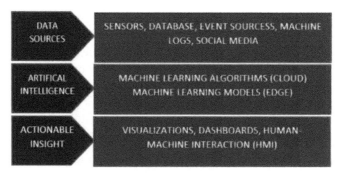

Figure 10.9   Edge computing architecture [13].

IoT devices collect information and transmit to the centralized system, which can have critical side effects and can overruled with local storage and processing in edge computing which are known as near-end devices (i.e., distributed gateways, etc.).

Edge computing network structure is given in Figure 10.10. There are three main components—cloud, local edges, and device edge, where edge

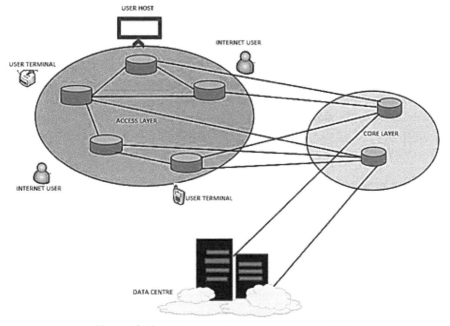

Figure 10.10 Edge computing network structure [13].

devices are end devices where actual data is collected or operations take place (i.e., industrial machine, sensors, cameras, etc.).

10.2.1 IoT and edge computing integrated framework

However, both IoT and edge computing have similar features, where IoT devices work as end-computing nodes of edge computing framework with enough processing and storage capacity. However, edge computing has capabilities to handle large amounts of data in terms of storage and processing which is collected from a large number of IoT devices and transmitted through gateways. In network-oriented services like IoT and cloud, privacy and security of the data is at most a requirement of an organization, which is far better in edge-based IoT frameworks as compared to centralized cloud-based IoT frameworks. Another issue with network-based frameworks is the unobservability issue where someone is using the services without letting others know that the service is in use. In the next, while transmission of data and use of services remotely is an open area to have cyberattacks, and with the help of edge computing, exposure of such openness can be minimized and attackers cannot know that two distinct objects are in communication. In IoT-based edge computing architecture, anonymity can be achieved where nodes can use the resources without revealing their identity. However, system security can be analyzed by analyzing different aspects [13] like confidentiality of the data and services used by end users, and availability of the services to the end nodes. Another aspect is integrity of the data and services, where data must not be tampered during transmission and privacy of the data and identity must be ensured. Access control also plays an important role in ensuring security and privacy needs of the data and end-user identity.

10.2.2 Pros and cons

IoT infrastructure has a number of risks which have to be addressed by the service provider or the institution which is owns the services. Risk mitigation of the IoT framework can be done by adopting certain policies like password policy; an ideal password policy has to be adopted to ensure the prevention of password crack (i.e., combination of various characters and symbols, etc.). Another communication must be encrypted whether it is inbound or outbound. A multilevel authentication system always helps us to ensure enhanced security, where after initial level access other different types of verification take place to minimize the risk incurred in the process. Nowadays, SMS-based or voice-based authentication codes are in practice to enhance the security. Sometimes, small hardware or software are also in practice to enhance the security of data. Similarly, a push notification mechanism is also in practice, where the same message turns up on user devices like mobile phones for validation of the identity.

10.2.3 Opportunities and challenges

We are in the digitized age where a number of IoT-based devices like sensors are collecting information from different environments and transmitting it in the network as raw data and later as processed information. IoT devices have their own communication mechanism among them and this digital revaluation is influencing human decision and interactions. Then, it is important to ensure a seamless human–machine interaction regardless at what rate the IoT devices are increasing. However, with the increasing use of IoT frameworks, we have increasing challenges also.

(i) **Privacy**: Privacy of the user and information both are important and somehow we have to achieve it for the data in transmission over the network and nodes in communication. At that time, they are very prone to the cyberattacks which can easily steal the private information of the node/person, and at the same time location and behavior also can be identified. An edge computing framework somehow overcomes this issue up to some extent, because local nodes store data and process it locally.

(ii) **Optimization metrics**: In edge computing at different layers, computation capabilities are incurred and it is pre-decided which layer will be doing what kind of operations. However, we have four different optimization matrices to ensure the optimality of the edge computing-based framework (i.e., bandwidth, computation, data and time latency, energy consumption and maintenance, operational cost, etc.).

10.3 Edge Computing Devices

Edge computing devices are used to process, collect data (i.e., sensor), network, storage, etc. The edge computing process takes place at the edge gateways which are beyond the edge of the network. In the edge devices, data is processed through fat client software which hardly transfers data. One of the important devices of the edge computing IoT-based framework is the server. Server is a powerful processing device placed at the edge of a network. These servers are physical and logically nearer to the data-collecting and information-utilizing devices. Edge computing server is similar to the centralized cloud server except its placement. Edge servers have capability to handle a lot of data in real time and disperse the decision or instruction to the designated devices/machines in real time to achieve real-time processing accuracy along with low computation and storage/transmission cost, at great security and privacy of the data and identity of nodes [13]. Edge servers can be placed anywhere, including regional edge or on-premises edge and its nature is also differing across the edges.

Edge hardware is expressed in different forms [14]:

(i) **Platform as a service (PaaS)**: In PaaS, we can deploy our applications on the cloud and need not manage the platform, which is automatically managed by cloud service provider (i.e., AWS, Azure, etc.)

(ii) **Software as a service (SaaS)**: In this service we can access applications which are already available on cloud platforms.

(iii) **Infrastructure as a service (IaaS)**: It is a very important service provided by the cloud platform where we have the flexibility to manage OS, applications, bandwidth, etc., as per our requirement.

10.4 Telcos and Edge Computing

With the increasing technological advancements, network bandwidth require-ment is also increasing. However, mobile cell networks are becoming complex day by day and classical optimization or deployment techniques are becoming ineffective. So, a new edge of computation and transmission mechanism over the virtual environment is required. Edge computing is the defining movement for next-generation network architectures and telcos operators. Telcos have rolled out the 5G services and shifted the services to cloud and virtual architectures, which are enhancing the processing power in a decentralized manner. However, edge computing is offering a vast platform to the telcos services even though before moving, it has to be decided which edge computing-enabled service is really fruitful for the telecoms sector. The existing operations for edge computing are AWS, Google Cloud, and Microsoft Cloud. In telecommunication, edge is the subset of multi-access edge computing and in edge architecture for telcos is able to position edge computing infrastructure encompassing network, storage and computing, and the virtualization of the resources is also one of the important technolog-ical advancement for telco industry. In the telco industry, edge computing is termed mobile edge computing (MEC), multi-access edge computing (MAEC) and it really processes at the edge which helps the telco operators to manage and deliver better services to customers. Across the infrastructure layers (i.e., access, transport data to the end nodes), telco industry is benefited with edge computing framework. However, it is a big question that the telco industry will be making edge computing frameworks commercially visible with increasing capital expenditure on 5G and fiber networks. Edge comput-ing is infrastructure as a service (IaaS) for the telco operators which enhances the processing speed and solves centralized data storage problems in an effi-cient manner at low cost. Edge computing offers different opportunities to the telco operators (i.e., new-edge capabilities-based solution and applications, distributed computing network, network performance, data privacy, security, etc.). A simple telco edge computing framework is given in Figure 10.11. In 5G technology, the edge nodes can be established on the basis of some factors.

(i) **Latency**: There are a number of applications which require 50 mil-liseconds latency (i.e., virtual reality, mission critical application, etc.). Therefore, the edge nodes must be as close to the end user as possible.

(ii) **Network topology**: On the basis of existing network topology at some certain places, the edge computing latency can be achieved. So, such points in the network have to be identified and edge nodes can be placed there.

(iii) Virtualization: Virtualization of the resources (either processing or storage) is another aspect which affects the placement of edge nodes in the network.

(iv) Maintenance cost: Maintenance cost is one of the important factors which affect the placement of edge nodes in the network. The edge node placement should be in such a way so the maintenance cost can be minimized.

(v) Availability: Availability of edge nodes is also a factor that influences their placement. It is possible at some places due to electricity or network issues the nodes may not be available for service. So, we have to place the edge nodes in such places of the network, where their availability can be ensured.

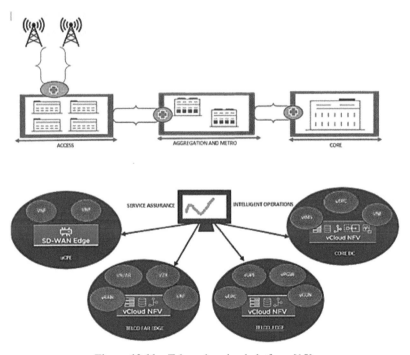

Figure 10.11 Telco edge cloud platform [15].

10.4.1 Telco network modernizations challenges

Modernization of the existing telco infrastructure is a big challenge to the operators. This includes network operation simplification, enhancing flexibility and availability of the services, reliance on data, time efficiency and at most scalability of services by reducing data and time latency and providing real-time service response. With the increasing use of virtualization of the resources workload can be optimized. Such optimization is very important for video streaming, cloud-based gaming, AI applications, and for business critical operations. Resiliency and availability of the services can be enhanced with the distributed nature of edge computing because failure of local nodes does not affect the availability of information of service at other locations because it can be backed up by another nearby local edge node. Data sovereignty is a very important aspect which has to be taken care of by telecom operators, where movement of the data across national boundaries is restricted and it has to be stored and processed locally. As we know a large amount of data the IoT devices are producing which has to be handled in real time, so the scalability of the operations can be enhanced with the help of the distributed nature of edge computing.

10.4.2 Multi-access edge computing [16], [17]

Multi-access edge computing (Figure 10.12.) is a future of telco and IT services with radio access cloud computing operations. Multi-access edge computing stores data and analyzes it at the local edge node and reduces the time and data latency to the end user. For new businesses, MEC have the potential to develop a wide range of services and applications. It helps the telco operations collect more and more information about the end user (i.e., interest, location, behavior, needs, etc.) to enhance the business and adopt new applications and solutions to enhance profits. Open radio network capabilities of the edge computing platform offers multi-tenancy and multi-service to the users with third- party authentication and allows processing and storage facilities as customized requirements. MEC gives ubiquitous cellular network coverage and enables IoT services and M2M services for smart cities, automobiles, energy sector, etc. However, security is another concern in MEC where third-party players can use and share the resources at great security and privacy.

A multi-access edge computing framework is given in Figure 10.12. MEC is a natural progression in data processing and computing and it improves the utilization of network resources. It allows better data processing, and drives efficiently as compared to simple cloud computing architecture. In edge computing, the distance of end user and data processing nodes has been reduced which decreases latency and allows time-efficient data storage and processing.

The standard designed networks of MEC help the business to disperse edge computing capacities. However, edge computing is already in the business models and MEC standards are still in the growing edge and it

Figure 10.12   Multi-access edge omputing [16], [17].

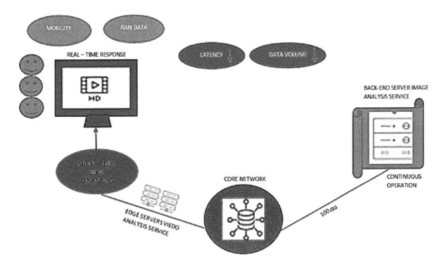

Figure 10.13   Simple MEC working [16], [17].

improves the network latency and efficiency. Edge computing solves the latency problem by processing and storing data on a nearby edge server to the end application. However, the MEC is better in terms of storage and processing and storage server. In Figure 10.13, simple working of MEC is given. MEC installed a radio access network at the data centers which reduces the latency and it transfers the load from front end server to the backend server to increase front end server availability for new tasks. However, in the 5G technology, MEC has a great scope because in 5G a faster data transmission and large storage capacity is required, and the processes must be time efficient.

10.4.2.1 Advantages of MEC

MEC has great scope in 5G technology. MEC has a great advantage because in 5G, a faster data transmission and large storage capacity is required, and the processes must be time efficient. There are some visible benefits of MEC are as follows [18]:

(i) **Latency**: So far it has been clarified that edge computing enables faster communication and data transfer over the network and can process the data with great time accuracy. The real-time insights can proactively monitor the IoT devices which help in proactive maintenance of the machines which later can be a cause of operational setbacks. So, in MEC the latency can be reduced which finally converts into profitable growth of a business.

(ii) **Security and reliability**: As we know, IoT has been an integral part of the social and industrial works. It has been involved in every area of business and it is also very clear to us that IoT devices are spread out remotely, which assess data from the environment and transmit it over the network to processing cloud units. This expansion of the IoT network and transmission of data through wired or wireless networks is security threat prone and weakens the reliability level of the systems. Edge computing is completely distributed in nature. The data collected by IoT devices is stored and processed on local nodes only. Only some specific data packet sends to the centralized cloud storage and processing. It minimizes the hacker's chances to attempt attack while data is in transmission and at the same time reliability factor of the system also increases because data is processing on local edge nodes, which are very close to the end nodes and if any edge node has failed due to any technical reason, the other neighboring edge node can accomplish that task.

(iii) **Scalability**: Another advantage of IoT-based edge computing is unlimited scalability of the IoT networks without consideration of existing storage capabilities and incurred cost.

(iv) **Device management**: It is another great advantage in MEC-enabled businesses to maintain devices at great efficiency and continuously monitor the real-time health of the devices, which helps to take proactive actions to save production setbacks.

10.5 Task Scheduling in Edge Computing

We are living on an advanced technological edge where technology is imparting in every aspect of life. IoT has been an essential ingredient of human society and the elaboration of 5G technologies has increased the mobile traffic and more and more heavy applications need high data transmission over the network. However, the high latency application requirements are directly affecting human health which has to be addressed from time to time. The cloud- based intensive computing tasks can easily meet up the latency requirements and this scenario can be efficiently uplifted by introducing edge computing framework with IoT, where intensive computing operations can be done at the edge nodes [20].

10.5.1 Scheduler-based edge computing

In scheduler-based edge computing, edge devices and servers are composed with edge scheduler, where tasks can be locally processed. The edge scheduler schedules the task to the edge nodes or the edge server as per the defined scheme.

The scheduling policies usually affect real-time computing, task-status, storage requirement and availability, network latency status, availability of edge server and nodes, etc., and the source device receives the computed results back through the scheduler. In this way, computation is evenly distributed to the edge nodes and servers, which realizes the collaborative processing of the tasks and it enhances the optimum utilization of computational resources. According to Figure 10.14, edge scheduler handles the task scheduled to edge nodes and servers. In another scheduling scenario, edge scheduler is composed of cloud center, edge server, and edge nodes where processing happens on edge nodes. In this scenario, the scheduler receives

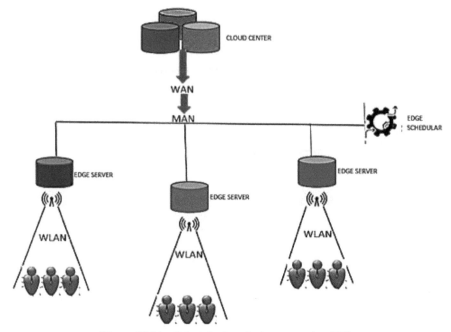

Figure 10.14   Scheduler-based edge computing [20].

all tasks and distributes them to the edge devices in an optimized way to achieve optimum utilization of the resources like bandwidth, computation power, etc. [20].

10.5.2 Computing task analysis

Before actual scheduling of the tasks, it has to be ensured that the task can be accurately scheduled on designated edge nodes or servers to meet the predefined objectives (i.e., time efficiency, energy consumption, cost, etc.). A task can be individually scheduled as a whole or it can split into different sub-tasks.

10.5.2.1 Local execution

Status of the edge resources like edge nodes, bandwidth, and servers are very important to determine if a task can be scheduled locally or not. If bandwidth is not sufficient to transmit the task to the centralized computing cloud, then it can only be computed on a local node. At the same time to achieve real-time decision- making the task has to be executed locally otherwise the time efficiency can be decreased due to the long way transmission of the data.

10.5.2.2 Full offloading
Before scheduling the task on the local edge nodes or on servers, we have to ensure the availability of the local server, sufficient bandwidth, and computation capability of the node. If the assessment findings are supportive and the local server can complete the designated task with great time efficiency and bandwidth is sufficient enough to transmit data, then the complete task can be scheduled to the local server.

10.5.2.3 Partial offloading
We know individual processing operations can be handled by the local edge nodes. However, distributed task handling is challenging work. Suppose we have to schedule a task which cannot be handled by a single edge node, there we have to identify other free edge nodes which can easily share the task. According to the computation and storage capacity of the edge nodes, we distribute the task to available nodes, later on, the resulting outcome is combined to pretend a single outcome.

10.5.3 Computing task scheduling scheme

A simple task analysis is given in Figure 10.15. After analyzing the edge computing resources, a synergistic task scheduling is needed where a competitive environment for the task is required due to the limitation of resources at edge node ends.

So far various algorithmic schemes have been proposed which are based on first-come-first-serve basis, min-max, max-min, time efficiency, time requirement, storage requirement, computation requirement basis, etc. Some standard heuristic task-scheduling algorithms are (GA, ant colony optimization, artificial immune algorithm, particle swarm optimization, simulated-annealing etc.). The algorithms which can achieve latency requirements, optimized resources allocation, and low energy consumption are expressed as minimal time delay. The time efficiency of the task is measured by combining the data transmission time over the network + processing time taken by the unit + result returning time to the initial node and the minimization of the time requirement at all stages is turn in quality of service (QoS). Minimal energy consumption, however, we talked about the time requirement, that is

Figure 10.15   Task analysis [20].

related to the energy consumption or we can say it is directly proportional. As the data transmission distance increases either as raw data transmission or result returning, the time requirement will also be increasing which increases the energy requirement in data transmission and at the same time the longer and shorter processing of the task also increases or decreases the energy consumption [20].

10.5.3.1 Task scheduling algorithm [21]

For the optimized resource scheduling we have to adopt distributed computing which definitely improves the processing performance, and minimizes the computation overhead. With the sensitive edge node, processing optimization can be achieved with low cost, less time, and great efficiency. Similarly, by minimizing task migration overhead we can easily minimize the energy consumption overhead and resource management cost. Such optimization is the guarantee of optimization of all task scheduling decisions. To achieve the

Input: T, S
Output: T and S
    Step1: Arange (T)
    Step2: for $event_k$ in T:
    Step3: Ser_k = getService($event_k$)
    Step4: for ser_i in S_k
    Step5: p_i = quote($service_k$, ser_i)
    Step6: n = select(ser_i)
    Step7: price ser_i = price ser_i * pow(ϑ, n)
    Step8 if price ser_i < $prices_{i-1}$
        bid($event_k$) = ser_i
    Step9: return ($event_i$, bid($event_i$))

Figure 10.16   Algorithm: Task scheduling algorithm [21].

real-time response demand of the scheduling in edge computing framework, needs a highly dynamic network which requires high computing power at dispatch centers. However, optimal utilization of the surrounding resources of edge nodes can help to achieve the requirement of low computation power and less energy consumption, reduced transmission time and distance, etc. A pre-decided schedule can abate the exodus of the computing resources and data in the edge computing and choice of the appropriate end node processing unit is also significant which gives the guarantee of originality of the work, time efficiency etc.

10.5.3.2 Bidding-model optimization and adaptive algorithm

Selection of the offloading nodes is a very important question, while a node wants to schedule a task and when the ratio of requesting nodes and offloading nodes is different then it becomes a critical question. A specialized optimization strategy can only answer to this question. Bidding model [21] is an answer of this question, where task scheduling takes place through the bidding process. It has two main actors; inviter and bidder. Inviter provides operational details (i.e., quotation, labor, etc.) and the bidder anticipates the lowest quoted bid. The same strategy is shown in Figure 10.16

It is a balanced scheduling in this model we have to arrange tasks in any order for quotation calculation. The balance scheduling objective is achieved by allocating a big task to a strong processing capability node. The scheduling

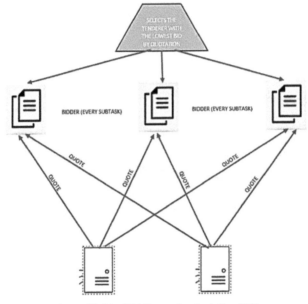

Figure 10.16   Bidding task scheduling [21].

is done through a mathematical formulation, however it is hard to schedule the task as it is confirmed to a specific node, because if some other bidder is making a query for the same node, increasing the latency and bandwidth occupancy [21].

10.6 Security and Privacy Issues of Edge Computing

However, edge computing is enhancing the use of IoT in development of new-age businesses. Thus, it has incurred various privacy and security issues. In edge-computing, the placement of the processing nodes keeps near to the users, which encourages the receivable of large volume private sensitive data and leakage of such sensitive data from local nodes can be a cause of serious problems. Traditionally privacy and security can be understood through availability, integrity, and confidentiality, in addition to this in a distributed environment this can be increased with some other factors like accountability, octave-categorization, auditability, preservation of privacy, trustworthiness, etc. However, auditability and accountability are directly related to trustworthiness which is highly concerned with authenticity.

The security requirements the edge computing based-IoT framework can be expressed as:

(i) Confidentiality ensures the genuineness of the data which is transmitted by the sender and received by the receiver.

(ii) Availability is another requirement of the edge computing framework, where we ensure that all the resources are available to the requesters as and when required.

(iii) Integrity is the next issue related to the security and privacy of the edge computing IoT frameworks. In this, we try to ensure the transmission of original information to the receiver, without any modification during transmission.

(iv) Non-repudiation is the issue where the sender or receiver denies that the communication information is not related to them. In the transmission both sender and receiver must not allow to deny the ownership of the information.

(v) Privacy preserving is to keep all the secret information private at both the sender and receiver ends to enhance the trust.

(vi) Authenticity is another security requirement in an age computing environment where the sender and receiver nodes must have specific identification to prevent unauthorized participation in the communication.

(vii) Attack detection and prevention is another security and privacy requirement, because we cannot stop transmitting data either at short distance or long distance through the network, that time cyberattackers can easily steal the data. So, we have to ensure it in the edge computing framework, that while transmission of data or even at storage of data, any unwanted suspicious activity can be detected as it encountered and prevented from the stealing of data and corrupting the data.

(viii) Reliability is the most important issue of any system, similarly, it is important in edge computing also where the nodes must be reliable in terms of time-efficient quality processing, stealing of information, and other security- and privacy-related threats.

10.6.1 Attacks classification

In this section, we are exploring some attacks which are threat to privacy and security of edge computing-based IoT framework [17], [22].

(i) **Software injection/malicious hardware**: Unauthorized software or hardware components can be suppressed by the attacker in our edge computing-based IoT network and this usually happens at the time of communication. Such malicious components can expose the critically secure processing information to the attacker and they can bypass the designated authentication process and steal the information. Similarly, hard injection attacks can be classified in different classes which include false replication of executing nodes, by assigning false identification to it. Hardware Trojan provides the illegitimate access to IT circuits to control the electronic circuit. Camouflage is another security challenge, where a counterfeit node inserts into the IoT network which behaves like a normal edge computing node in operations like data sharing, receiving, processing, storing data, redirecting the signals, transmitting data packets, etc.

(ii) **Flooding**: Intentionally the attacker transmits the counterfeit message in the network to exhaust the network capacity. So, the network should not respond to the authorized clients.

(iii) **Denial of service (DoS)**: In distributed DoS attack, battery draining, sleep deprivation, and outage attack are most common. However, such attacks mainly jam the signal transmission.

(iv) **Tampering**: This can happen when the physical access of the edge nodes or other IoT devices is available to the attackers and private even encrypted information can also be extracted, physical electronic circuits can be tampered for further data stealing or for miscommunication.

(v) **Sniffing**: In this attack, attackers anticipate the authentication-related data (i.e., password, usernames, etc.) privately from the communication link and after getting authentication information they can steal secret data from the network nodes.

(vi) **Non-network side-channel**: Even the ideal nodes of the network can reveal the secret data, that is, acoustic data from the medical devices can reveal the private medical data of the patient.

(vii) **Routing information**: During communication, attackers drop data packets to redirect the routing information. This happens by inserting malicious nodes in the network.

(viii) **Forgery**: A fraudulent information packet injects in the network to pretend that the system has failed, through a communication link. Such packets are inserted through malicious data packets, modified data packets, and by replicating previous packets.

(ix) **Unauthorized control access**: In edge computing-based IoT networks, edge nodes communicate with each other and transmit data also. Somehow, an attacker can have access to a single edge node only, and then it can control the whole network.

(x) **Inessential logging**: From the insecure log files authentication data can be leaked. That is why most of the time log files are kept in encrypted format.

(xi) **Attack through I/O devices**: Here attackers use mobile botnets, IoT malwares, and ransomware which are security concerns for both application and edge node.

For the security and privacy concern we need to have a strong counter mechanism to prevent such security threats. To counter the malicious injections some effective techniques are in use. Hardware Trojans can be detected through the side channel analysis. This detection happens through temperature testing, power consumption testing, timing efficiency testing, etc. At the same time malicious application detection is done through behavioral analysis of the edge nodes. A policy to identify violating behavior of the node is also another level of counter mechanism to identify the attack. In such a policy-based counter mechanism, we ensure that the standard rules are followed or not. The update of the security framework is also very important to overcome the security threats. By creating a list of the trusted nodes for sensitive data transmission, the reliable transmission can be ensured. Even though the need for an intrusion detection system is very high to mitigate the security threats. Through intrusion detection systems, we can monitor the network behavior, suspicious events can be reported. A cryptographic countermeasure is most important to secure the data either on the transmission or at storage. The logs of the application can be stored in encrypted forms and the data over the network also can be transmitted in encrypted format. In the edge-based IoT networks, the decentralization of the processing and storage unit is a very important feature which data need to travel for long distances and can be analyzed and utilized at local nodes [17], [22].

10.7 Conclusion

As we know, IoT enhances the communication among different digital assets (i.e., sensors, devices) and helps to capture required data and transmit it to processing centers. However, IoT devices generate a volume of data at the end-user level and it needs to be processed in a short span of time on the cloud. In this chapter, the concept of IoT and edge computing are explored with how the new technologies are enhancing the living standards of the society and enhancing businesses. Thus, different applications and use cases of the edge computing-based IoT framework are discussed. Then, various challenges and security issues are discussed along with various cyberattacks and their countermeasure mechanism. However, it is true that new-age technologies like IoT, edge computing are drastically changing the traditional business strategies and helping to improve the living standard of human society. Apart from these benefits they are bringing various challenges like privacy breach of personal and secure information. Over-dependency and utilization of technology is also harmful to nature and human society. Finally, we need to be careful enough to utilize the new-age technologies carefully to enhance the qualitative achievements and minimize the negative aspects.

References

[1] Goyal S. et. al, 'Precedence & Issues of IoT based on Edge Computing', in IEEE 9th Int. Conf. on Commu. Systems and Network Technologies (CSNT), pp. 72-77, 2020.
[2] Alrowaily M. et. al, 'Secure Edge Computing in IoT Systems: Review and Case Studies', in IEEE/ACM Symposium on Edge Computing (SEC), pp. 440 – 444, 2018.
[3] https://www.designtechproducts-ptc-iiot.com/articles/what-is-edge-co mputing.
[4] Sharma M. et.al, 'Functional Framework for IoT-Based Agricultural System', in Internet of Things and Analytics for Agriculture, Vol.3, pp.1-27, 2021.
[5] Varghese B. et.al, 'Challenges and Opportunities in Edge Computing', in IEEE Int. Conf. on Smart Cloud (SmartCloud), pp.20-26, 2016.
[6] Sharma M. et.al, 'Intelligent Agro-Food Chain Supply, Internet of Things and Analytics for Agriculture', Vol.3, pp. 65-91, 2021.
[7] https://www.manufacturingtomorrow.com/article/2021/02/10-fundame ntal-computer-hardware-needs-for-edge-computing/16443.
[8] https://phoenixnap.com/blog/edge-computing-vs-cloud-computing.
[9] https://www.i-scoop.eu/industry-4-0/.
[10] https://www.redhat.com/en/topics/cloud-computing/cloud-vs-edge.
[11] Qian Liu et. al, 'Cloud, Edge, and Mobile Computing for Smart Cities', in Urban Informatics, pp. 757-795, 2022.

[12] Lin J. et.al, 'An Edge Computing Based Public Vehicle System for Smart Transportation', in IEEE Trans. on Vehicular Technology, Vol. 69, No.11, pp. 12635-12651, 2020.

[13] https://www.onlogic.com/company/io-hub/what-are-edge-servers/

[14] https://stlpartners.com/articles/edge-computing/what-is-an-edge-server/

[15] https://telco.vmware.com/content/dam/digitalmarketing/vmware/en/pdf/solutions/vmware-ebook-telco-edge.pdf.

[16] Taleb T. et.al, 'On Multi-Access Edge Computing: A Survey of the Emerging 5G Network Edge Cloud Architecture and Orchestration', in IEEE Commu. Surveys & Tutorials, Vol.19, No.3, pp. 1657-1681, 2017.

[17] Alwarafy A. et. al, 'A Survey on Security and Privacy Issues in Edge-Computing-Assisted Internet of Things', in IEEE Internet of Things Journal, Vol.8, No.6, pp. 4004-4022, 2021.

[18] https://www.avsystem.com/blog/multi-access-edge-computing/

[19] Liao J. X. et.al, 'Resource Allocation and Task Scheduling Scheme in Priority-Based Hierarchical Edge Computing System', in 19th Int. Sympo. on Distributed Computing and Applications for Business Engineering and Science (DCABES), pp. 46-49, 2020.

[20] Shichao Chen et al., 'Recent Advances in Collaborative Scheduling of Computing Tasks in an Edge Computing Paradigm, sensors', in MDPI, pp.1-22, 2020.

[21] Zheng Shi et al., 'Multi-node Task Scheduling Algorithm for Edge Computing Based on Multi-Objective Optimization', in Journal of Physics: Conference Series, Vo.1607, pp.1-13, 2020.

[22] Yahuza M. et.al, 'Systematic Review on Security and Privacy Requirements in Edge Computing: State of the Art and Future Research Opportunities', in IEEE Access, Vol.8, pp. 76541-76567, 2020.

11

Low-energy Network Protocols

Hiren Kathiriya[1], Arjav Bavarva[2], and Vishal Sorathiya[3]

[1]Department of Electronics and Communication, R K University, India
[2]Department of Information and Communication Technology, Marwadi University, India
[3]Faculty of Engineering and Technology, Parul Institute of Engineering and Technology, Parul University, Gujarat, India
E-mail: hiren.kathiriya@rku.ac.in; arjav.bavarva@gmail.com; vishal.sorathiya9@gmail.com

Abstract

IoT and Industry 4.0 are the integration of sensors, processors, and power supply units, which can be implemented using numerous networking technologies, for instance, wireless sensor networks (WSNs) as well as low-power wide area networks (LPWAN), and low-power wireless personal area network (LoWPAN) that contains sensors and actuators. Wireless sensor nodes gather scalar information like temperature, speed, location, and humidity from predefined areas and forward to the sink using single hope or multiple hopes. A sink, coordinator, or gateway is the intermediate device that connects IP-less networks with IP-enabled networks. Predefined areas can be monitored and controlled from a remote place. The communication between predefined areas and the real physical world provides several applications which can be used for IoT and Industry 4.0. This chapter contains a variety of technologies, algorithms, and innovative concepts used to design low-power, energy-efficient Industry 4.0 protocols. ZigBee, thread, Z-Wave, and EnOcean are popular low-energy protocols used to implement low-energy wireless personal area network technologies. Various technologies like LoRa/LoRaWAN, Sigfox, NB-IoT, etc., are used for LPWAN. IP-based lightweight protocols include 6LoWPAN, 6Lo, 6 TiSCH, RPL, and CoAP. Routers and aggregators that use energy-efficient routing and data aggregation use techniques such as scheduling, listening after sending data, channel sampling, etc. Protocols used for energy harvesting are also in demand to design self-powered sustainable WSN with the help of energy-neutral operation (ENO). Some protocols are designed along with low-energy consumption, providing solutions for encoding, encryption, security, and management. This chapter also analyzes various low-energy network protocols used in various technologies to implement IoT and Industry 4.0.

Keywords: Low-energy protocols, Industry 4.0, IIoT protocols, IIoT network architecture.

11.1 Introduction

The introduction of steam and water to manufacturing led to Industry 1.0. This led to a more efficient, faster, and easier production process. Industry 2.0, also known as "The Technological Revolution," was superior in electrical technology, allowing even more production by sophisticated machines. Industry 3.0 was the era of computers and the internet and the usage of electronics, communication, and information technology. As a result, more automation became possible in production and the assembly line. Industry 4.0 is beyond automation. Autonomous systems with smart machinery, storage devices, intelligent protocols, and algorithms are capable of exchanging information and taking proper decisions as per the situation. Fully automated and smart industries are possible with the help of Industry 4.0. Wireless sensor networks (WSN) and industrial internet of things (IIoT) are responsible for Industry 4.0. However, the execution of Industry 4.0 brings many challenges, and much work must be done to implement the Fourth Industrial Revolution [1].

Industry 4.0, or the Fourth Industrial Revolution, is gaining momentum. The transformation of industries through innovative production strategies using IIoT acquires real-time data, uses machine learning algorithms in data processing, enables the system to self-determine, and provides interaction across the entire system [2]–[4]. WSNs are used to monitor and control various parameters. IIoT and WSN play vital roles in implementing Industry 4.0. However, practical execution incorporates many challenges in energy efficiency, security, and data management and analysis.

Provision of regulated and constant power supply to the sensor nodes is complicated for many applications of IIoT networks and networks deployed in mountains or mining areas. Energy can be harvested from the available

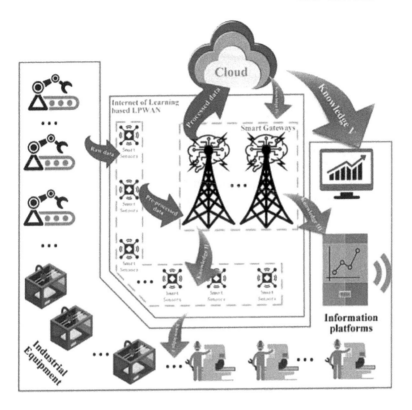

Figure 11.1 Concept of IIoT network [9].

external energy sources [5], but then again, existing protocols and algorithms must be revised significantly for network layers, as the energy levels of the nodes keep varying. Many important decisions related to the network, for instance, cluster formation, sleep cycle, data transmission and reception, are taken based upon the accessible energy of the node. Based upon the potential energy harvesting model, it is expected that an improved energy harvesting policy will be planned to improve the energy efficiency of the network. The proposed protocol for self-powered wireless nodes will show better performance with lower power consumption. IoT is a system embedded with sensors, processing units, storage devices, and software that can exchange the data and information by connecting with other devices without human interaction. IoT can be interconnected with industrial instruments and other networks for industrial applications like production, manufacturing, and energy management [6]–[8].

Figure 11.1 depicts the stream of the IIoT; the whole track is designed for the process of data gathering to critical decision-making. Data collection is usually conducted via a sensing network in wireless sensor networks. The flow is routed to a processing network after the collected data has been processed. The processing engine plays a crucial role in understanding and translating the composed data into understandable formats for a better understanding. Finally, the data is transmitted via various communication protocols to the gateway node, where critical decisions and corrective actions are made. It is designed to act as a closed-loop system to increase efficiency and the ability to make better decisions within the system.

11.2 IIoT

After the year 2013, the usage of IoT-connected devices and revenue of IoT gradually increased worldwide [10]. Over the past decade, the IIoT concept has been adopted by North America, Asia-Pacific, and European countries. In 2019, the United States, Asia-Pacific, and Europe were responsible for global IoT spending at 35.7%, 27.3%, and 23%. By 2024, Europe expects global IoT spending to grow by 24%, while North America and Asia-Pacific will experience lower annual growth of 11% and 13.2%.

India is one of the most emerging countries in manufacturing enterprises and has more than 1.37 crores of manufacturing enterprises. Most of these manufacturing enterprises are at the second or third stage in connection with the industrial transformation. The Government of India also promotes the industrial revolution through SAMARTH Udyog Bharat 4.0. India's Ministry of HI & PE government has launched the Industry 4.0 initiative SAMARTH Udyog Bharat 4.0.

Industrial automation using IIoT increases efficiency, minimizes errors, reduces costs, and improves production and safety. Industrial machinery can be monitored and controlled from remote places, update the functionalities on human–machine interface software, analyze the machine maintenance, manage the data from multiple locations, and optimize industrial actions; all these are possible using IIoT [11]–[13].

11.2.1 Network architecture

The architecture of IIoT is categorized into four layers named device, network, infrastructure, and application, as shown in the Figure 11.1. Sensors and actuators come under the device layer. Various sensors are used to sense the environment, such as gas, temperature, humidity, motion, etc. Actuators control the mechanism or system based on the decision or operated by the controller either from a local or remote place. The network layer is responsible for connecting devices with the gateway, the coordinator, or the sink. Different technologies can create connectivity like near-field communication, Ethernet, wi-fi, Bluetooth, software-defined networking, etc. The gateway is the central device where data acquisition takes place. Data processing, management, sorting, analysis, and decisions are taken at the infrastructure layer, especially for fully automated systems [14], [15]. The gateway should be connected with the user through an application layer using an internet protocol-based network. Applications are the user interface to monitor and control the sensors/actuators connected at the device layer if needed. Software upgrades, quality control, safety, and security are significant application layer concerns.

11.2.2 Protocols

Protocols used to realize an open and interoperable IIoT are classified into polling-based and event-based protocols. In communication perspective, the IoT platform consists of transmission control protocol/internet protocol (TCP/IP) and standard protocols, widely accepted, mandated, the fixed procedure for completing the task. Several protocols like advanced message queuing protocol (AMQP) and message queuing and telemetry transport (MQTT) as well as constrained application protocol (CoAP), extensible messaging and presence protocol (XMPP), and JavaScript object notation (JSON) are considered as standard protocols for IIoT. MODBUS TCP, a pooling-based protocol, can be used on application layers for industrial applications with an optimized message structure. HyperText Transfer Protocol (HTTP), polling-based, cannot be used for IIoT applications as it uses an asynchronous communication request-response model. XMPP, a byte-oriented message protocol, is not advisable for IIoT applications because it increases the message size. MODBUS has small-sized data units and is suitable for monitoring, control, and automation-related applications compared to XMPP. MQTT is designed for machine-to-machine conversation, which can coexist with MODBUS TCP, a request-based protocol used for monitoring and control purposes. CoAP also works on application layer IPv6 network infrastructure, which uses an exponential back-off mechanism [12].

Researchers have compared various protocols in terms of round-trip time (RTT), delay, jitter, and energy consumption for various applications. MQTT is one of many TCP-based message-oriented protocols, while CoAP is a UDP-based protocol. Thus, MQTT is more reliable. Results show that MQTT has lower RTT compared to CoAP [16]. MODBUS TCP performs well for IIoT applications as it uses transport layer security, and security in IIoT applications is a prime concern. MQTT parallelly works with MODBUS TCP to get the advantage of both protocols. MQTT performs well for M2M communications in terms of three classes of quality of services messaging classes: exactly once, fire and forget/unreliable, and at least once, while MODBUS TCP performs better for industrial functionalities [17]. Though MQTT is one of the most popular IIoT application protocols, it does not provide real-time services. The extended MQTT protocol known as real-time MQTT (RT-MQTT) specifies actual time requirements with the help of software-defined networking [28]. Low-latency network can also be implemented using named data network [18], [19].

11.2.3 Testbed

Experimental setup for the IIoT has been done using hardware and software modules [17]. Various hardware supports IIoT applications like Raspberry Pi 3, Raspberry Pi 2, ESP 8266, Tessel 2, Intel Edison board, Particle Photon, and Arduino Uno. Drivers for sensors, radio communication, timers, and input/output interface are installed over hardware devices. Operating system (OS) is system software that manages hardware and software utilization. IoT OSs are embedded operating systems specially designed to perform over limited constraints and low-power small hardware modules. Few OSs have special libraries that support programmers in making programs small and manageable. TinyOS, OpenWrt, RIOT, FreeRTOS, and Contoki NG are open sources and widely used OSs for IoT devices. OSs like Embedded Linux is also freely available, while Mbed OS and Windows 10 IoT are well-known for their high-grade security. User applications run on OS through middleware. Middleware is a bridge between OS, database, and applications. OpenIoT is a famous open-source middleware that establishes communication between heterogeneous sensors and applications through the cloud.

OpenIoT and Open IoT-VDK are ready to use packages of OpenIoT. ThingsBoard is open-source which can be used standalone or in a cluster. It is widely used in IoT project development, management, and scaling IoT applications. RabbitMQ and Mosquitto are popular brokers, an intermediary program module that works as a protocol translator between sender and receiver. RabbitMQ is a message broker that is freeware and open access that

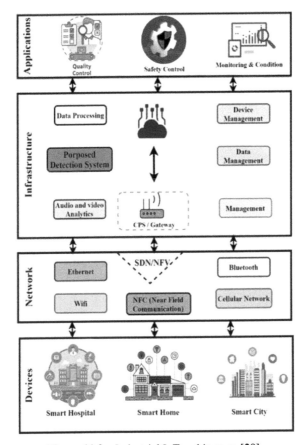

Figure 11.2 Industrial IoT architecture [20].

supports AMQP, MQTT, and HTTP while Mosquitto is a lightweight broker for MQTT [17].

11.2.4 Criteria for implementation

IIoT comes with many challenges. Essential criteria must be kept in mind before we design IIoT applications. The primary criteria are communication, technology, job, privacy and security, legal regulations, and culture, further subdivided into 26 sub-criteria [9]. Multiple-criteria decision-making processes are employed to categorize criteria, such as the analytic hierarchy process and the analytic network process. In addition, weight factors are allotted to main and sub- criteria to set the priorities and importance. It helps to evaluate the difficulties in IoT multi-criteria decision-making implementation.

11.3 Low-energy Technologies

It is becoming viable to develop a new generation of sensorization services due to the proliferation of cost-effective communication devices. These services have great potential to be very valuable in industrial applications. But the solutions created so far to manage such a system do not totally match the severe needs of commercial networks, specifically those about energy economy and dependability. The low-power wide area networking (LPWAN) paradigm, a revolutionary networking paradigm, will be discussed in detail in this article (LPWAN). LPWAN-based solutions, which use cellular-type architecture, are meant to solve the consistency and efficiency issues that long-term industrial networks are experiencing by resolving. Thus, the major leading LPWAN choices are examined in length, with the merits and downsides of each solution being identified and described in detail. The current status of deployment of these technologies in Spain also is examined as part of the research. However, even though they are still in the early stages of development, low-power wide area networks (LPWANs) provide a feasible option for expanding upcoming industrial Internet of Things networks and services.

LPWAN is a new networking paradigm that has recently evolved to close the gap in deploying congested M2M networks [21], [22]. LPWAN is a low-power wide area network that transmits data over long distances at low power. The deployment of highly scalable systems, which are often performed in an automated manner and include low-cost edge devices with minimum energy use, is one of the most significant components of these systems. In Figures 11.1 and 11.2, you can see an illustration of the design of a typical LPWAN network. If we consider the network architecture of cellular networks, it is nearly identical, with one or a sequence of base stations enabling edge devices to be directly connected to the backhaul network and, subsequently, to the cloud, where the data is gathered made ready for user access. It is crucial to note that the edge-network architecture utilized by conventional WSNs differs significantly from the design used by traditional WSNs in terms of performance. The end nodes interact directly with the base station rather than building a local network. A gateway is used to transfer the obtained data outside of the network. This approach reduces network administration

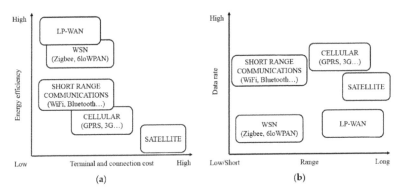

Figure 11.3 (a) Coverage and data rate in IIOT technology, (b) Terminal, connection cost and energy efficiency in IIOT enabling technologies **[10]**.

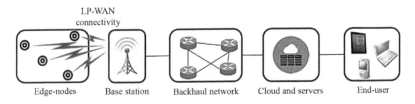

Figure 11.4 Network architecture in low-powered WAN network **[10]**.

complexity while also lowering the total energy consumption since routing functions are not performed in this configuration [10].

The continual rise of Industry 4.0 has launched tremendous openings and scopes for engineers to increase the effectiveness of equipment, which is just the beginning of what lies ahead for them. However, because of the unpredictable nature of the communication path, many industry managers remain leery of utilizing the internet to control their equipment, despite technological advancements. As long as thorough verification of organizations is performed, and trust is created between them, the use of the internet to manage industrial operations has the potential to become generally accepted in the future. However, traditional systems are impossible to be applied to network devices with limited resources due to their inherent security issues and other complications. Because of this, the Make-IT protocol offered a robust mutual authentication and secret key exchange mechanism to circumvent the drawbacks of current approaches. Using a range of cryptographic operations such as hashing, ciphering, and so on, it has offered secure mutual authentication and secret key exchange across various organizations to prevent unauthorized access to sensitive information in the past. Performance and security assessments of the proposed work demonstrate that it is both more energy efficient (since processing and communication are inexpensive) and more resistant to attacks when compared to existing systems [23]. Energy harvesting wireless sensor networks, also known as EH-WSNs are generally recognized as one of the essential empowering technologies for the growth of the internet of things. EH-WSNs are a wireless sensor network that harvests energy from the sun (IoT). In part, this is because most existing research on EH-WSN does not consider the relationship between energy state and data buffer limitations. As a result, they cannot address the issues of low-energy efficiency and long end-to-end latency associated with standard wireless sensor networks (WSNs). Considering the above concerns, this work provides a totally new greedy strategy-based energy-efficient routing protocol based on greedy optimization. As part of the system modeling process, we first construct an energy assessment model that considers all of the elements that impact energy harvesting, consumption, and classification to establish the energy status of each node and the overall energy state of the system.

A communication range judgment model based on channel feature information is then developed to determine the transmission area of nodes. Finally, a reception state adjustment mechanism must be developed to combine the two models outlined above. When altering the data reception status of nodes on a network, it considers the buffer occupancy and the MAC layer protocol used by the network. Based on this information, a greedy strategy-based routing technique is recommended [24]. Long-range wide area network (LoRaWAN) is one of the intense research topics in IIoT. There are many applications where LoRaWAN gateways are not connected with the grid, and due to the limited power storage, network lifetime becomes a major aspect. Researchers proposed LoRaWAN energy-efficient communication protocols to improve the overall lifetime of the network [25].

11.4 Self-powered Network

11.4.1 Comparison between energy harvesting WSNs and battery-operated WSNs

A battery-powered sensor node can operate indefinitely until its total energy is depleted. Therefore, the prime objective of existing network protocols is to save energy and extend the duration of the network lifetime. Such protocols cannot be implemented for energy harvesting WSNs because each node can operate unless ambient power is available, making the balance between power consumption and power harvested the primary goal. Table 11.1 summarizes the comparison of battery-powered WSN and energy harvesting WSN.

The total energy decreases during the lifetime of a battery-powered node, which means that the sensor node can function until its energy level approaches an inoperable level. An energy harvesting WSN nodeâĂŹs energy characteristics differ from those of a battery-powered sensor node. Utilizing energy harvesters, EH-WSN nodes can replenish their energy [26]. Nevertheless, as the energy harvesting rates achieved by the EH-WSN devices available on the market today are much lower than the power consumption needed for node operation (sensing, processing, and communication), The node can only operate when a certain amount of energy has been accumulated in a storage device [11, 26].

11.4.2 Energy-harvesting sources for sensors

It is shown in Figure 11.5 how mechanical energy can be harvested from mechanical systems and used to power energy harvesting WSNs. Based on the types of energy, these sources can be classified as light, electricity, heat, and motion, which are the results of mechanical systems and their surroundings.

Table 11.1 Comparison between energy harvesting WSNs and battery-powered WSNs **[11]**

	Battery-powered WSNs	Energy harvesting WSNs
Prime objective	Because batteries cannot be replaced, maximize life at the cost of output and latency	Considering the energy harvesting rate and the ability to replenish the energy, maximize throughput and minimize delay
Protocol development	Sleeping and wake-up times can be precisely determined	Schedules of sleep and waking up are more difficult to predict and assign
Prototype of energy	The energy prototype is well-comprehended	A variety of factors affect the rate of energy harvestings, such as location, time, and equipment type

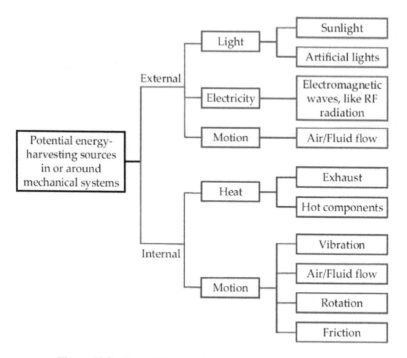

Figure 11.5   Potential energy-harvesting sources for sensors [27].

11.4.3 Energy stored by a supercapacitor

$E = \frac{1}{2} \times CVV$ is the formula for calculating the energy stored in a capacitor. So, in our case for a 5.5 V, 1.5 F, the energy stored by the capacitor, when charged fully, will be

$E = (1/2) \times 1.5 \times 5.52$

$E = 22.69$ Joules

We can now use this value to calculate how long the capacitor can power things if we need 500 mA at 5 V for 10 seconds. The energy needed for this device can then be calculated using the formula energy = power × time. Again, P = VI is used to calculate power here, so the power for 500 mA and 5 V is 2.5 Watts.

Energy = 2.5 × (10/60 × 60)

Energy = 25 Joules or 0.00694 Watt-hour

Let's assume that a remote sensing application transmits active Bluetooth low- energy data once every 2 seconds during its working life. The rest of the time, the application has remained in sleep mode (wake-up from a pin) or has not transmitted or received data as shown in Figur 11.6.

Power = 15 mW (Tx), 0.3 μW (deep sleep)

Tx Energy = (0.015) × (0.007/3600) = 3 × 10^{-8} Wh

DS Energy = (0.3 × 10^{-6}) × (1.993/3600) = 1.66 × 10^{-10} Wh

Total Energy = 54.3 μWh (1 hour) = 0.2 joule

So, if the supercapacitor is fully charged, it can power BLE for 125 hours. (22/0.2 = 110 hours)

11.4.4 Proposed circuit

11.4.5 BLE–RSL10 from ON Semiconductor

Specifically designed for IoT applications, the RSL10 is an ultralow-power, multiprotocol 2.4 GHz radio with a highly flexible interface. In addition, due to its Arm Cortex-M3 processor and LPDSP32 DSP core, the RSL10 can support Bluetooth low energy and proprietary 2.4 GHz protocols without sacrificing power consumption.

11.4.6 Key features of BLE-RSL10 [17]

- Rx sensitivity (Bluetooth low-energy mode, 1 Mbps): −94 dBm
- Data rate: 62.5–2000 kbps
- Peak Rx current = 5.6 mA (1.25 V VBAT)
- Peak Rx current = 3.0 mA (3 V VBAT)
- Peak Tx current (0 dBm) = 8.9 mA (1.25 V VBAT)
- Peak Tx current (0 dBm) = 4.6 mA (3 V VBAT)
- Arm Cortex-M3 processor clocked at up to 48 MHz

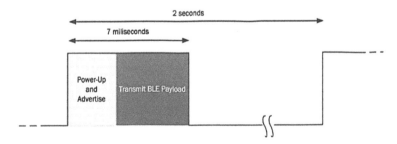

Figure 11.6 Duty cycle power considerations for Bluetooth low-energy technology **[28]**.

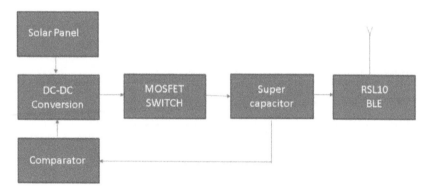

Figure 11.7   Solar-powered wireless sensor node using BLE.

Figure 11.8   Supercapacitor charging circuit.

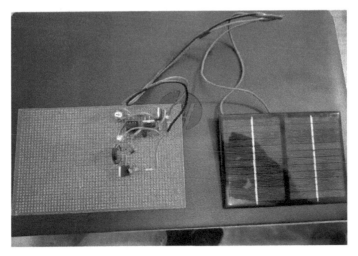

Figure 11.9 Supercapacitor charger circuit using solar panel.

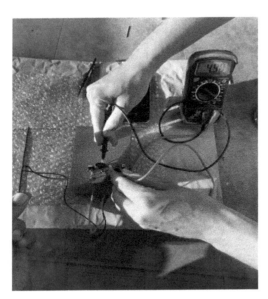

Figure 11.10 Supercapacitor charger circuit using solar panel.

- LPDSP32 for audio codec
- Supply voltage range: 1.1–3.3 V
- Current consumption (3 V VBAT):
  - Deep sleep for IO wake-up: 25 nA
  - Deep sleep with 8 kB RAM retention: 100 nA
  - Audio streaming with 7 kHz audio bandwidth: 0.9 mA RX, 0.9 mA TX
- 384 kB size of flash memory
- Highly integrated system-on-chip (SoC)

11.4.7 Lifetime improvement of WSN

- Based on previous residual energy information, we propose a technique to extend the lifetime of mobile or fixed wireless sensors by using this chapter.
- In this way, the energy consumption of the sensor nodes is optimized while maintaining data throughput.
- Scheduling nodes can determine activity levels to communicate with sinks at specific fixed times or frequency slots during sensory measurements recorded.
- A solar-powered sensor node is anticipated to recharge sensor batteries or supercapacitors.

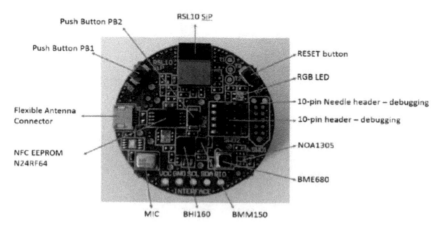

Figure 11.11 BLE–RSL10 from ON Semiconductor **[29]**.

- In comparison to non-optimized networks and idealized greedy algorithms, the proposed optimization approach extends the lifetime of sensor nodes based on residual energy information [30], [31].

11.4.8 Energy model for WSN

Almost all clustering algorithms for WSNs use the presented radio energy model as the de facto standard [17], so this model is considered for energy consumption during different activity as shown in Figure 11.12.

Eqn (11.1) shows the energy spent in data transmission:

$$E_T(l, d) = \begin{cases} E_{\text{elec}} \times l + E_{\text{fs}} \times l \times d^2 \text{ if } d < d_0 \\ E_{\text{elec}} \times l + E_{\text{mp}} \times l \times d^4 \text{ if } d \geq d_0 \end{cases}, \qquad (11.1)$$

where d and l are the data size and distance measurement value, respectively. The energy used on receiving purposes is expressed in eqn (11.2).

$$E_R = E_{\text{elec}} \times l, \qquad (11.2)$$

where E_{elec} designates the amount of energy required to drive the electronic circuitry, and E_{fs} and E_{mp} denote the energy required for free space and multipath amplifiers, respectively. Eqn (11.3) depicts the distance threshold value (d_0):

$$d_0 = \sqrt{\frac{E_{\text{fs}}}{E_{\text{mp}}}}. \qquad (11.3)$$

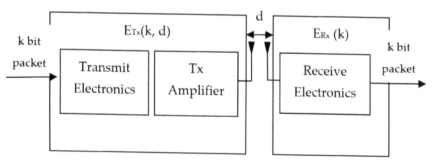

Figure 11.12 Model presentation for radio energy [31].

11.4.9 Lifetime optimization in a mobile WSN with solar energy harvesting

The majority of each sensor node's energy is expected to be expended throughout communication events with the sink node. It is to be noted that the word "activity level" refers to the amount of time or frequency with which these activities are carried out. Each frame of the communication process is divided into time or time-frequency slots. It is determined by the activity levels how many slots in each frame each sensor is permitted for communication. The appropriate activity levels must be calculated dynamically at the sensor nodes to maximize the network lifetime due to node mobility [16].

11.4.10 Time-slotted approach

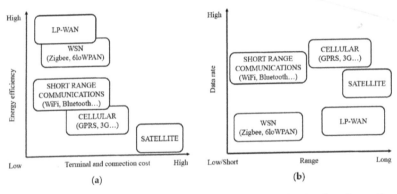

Figure 11.13 Improvement in the mobile sensor nodes activity levels using a time slot technique [16].

11.4.11 Optimization algorithm

Algorithm: Optimization of sensor activity levels.

Input data:
   Amount of sensor nodes, N,
   Optimization span, F,
   Batteries at their initial charge, $s1(1)$,
   Estimated consumption of energy in first F frame, $B(1)$, for $k = 1, 2, \ldots$
(Until the death of the network),
   Calculate $X(k)$ by solving the optimization problem. Based on the information on actual residual energies and recharge at sensor nodes, compute

matrix $B(k + 1)$. The sensor nodes should be assigned time or time-frequency slots based on the activity levels in $X(k)$ [33].

11.5 Simulation of WSN Network and Results

11.5.1 Simulation parameters

- Number of optimization events (or runs), $K = 80$
- Each optimization event spans the number of frames, $F = 5$
- Number of sensor nodes, $N = 10$
- An initial maximum stored energy for a sensor node, $S_{max} = 10$
- Energy of death, expressed as a fraction S_{max}, $Sd_r = 0.05$
- Fraction of maximum recharge value S_{max}, $r_r = 0.25$
- The activity level matrix is defined by
  $X(k) = [\mathbf{x}_1(k), \mathbf{x}_2(k), \ldots \ldots \ldots \mathbf{x}_f(k)]$
- Maximum energy consumption matrix
  $B(k) = [\mathbf{b}(k), \mathbf{b}_2(k), \ldots \ldots \ldots \mathbf{b}(k)]$
- Matrix of residual energy
  $S(k) = [\mathbf{s}_1(k), \mathbf{s}_2(k), \ldots \ldots \ldots \mathbf{s}_{f+1}(k)]$
- Matrix of recharge
  $R(k) = [\mathbf{r}_1(k), \mathbf{r}_2(k), \ldots \ldots \ldots \mathbf{r}_f(k)]$
- The residual energies
  $\mathbf{s}_{f+1}(k) = \mathbf{s}_f(k) - \mathbf{b}_f(k) \circ \mathbf{x}_f(k) + \mathbf{r}_f(k)$

11.5.2 Simulation results

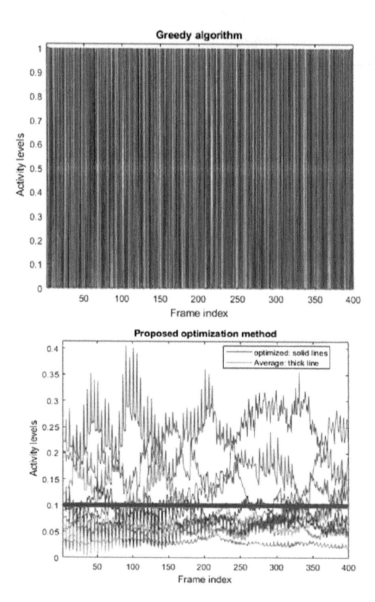

Figure 11.14   Average activity levels.

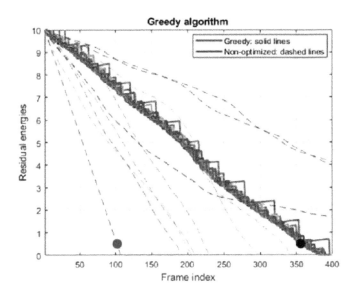

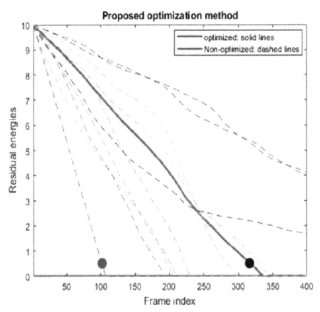

Figure 11.15   Normalized residual energies.

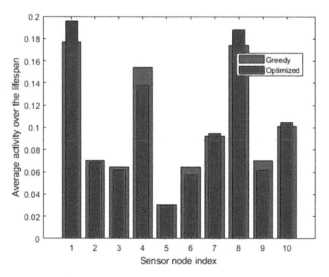

Figure 11.16 Average activity over life span.

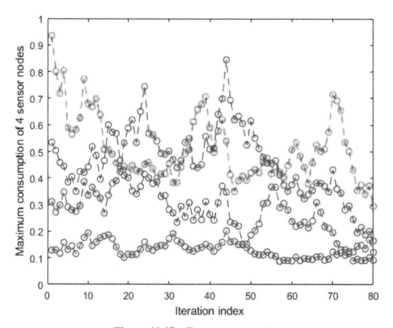

Figure 11.17 Energy consumption.

11.6 Future Scope

In the revolutionary era of Industry 4.0, the manufacturing sector has many opportunities and challenges. However, it is hard to attain the goals of Industry 4.0, and it is likely to take years to realize. Currently, Industry 4.0 is a vision for the future, as it involves many aspects and faces many types of difficulties and challenges, including technological challenges like the development of sustainable and reliable IIoT. Energy harvesting technology for the industrial internet of things can perform key role to reduce burden of powering wireless sensor nodes as using the main power and batteries is mainly restricted due to the small sizes of many devices and the fact that these devices are installed in hard-to-reach areas, where regular battery maintenance is impractical and very expensive. Energy harvesting techniques play a critical role in increasing the device lifetime by providing a sustainable way to recharge the batteries.

Moreover, an energy-efficient protocol supports the system to consume the least energy for its operation. Henceforth, a self-powered wireless sensor network and energy-efficient protocol are helpful to power up many industrial IoT applications. Therefore, a proposed research work aid in developing sustainable and energy-efficient protocol for wireless sensor networks with an extended life span and contributes to the fourth generation revolution of manufacturing sectors.

11.7 Conclusion

Industry Revolution 4, also known as Industry 4.0 was possible because of research and development toward advanced technologies like IoT and WSN. IIoT has many applications though many challenges have to be overcome for successful implementation. Security and energy consumption reduction are the primary concerns. The network architecture of IIoT has been explained layer-wise and mainly focused on protocols. MODBUS TCP and MQTT perform well in many protocols compared to others. Experimental results also show that combining both these protocols is also possible. Various hardware components, OS, and middleware, are also compared, which will help identify the best options for the testbed design and implement any application. Implementation criteria are also discussed and divided into various categories and subcategories, and different priorities are set according to their weight calculation. The low-power energy model has been discussed along with its protocols. Traditional protocols were designed with limited energy levels. We propose a self-powered network where nodes' energy keeps changing, and hence hardware and protocols need to be changed. Hardware design for self-powered sensor nodes is proposed while the proposed protocol is verified using simulation.

References

[1] K. Wójcicki, M. Biegańska, B. Paliwoda, and J. Górna, "Internet of Things in Industry: Research Profiling, Application, Challenges and Opportunities—A Review," Energies, vol. 15, no. 5, p. 1806, Feb. 2022.

[2] T. L. J. Phan, I. Gehrhardt, D. Heik, F. Bahrpeyma, and D. Reichelt, "A Systematic Mapping Study on Machine Learning Techniques Applied for Condition Monitoring and Predictive Maintenance in the Manufacturing Sector," Logistics, vol. 6, no. 2, p. 35, May 2022.

[3] J. Rosenberger et al., "Deep Reinforcement Learning Multi-Agent System for Resource Allocation in Industrial Internet of Things," pp. 1–23, 2022.

[4] G. P. Tancredi, G. Vignali, and E. Bottani, "Tools for Anomaly Detection : An Application to a Food Plant," 2022.

[5] Y. Zou et al., "A High-Performance Flag-Type Triboelectric Nanogenerator for Scavenging Wind Energy toward Self-Powered IoTs," Materials (Basel)., vol. 15, no. 10, p. 3696, May 2022.

[6] V. Fernandez-Viagas and J. M. Framinan, "Exploring the benefits of scheduling with advanced and real-time information integration in Industry 4.0: A computational study," J. Ind. Inf. Integr., vol. 27, no. July 2020, p. 100281, 2022.

[7] M. Waters et al., "Open Source IIoT Solution for Gas Waste Monitoring in Smart Factory," Sensors, vol. 22, no. 8, p. 2972, Apr. 2022.

[8] J. Rymarczyk, "The Change in the Traditional Paradigm of Production under the Influence of Industrial Revolution 4.0," Businesses, vol. 2, no. 2, pp. 188–200, Apr. 2022.

[9] J. Qin et al., "Industrial Internet of Learning (IIoL): IIoT based pervasive knowledge network for LPWAN—concept, framework and case studies," CCF Trans. Pervasive Comput. Interact., vol. 3, no. 1, pp. 25–39, Mar. 2021.

[10] R. Sanchez-Iborra and M. D. Cano, "State of the art in LP-WAN solutions for industrial IoT services," Sensors (Switzerland), vol. 16, no. 5, 2016.

[11] P. Fraga-Lamas, T. M. Fernández-Caramés, and L. Castedo, "Towards the internet of smart trains: A review on industrial IoT-connected railways," Sensors (Switzerland), vol. 17, no. 6, 2017.

[12] L. C. Souza, E. R. Neto, E. S. Lima, and A. C. S. Junior, "Optically-powered wireless sensor nodes towards industrial internet of things," Sensors, vol. 22, no. 1, 2022.

[13] J. A. Kaw, S. Gull, and S. A. Parah, "SVIoT: A Secure Visual-IoT Framework for Smart Healthcare," Sensors, vol. 22, no. 5, p. 1773, Feb. 2022.

[14] M. A. Almaiah, A. Ali, F. Hajjej, M. F. Pasha, and M. A. Alohali, "A Lightweight Hybrid Deep Learning Privacy Preserving Model for FC-Based Industrial Internet of Medical Things," Sensors, vol. 22, no. 6, p. 2112, Mar. 2022.

[15] T. Gkamas, V. Karaiskos, and S. Kontogiannis, "Performance Evaluation of Distributed Database Strategies Using Docker as a Service for Industrial IoT Data: Application to Industry 4.0," Information, vol. 13, no. 4, p. 190, Apr. 2022.

[16] D. Silva, L. I. Carvalho, J. Soares, and R. C. Sofia, "A performance analysis of internet of things networking protocols: Evaluating MQTT, CoAP, OPC UA," Appl. Sci., vol. 11, no. 11, 2021.

[17] S. Jaloudi, "Communication protocols of an industrial internet of things environment: A comparative study," Futur. Internet, vol. 11, no. 3, 2019.

[18] M. A. P. Putra, D. S. Kim, and J. M. Lee, "Adaptive LRFU replacement policy for named data network in industrial IoT," ICT Express, no. xxxx, 2021.

[19] E. Shahri, P. Pedreiras, and L. Almeida, "Extending MQTT with Real-Time Communication Services Based on SDN," Sensors, vol. 22, no. 9, p. 3162, Apr. 2022.

[20] Z. Hussain, A. Akhunzada, J. Iqbal, I. Bibi, and A. Gani, "Secure IIoT-Enabled Industry 4.0," Sustainability, vol. 13, no. 22, p. 12384, Nov. 2021.

[21] X. Xiong, K. Zheng, R. Xu, W. Xiang, and P. Chatzimisios, "Low power wide area machine-to-machine networks: Key techniques and prototype," IEEE Commun. Mag., vol. 53, no. 9, pp. 64–71, Sep. 2015.

[22] F. Wang and J. Liu, "Networked wireless sensor data collection: Issues, challenges, and approaches," IEEE Commun. Surv. Tutorials, vol. 13, no. 4, pp. 673–687, 2011.

[23] K. Choudhary, G. S. Gaba, I. Butun, and P. Kumar, "Make-it—a lightweight mutual authentication and key exchange protocol for industrial internet of things," Sensors (Switzerland), vol. 20, no. 18, pp. 1–21, 2020.

[24] S. Hao, Y. Hong, and Y. He, "An Energy-Efficient Routing Algorithm Based on Greedy Strategy for Energy Harvesting Wireless Sensor Networks," Sensors, vol. 22, no. 4, pp. 1–24, 2022.

[25] K. Banti, I. Karampelia, T. Dimakis, A. A. Boulogeorgos, T. Kyriakidis, and M. Louta, "LoRaWAN Communication Protocols : A Comprehensive Survey under an Energy Efficiency Perspective," pp. 322–357, 2022.

[26] F.-Y. Tsuo, H.-P. Tan, Y. H. Chew, and H.-Y. Wei, "Energy-Aware Transmission Control for Wireless Sensor Networks Powered by Ambient Energy Harvesting: A Game-Theoretic Approach," in 2011 IEEE International Conference on Communications (ICC), 2011, pp. 1–5.

[27] A. D. Ball, F. Gu, R. Cattley, X. Wang, and X. Tang, "Energy harvesting technologies for achieving self-powered wireless sensor networks in machine condition monitoring: A review," Sensors (Switzerland), vol. 18, no. 12, p. 4113, Nov. 2018.

[28] D. M. E. Csikes, "RSL10: Ultra-Low-Power Bluetooth," Design Spark, 2018. [Online]. Available: https://www.rs-online.com/designspark/ultra-low-power-bluetooth. [Accessed: 28-May-2022].

[29] onsemi, "RSL10-SENSE-GEVK (and RSL10-SENSE-DB-GEVK) User Guide," 2019.

[30] C. M. Angelopoulos, S. Nikoletseas, T. P. Raptis, C. Raptopoulos, and F. Vasilakis, "Efficient energy management in wireless rechargeable sensor networks," in MSWiM'12 - Proceedings of the 15th ACM International Conference on Modeling, Analysis and Simulation of Wireless and Mobile Systems, 2012, pp. 309–316.

[31] P. K. Mishra and S. K. Verma, "Ffmcp: Feed-forward multi-clustering protocol using fuzzy logic for wireless sensor networks (wsns)," Energies, vol. 14, no. 10, p. 2866, May 2021.

[32] Y. Chen and Q. Zhao, "On the lifetime of wireless sensor networks," IEEE Commun. Lett., vol. 9, no. 11, pp. 976–978, Nov. 2005.

[33] D. A. Guimarães, L. J. Sakai, A. M. Alberti, and R. A. A. de Souza, "Increasing the lifetime of mobile WSNs via dynamic optimization of sensor node communication activity," Sensors (Switzerland), vol. 16, no. 9, p. 1536, Sep. 2016.

12

Consensus Algorithms in Blockchain for an Efficient and Secure Network

Medini Gupta[1], Sarvesh Tanwar[2], Niranjan Lal[3], and Ritvik Pawar[4]

[1,2]Amity Institute of Information Technology, Amity University, Uttar Pradesh, India
[3]Computer Science and Engineering, SRM Institute of Science and Technology, India
[4]Mahatma Gandhi University, Meghalaya, India
E-mail: guptamedini642@gmail.com; s.tanwar1521@gmail.com; niranjan_verma51@yahoo.com; ritvik.charly@gmail.com

Abstract

Cryptocurrencies created a trendsetter after the emergence of Bitcoin in the last decade. Blockchain technology is the backbone behind cryptocurrencies and gave rise to its usage in multidisciplinary fields. The future of payment with cryptocurrencies might lead to the end of manual cash. Blockchain features such as encryption, peer to peer (P2P), transparency, and timestamp make an immutable ledger to overcome the problem of a centralized system with a decentralized and trustworthy approach. Consensus algorithms are an essential component responsible for the secure flow of the network on the distributed ledger. The security and poor performance of consensus algorithms can cause hindrance for a wide range of adoption of blockchain. Drawbacks of blockchain can be addressed with a better understanding of how each consensus algorithm functions. There are various consensus algorithms according to different types of blockchains. Selection of the most appropriate consensus algorithm will improve the network performance and provide better outcomes. The analysis of varying consensus algorithms about specific metrics has been discussed. In this chapter, we will go through an introduction to blockchain, the emergence of cryptocurrencies, related work done in this field, the role of consensus algorithms, comparative analysis of existing consensus algorithms, limitations of each consensus algorithm, conclusion, and future scope.

Keywords: Blockchain, cryptocurrency, proof of work, performance, consensus algorithm.

12.1 Introduction

Satoshi Nakamoto introduced the concept of cryptocurrencies to the world in 2008 by publishing a whitepaper on "digital currency" and mining the foremost blockchain cryptocurrency. Blockchain is an immutable ledger to keep track of digital commodities on a peer-to-peer network [1]. Cryptochromes have come out as a widely accepted application of blockchain. Consensus algorithms are responsible for establishing a secure and integrated network on a blockchain by involving all entities to agree on specific agreements. While using cloud computing to store confidential information, one needs to apply AWS. With blockchain's timestamp and smart contract mechanism, the risk of medication of data by a third party gets omitted. P2P is the backbone of a decentralized system. The P2P model in blockchain enables cryptocurrencies to be transmitted universally without including intermediaries. Each node on the network stores a duplicate copy of the distributed ledger and performs verification on the rest of the nodes to maintain transparency. When data in one system gets compromised, all the network nodes get alert and appropriate steps are taken to address the issue [2] in contrast to a bank, which is based on a centralized architecture, where records are kept privately and are handled by the bank.

Blockchain refers to the collection of records termed as blocks. Data is stored in a chronological sequence. To ensure data privacy, cryptographic techniques are deployed to encrypt the information and guarantee end-user privacy. Cryptography to secure data confidentiality is achieved through making use of two keys: private and public keys. Each node carries both keys to create a digital signature. Every transaction is authorized through the owner's digital signature [3]. P2P is a network of individuals with equal power with the same data copied in different systems on the same web. The centralized system is prone to one-point failure, where the collapse of one system disrupts the functioning of the entire network.

P2P is principled in a distributed environment, which makes it more efficient, fast, and impervious to cybercrimes. Smart contracts are computer programs executed automatically when specific conditions are fulfilled. A smart contract on blockchain is very similar to a manual agreement in the physical world, which states some conditions signed by both buyer and the selling party during the purchase of any asset. Due to encryption, there is no question about the transparency and trust of the smart contract. As soon as predefined conditions are achieved, the smart contract activates the next course of action because this requires no manual work, which often contains errors and consumes considerable time.

12.1.1 Motivation

Blockchain is a distributed ledger that stores a large amount of information in a secured environment while maintaining privacy and transparency of information and without including a mediator to verify the record [4]. Blockchain gains these features through the successful implementation of consensus algorithms. These are approaches for decision-making among peers to reach a common verdict and to regulate equity and fairness in the virtual world. Byzantine fault tolerance is a failure problem that mainly occurs in distributed systems. This problem led to the development of rules and regulations for blockchain. The main challenge in the Byzantine problem was the difficulty to reach on a joint agreement. As a single failure occurs, nodes cannot agree on a common consensus. We will learn about the Byzantine problem in the upcoming section [5]. Blockchain consensus protocols do not come across any such issue. Their target is to reach a particular goal anyhow. Thus, blockchain consensus protocols are more fault-free and resilient than Byzantine. A consensus algorithm is a better option for getting solutions related to contradictory output in a distributed architecture [6].

12.2 Literature Review

Ferdous *et al.* [7] proposed an approach for consensus protocols indicating various parameters contributing to a better future. Consensus protocols are divided into two categories: incentivized and non-incentivized, which are further divided into different categories. A systematic overview of varying blockchain consensus algorithms is discussed. Cryptocurrencies and public blockchains use incentivized protocols and require stimulus for nodes to influence them so that they actively work on them. The non-incentivized protocol does not incentivize the required behavior because nodes are trusted. Private blockchains commonly use these algorithms.

Samy *et al.* [8] provided a reliable solution for the consensus algorithm by modifying the algorithm base on voting in the Byzantine fault tolerance problem that gives high throughput. The modified algorithm solved the scalability issue in a private blockchain. It has also addressed the prerequisite of consensus algorithm to be fault free. Blockchain gadget—Hyperledger Caliper is used to examine the performance parameter of an open-source platform. Consensus algorithms are regulations responsible for maintaining equality on the network. Nodes must come to a big statement to update any data in the blockchain.

Frikha *et al.* [9] proposed a hybrid procedure for hardware/software models and developed proof of work (PoW) consensus, one of the widely accepted protocols in an immutable ledger. Ethereum blockchain is used to validate the consensus algorithm by implementing Keccak-256 as a hashing algorithm. Better result in decreased execution time and reduced power consumption up to 255% compared with Nvidia Maxwell GPUs and discussed embedded technologies of robotic integration handled by blockchain. A secure, error-free, and authenticated consensus algorithm ensures the successful functioning of the blockchain.

In their review paper, Chaturvedi *et al.* [10] mentioned the most popular consensus algorithms in the decentralized ledger. The paper analyzed consensus algorithms' characteristics, usage, and performance over the network. Hashing techniques act as an added layer to enable security. Records are spread among every node in distributed blockchain architecture. Scalability guarantees the trouble-free working of blockchain-related throughput even when there is huge demand. Proof of work takes a considerable time to mine an additional block. Litecoin cryptocurrency requires minimal cost or zero cost that deploys proof of work. Another algorithm, proof of stake (PoS), is a substitute for proof of work, requiring much less power consumption. Proof of elapsed time has low mining time compared to proof of work, thus utilizing minimal resources.

Wang *et al.* [11] in their review paper, compared the existing blockchain consensus algorithms and discussed the pros and cons of each of them with future scope in upcoming years. PoW and PoS are trending algorithms to develop a more transparent smart contract in the current scenario. PoS utilizes meager resources, and block creation time is also low compared to cryptocurrencies based on PoW. The risk of security breach exists in PoS due to human intervention and detailed execution guidelines. PoW is a decentralized algorithm with simple implementation. High investment is made in the security aspect of PoW; thus, it is a highly secure protocol. Nodes only need to solve the numerical problem, and no extra details are required to reach a particular consensus. PoW utilizes more resources because of the complex mining scheme.

Fu *et al.* [12] proposed a unified consensus protocol. Accountant selection is the initial step of the protocol. The first step consists of block generation, all the nodes and transactions act as input, and a new block is received as output. In the second step of adding a block in the ledger, each node has a Xerox copy of the immutable ledger. Once the node gets a new block, then the node verifies the concerned miner and block. Block transaction and header are verified during block verification. The block header consists of hash values of all the blocks. In the final step, the confirmation transaction takes place based on the type of node each blockchain holds. Present blockchain acts as an input, and confirmed transactions are considered output. Different modes of consensus protocol: leader-based mode, voting-based mode, and committee-voting-based mode are discussed. A comparison of three distinct methods of consensus protocol concerning a list of consensus protocols is made.

Zhu [13] examined and demonstrated the practical implementation of Hyperledger, Bitcoin, and Ethereum. Various consensus protocols differ in fault-free nature, resource complexity, and flexibility. The regulation, uniformity, performance, and ease of use, are a few parameters where more research is required. Bitcoin was the primary digital currency for proof of work, and SHA-256 is used in hash cash PoW.

Shukur *et al.* [14] worked on consensus algorithms for public blockchain, the notation behind the development of consensus algorithms, and research disparity in this field. Related work was done on 25 consensus protocols collected from 36 research publications and divided into public, consortium, and secret immutable ledger consensus algorithms. Tindermint algorithm is developed to discourse the environmental, speed, and flexibility challenges of PoW. Delegated PoS is introduced for bit shares. Shareholders choose nodes that act as witnesses for creating blocks for a particular period and keep some of them as a backup. Nodes will build the union in a round-robin design, and those who fail to do so in a given time are exchanged by those nodes which are kept back.

Kaur *et al.* [15] proposed an architecture based on permissioned blockchain concerning present challenges faced with electronic health records (EHRs). Different algorithms define a transparent and immutable system flow to share safely, query and store EHR involving physicians, patients, and hospitals. CouchDB is a database that not only searches through keys but data values are also involved in searching. Hyperledger Caliper is a tool that has measured performance based on particular metrics. The preliminary outcome showed that adding a large number of peers to the simulation resulted in reducing the throughput.

In their review paper, Musilek *et al.* [16] worked on distinct types of distributed ledgers focusing more on blockchain consensus algorithms. They reviewed more than 100 consensus mechanisms from various publications. An extensive classification of blockchain consensus protocols is being proposed based on building blocks. Consensus protocols in the same category represent features in terms of functionality, advantages, and challenges.

Yao *et al.* [17] discussed the workflow of blockchain consortium algorithms. Six metrics were taken into consideration to examine eight consortium algorithms. Fault tolerance, scalability, performance, asset utilization, efficiency, and decentralized nature are the metrics. Their research allows stakeholders to understand the present scope of blockchain consensus features. Proper decentralization, full-proof security and scalability are critical requirements for the appropriate deployment of blockchain in various fields.

Alhejazi *et al.* [18] proposed a novel architecture for the weighted majority algorithm of blockchain. Practical implementation of algorithm with Java is mentioned. This algorithm identifies the anomalies in the IoT platform and improves the security of the immutable blockchain. Depending on the performance, specific weight is allocated to each voter. When voter fails to vote in the proper manner, then there is a reduction in voters' weight by a particular ratio. The algorithm's overall performance is examined using five scenarios.

Gürcan [19] presented their review paper on proof of work based solely on feedback. The higher the number of participants, the more positive the input will be. Tampering of data is caught by generating hashes. The length of the hash generated will be same as that of the original data. Hash is only applied to validate whether any changes are mad mob original data. Mutual consciousness is present among the entities. They are not necessary to learn whether other entities are available on the blockchain. Proper sequencing exists in the system. When a specific condition is fulfilled, the next course of action takes place.

Wang *et al.* [20] worked on improving food traceability by proposing a blockchain consensus protocol based on food credit and clustering. Blockchain is a tamper-proof ledger that plays a crucial role while registering for digital products and securing finance without taking consideration from third parties. Implementation of blockchain in food traceability can be done by detecting fault tolerance and properly tracking assets. The proposed blockchain protocol has effectively brought down the technical challenges of food traceability and ensured food security.

Wu *et al.* [21], in their review paper, mentioned about principal technology behind blockchain and background work done on consensus algorithm. Classification of blockchain based on permissions granted and access mode is done. Extensive analysis of features of blockchain with comparison is discussed. Consensus protocols remain an essential part of distributed ledgers despite of type of blockchain being used. The long duration of consensus time, too much delay in large transactions, and low throughput are a few shortcomings of consensus protocols.

12.3 Emergence of Cryptocurrencies

Cryptocurrencies are electronic assets that are momentary units in digital space. As the name suggests, cryptocurrencies deploy effective cryptographic techniques to secure the medium of exchange. Conventional currencies have their values fixed by the bank. It is not the same case with digital currencies [22]. The importance of cryptocurrencies depends on the number of transactions, and they are not under the control of any central surveillance. Cryptocurrencies are decentralized by nature. Prominent business solutions can be delivered by digital currencies where trust and fraud detection form the base of establishment, from handling manufacturing to protecting the secure delivery of goods and services. Mathematical equations of subsequent blocks need to be calculated to update the existing partnership, which takes much time and requires enormous numerical calculations. Additional blocks can be added, but previous records cannot be removed permanently, leading to a robust environment. In the last decade, the world saw the emergence of around 1000 cryptocurrencies. Bitcoin was the initial digital cash introduced back in 2008, and then it started the era of crypto space. Figure 12.1 represents the widely popular cryptocurrencies in 2022.

Bitcoin—In January 2022, 882 million bitcoins were circulated, and it is on the top list of the most acceptable cryptocurrencies. Miners solve complex arithmetic puzzles to agree on transactions that form a single block [24]. On successful completion of block creation, miners are reinforced with bitcoins. Nodes can buy, trade, or sell virtual tokens through cryptocurrencies. All the bitcoin transactions are openly available, which makes records transpirable and lower illegal activities. At present, the bitcoin supply has a limitation of

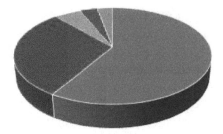

■ Bitcoin ■ Ethereum ▪ Tether ■ Solana ■ Cardano

Figure 12.1   Rise of top 5 cryptocurrencies in January 2022 [23].

21 million. E-cash delivers an alternative payment gateway independent of central authority [4].

Ethereum—Initially released in 2015, it gained acceptance with smart contract functionality. It works on the cross-platform operating system, written in multiple programming languages—Rust, Java, Go, and C#. Ethereum Request for Comments 20 (ERC-20) are commutable tokens on this platform that can be exchanged with other smart contract tokens. Virtual assets that are developed on decentralized ledger is termed as tokens. Details about transferring tickets from one node to another, getting account details, and knowing the availability of permits are functionalities undertaken by ERC-20. Ethereum, or ether, is the second most popular cryptocurrency after blockchain. Market analysis of Ethereum from its release in 2015 till January 2022 is discussed in Table 12.1. Various indicators, such as the cost of Ethereum, transactions per day, mining profitability, and market dominance, are considered. Table 12.1 represents the analysis for Ethereum in the period 2015–2022. The result discloses that ETH cost $1 in 2015 when it was first released, and currently, it is $3210 in January 2022. Ether number of daily transactions was $30,483 in April 2016, which rose to $1,174,849 in January. Mining profitability was $0.05 in 2017 and $0.5 in 2022. Presently ether's market dominance is $18% compared to $2.4% in 2015.

Tether—This cryptocurrency operates on the Ethereum blockchain, and Tether Limited grants tokens. Its market capitalization is $78 billion. It can be retrieved for a dollar and termed a stablecoin because it is a hard cryptocurrency. Merchants make use of tethers as a replacement for dollars.

Solana—Solana is a permissionless blockchain platform that is written in Rust. Its internal cryptocurrency is SOL, which is used to perform transactions. Implement proof of stake as a consensus algorithm and implies

Table 12.1 Analysis of Ethereum in the years 2015–2022 [25]

Parameters	Unit	August 2015	April 2016	October 2017	June 2018	November 2019	November 2020	March 2021	January 2022
Cost of Ethereum	USD	1	8	309	429	154	528	1842	3210
No. of transactions per day	USD	2,494	30,483	505,611	639,775	857,371	1,240,072	1,160,559	1,174,849
Mining profitability	USD	0.08	0.1	0.05	0.04	0.01	0.04	0.12	0.5
Market dominance	USD	2.47%	8.36%	16.06%	17.76%	8.01%	11.58%	11.19%	18.05%

verification of history to improve efficiency. Being censorship-resistant allows the network to remain open for executing applications and providing endless transactions. Solana underwent a high rise in the market capitalization of $52 billion in 2022.

Cardano—With a market rise of $44 billion, Cardano uses proof of stake. Metaverse, non-fungible, and decentralized transactions are recent growth on Cardano. It is designed on principles of past cryptocurrencies such as blockchain and Ethereum to facilitate scalability and security challenges.

12.4 Role of Consensus Algorithms

Decisions are made mutually in the consensus algorithm; if the node assumes that Jack has 40 coins, node Y assumes that Jack has 30 coins, and node X thinks that Jack owns 50 coins, then it is not considered consensus. Here comes the role of the consensus algorithm in the blockchain. Consensus algorithm is a mechanism to reach a common acceptance regarding any decision in a decentralized ledger. It provides reliability by developing a source of trust among all the nodes, thereby maintaining secure infrastructure [26]. Information on the blockchain should be authentic. The cooperative agreement is achieved when the blockchain is updated in real. Incentives are granted to miners once they finish the mining process successfully, thus aligning with the economic incentives of peers [27]. Blockchain is equitable and fair technology with its decentralized, timestamp, and open-source features. Consensus algorithms ensure the consistent working of blockchain infinity times, even during cyber risk or system failure.

Blockchain remains fault-tolerant and reliable, making it one of the most emerging technologies in this era. In a centralized architecture, a single user has authority over the network, and they can do data manipulation according to their wish. Table 12.2 provides attributes of a good consensus algorithm. Comparison analysis of different consensus algorithms is discussed in Table 12.3. There doesn't exist any proof of supervision for finalizing consensus. Participants' balances of cryptocurrencies are stored on the ledger. It is a must for all the nodes to maintain an exact copy of the ledger. Public key cryptography, or modern cryptography, secures users' privacy regarding their digital coins [10].

12.4.1 Proof of work

PoW was an initial procedure to evaluate blocks and cryptocurrencies on the blockchain, and decentralization played a significant role in eradicating the involvement of momentary institutes. Cryptocurrency works on blockchain, and every block consists of a unique hash value. To add a block to the ledger, miners have to generate a new hash that should not be greater than the hash of that block. Mining is similar to earth; miners gain rewards from cryptocurrency when they fulfill the target. Mining devices are used to quickly calculate various arithmetic computations [28]. The miner who completes the target first gets the opportunity to amend blocks on the ledger and receives cryptocurrency in return [7]. PoW provides a competitive environment to verify the records, and miners are uplifted to gain as much as possible due to financial value.

PoW was first implemented in bitcoin to ensure confidentiality on the blockchain. To earn bitcoin, security verification is done by generating a hash for every block. The SHA-256 algorithm is used in bitcoin that produces a hash of 64 characters. The working of PoW is explained in Figure 12.2. Various miners simultaneously compete to calculate the required hash for the block. Transaction fees and coins are given to miners through bitcoin rewards. Proof of work takes 10 minutes to add a new partnership [11].

Table 12.2   Attributes of good consensus algorithm

Features	Explanation
Equalitarian	Equal opportunity and weightage are granted to each vote cast by nodes.
Security	Valid outcomes are achieved by nodes based on established rules.
All-inclusive	Guarantees collaboration of every node for updating the ledger

Table 12.3 Comparative analysis of different consensus algorithm

Consensus algorithm	Year	Advantages	Issues	Use cases
PoW	1993	Highly decentralized Frequently implemented to date Openness to joining the network	Wastage of resources Low scalability More power utilization	Bitcoin Ethereum Litecoin
PoS	2011	Consumes less energy The simple procedure of staking virtual resources. Eco friendly	Prone to security attacks Complicated implementation Reputed validators rule on the network	Cardano Solana Avalanche
DpoS	2015	Highly scalable Great performance Selection based on voting Less hardware	Less decentralized Necessary participation Cartels can lead to malicious activities	EOS Lisk Tezos
PoB	2012	Sustainable Decreases power utilization It eliminates the need for hardware Look for long-term goals	Wastage of resources Require more testing to determine the efficiency Slow procedure Complicated verification	Slimcoin
PoC	2013	Requires standard hardware Saves more energy with a hard drive More decentralized	Prone to malware Less adaptable Resource wastage Security breaches	Burstcoin, Chia, SpaceMint
PoET	2016	Based on random selection Encourages equality in terms of participation Scalable Cyber-attack resistant	Not promote decentralization Low openness to joining the network	Hyperledger

When miners can quickly generate the hash, then computations become harder. If miners take too much time to solve the numeric problem, the calculation becomes more accessible.

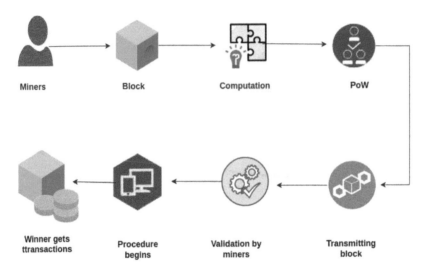

Figure 12.2   Working of PoW.

12.4.2 Proof of stake

Proof of Stake was introduced in 2011 as a substitute for proof of work. By reducing the size of computational numerals that authenticates the records and transactions, PoS evaluate the blocks based on the total stake of network participation. Both PoW and PoS have a common aim to validate transactions accurately. Still, instead of generating a new hash in PoW, machines of digital coin owners are taken into consideration for validating blocks through PoS. Miners keep their virtual resources which include tokens, at stake. The workflow of PoS is shown in Figure 12.3. Owners who have put their resources are termed as validators. Few validators are chosen arbitrarily to perform block validation. Randomly miners are selected to mine the block, unlike setting a computational problem in PoW.

Ethereum requires 32 ETH to be put on stake by miners before converting them into validators. Multiple validators verify the block, and once it is validated, it is often finalized to be appropriate. Distinct methods are deployed in the PoS consensus algorithm, which differs from system to system. Sharding is used in Ethereum to validate blocks through PoS. Sharding is the procedure that divides the ledger horizontally, thus spreading the weight and creating shards. Utilizing shards expands the transactions per second and brings down network overload. Security concerning 5% of controlling power is a topic of

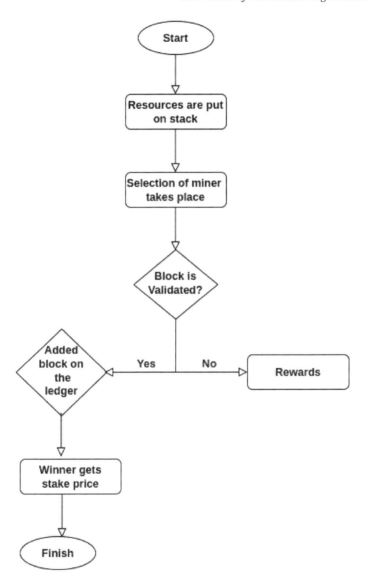

Figure 12.3   Working of PoS.

concern. Fifty-one percent attacks occur when hackers have control of 51% of records on the ledger and take advantage of the majority to update the blockchain.

12.4.3 Delegated proof of stake (DPoS)

Delegated proof of stake consists of a voting system. Nodes do not race on computational consumption. Voting for the computers takes place where delegate votes for the system they consider most appropriate to successfully manage the network system. DPoS is fast and much more power efficient than Proof of Work as a new block is generated every next half second as compared to PoW [29]. To motivate many user engagements, digital tokens are produced as reinforcement. Similar to a parliament system in a democratic government, when delegates fail to fulfill the responsibility assigned to them, such as appending the ledger on a regular interval, the removal of delegates takes place, and new representatives are appointed. It depends on whether users have to trust the node because if delegates cannot validate records appropriately or do not provide enough power to the network, users can draw out their votes.

Digital coins are put at stake to get involved in the DPoS consensus algorithm. Delegates that are elected are the ones who perform validation of records and create a new block. Delegates that got elected are termed, block producers. All coins are staked on a centralized staking ground, and virtual coins are linked with a particular delegate. Records are stored on multiple nodes; crypto-hash functions carry out validation procedures. Instead of handling transactions by the standalone entity, transactions are spread everywhere on the network. Delegates are selected on a reputation basis. Users who carry a minimum of one coin with a DPoS mechanism can vote for the nodes they consider qualified for validating transactions.

12.4.4 Byzantine fault tolerance (BFT)

This consensus algorithm came into existence to provide a solution for unreliable nodes. When inconsistent details regarding transactions are shared by members of the network, then it leads to the downfall of the blockchain, and no supervision authority could take a step to dismiss it. Proof of stake requires an ultimate solution and proof of work utilizes Byzantine fault tolerance with its powerful processing. The system works correctly even if some nodes show malicious or fraudulent behavior. This consensus mechanism is taken from Byzantine generals problem. To reach this algorithm, nodes have to work collectively.

12.4.5 Proof of burn (PoB)

Here, "burn" in proof of burn refers to the concept of coin burning. Miners eliminate virtual coins from circulation to end their availability for further trading. Virtual coins are transferred to the unapproachable but verified address, whose private key cannot be accessed. Practical execution of PoB also includes PoW and PoS along with core PoB. Only those cryptocurrencies with an excess supply come under the PoB mechanism. Mining rigs are a single computer system allocated for mining digital currencies. Mining rigs are used in this process, unlike a PoW system where multiple nodes are used for block creation. Miners have a high level of power if they burn more coins. Developers invest in large-term goals in exchange for a short duration of loss. To become validators and mine new blocks, developers have to burn more coins to get selected randomly in this process.

12.4.6 Proof of capacity (PoC)

Miners devote a specific amount of hardware space for storing solutions for cryptarithmetic hashing queries. More amount of space a miner holds, the greater the possibility of them getting randomly selected for gaining rewards and mining blocks on the network. Issue of more power consumption in PoW and stockpile problem of PoS mechanism is addressed through proof of capacity. Vacant space on the miner's hard disk is used to mine the provided cryptocurrencies. Expected solutions are stored on a hard disk even before the commencement of the mining process. The capacity to hold the computational key depends on the space allocated in the hard drive. Initially, the value is created by using repeated steps of hashing data. Each nonce consists of 8192 values beginning from 0 to 8191. Then scoops are made by combining the first two numbers and the following two numbers in a row. The mining procedure occurs in the next step, where miners calculate the scoop number. If the miner obtains 60 as a scoop number, it will move to a nonce number of 1 and apply that scoop information to find a final value.

12.4.7 Proof of elapsed time (PoET)

Permissioned blockchain mainly uses proof of elapsed time to determine who will create the next block on the network. Provides a fair chance to all the participants, based on the lottery system where participants get equal opportunities to win the rewards. This mechanism produces a random wait time for every validator on the ledger. Validators wait for the random duration of time as allocated to them and that wait time is added in their block. With a minimum wait time, the validator will be declared the winner for mining the block and appending the blockchain. Network energy is increased because nodes are put to sleep for that particular time. Various security checksum is deployed to ensure that validators get their wait time randomly instead of intentionally choosing a wait time of a short duration to win the mining block. Authentication of miners who have completed their wait time with fair means is also verified.

12.5 Limitation of Consensus Algorithms

Finalizing an appropriate consensus algorithm is a significant task; if chosen correctly, various vulnerabilities available in blockchain solutions can be handled effectively. All the consensus protocols have a similar aim to govern the functionality of nodes in a decentralized structure. Although it has a shared motive, it varies in terms of the route to reach a destinated consensus. The perfect consensus algorithm does not exist till now; it is engrossing to learn how much development has been done and adapted itself according to changing demands of algorithms.

12.5.1 Failure of consensus

Implementing an invalid consensus protocol in the industry could lead to consensus failure. Some fragments of nodes cannot participate in the procedure, and their unavailability leads to an inappropriate number of votes [30]. Error-free and expected results are not achieved.

12.5.2 Ledger forks

A ledger fork is a cyber threat that occurs when an inappropriate blockchain consensus algorithm is considered. A single chain gets divided into more than two chains [18]. Multiple isolated nodes are created, and the ledger starts functioning unforeseeably.

12.5.3 Substandard performance

Network portioning and nodes get defected, which cause the downfall of any business solution. Performance is a critical factor for measuring the success rate of implementation. Peers come together to build a sense of trust over a decentralized architecture. Performance is calculated in terms of the minimum tie taken by the transaction to get validated [31].

Scalability is the characteristic that determines the potential of the network to remain functional even though the rise in nodes and a surge in transactions occur [3]. Network latency detains the procedure of broadcasting data on the ledger, which in the long run, diminishes the performance of the network.

12.6 Conclusion

Cryptocurrency is the mode of payment in the virtual world developed based on cryptography. Blockchain involves many nodes to deliver overall security on the network, eliminate a single point of failure, and improve the ability against cyber vulnerabilities. Cryptocurrencies will be more effective in digital payment despite fluctuating values. Charges of cross-border payments will be directly affected due to high transactional costs. Bitcoin has been demonstrating its power since the time of inception. Shortcomings of the centralized platform include remaining under regulations [14]. Not only machines and networks are unpredictable, but the intention of miners controlling the network also plays a crucial role in providing a secure environment. BFT algorithms lack integrity and authorization mechanism, which ensures that attackers would not be able to manipulate the information unless proper verification. Controlling 51% of nodes leads to severe damage; thus consensus algorithm makes sure about consistency among the peers but also provides a mechanism for selecting the right decision collectively.

References

[1] S. Nakamoto, 'Bitcoin: A peer-to-peer electronic cash system', Decentralized Business Review, pp.1-9, 2008.

[2] M. Gupta, S. Tanwar, S. Badotra, A. Rana, 'A Systematic Review on Blockchain in Transforming the Healthcare Sector', Transformations Through Blockchain Technology, pp.181-200, 2022.

[3] P. Datta, S. Tanwar, S. N. Panda, A. Rana, 'Security and Issues of M-Banking: A Technical Report,' 8th International Conference on Reliability, Infocom Technologies and Optimization (Trends and Future Directions), IEEE, pp. 1115-1118, 2020.

[4] S. Tanwar, T. Paul, K. Singh, M. Joshi, A. Rana, 'Classification and Imapct of Cyber Threats in India: A review', 8th International Conference on Reliability, Infocom Technologies and Optimization (Trends and Future Directions) (ICRITO), IEEE, pp. 129-135, 2020.

[5] P. Datta, S. N. Panda, S. Tanwar, and R. K. Kaushal, 'A technical review report on cybercrimes in India,' International Conference on Emerging Smart Computing and Informatics (ESCI), IEEE, pp. 269-275, 2020.

[6] L. Kakkar, D. Gupta, S. Saxena, S. Tanwar, 'IoT Architectures and Its Security: A Review', Second International Conference on Information Management and Machine Intelligence, Lecture Notes in Networks and Systems, pp. 87-94, 2021.

[7] M. S. Ferdous, M. J. M. Chowdhury, M. A. Hoque, A. Colman, 'Blockchain consensus algorithms: A survey, arXiv preprint arXiv:2001.07091, 2020.

[8] H. Samy, A. Tammam, A. Fahmy, and B. Hasan, 'Enhancing the performance of the blockchain consensus algorithm using multithreading technology,' Ain Shams Engineering Journal, 12(3), pp. 2709-2716, 2021.

[9] T. Frikha, F. Chaabane, N. Aouinti, O. Cheikhrouhou, N. B. Amor, and A. Kerrouche, 'Implementation of blockchain consensus algorithm on embedded architecture,' Security and Communication Networks, 2021.

[10] S. Kaur, S. Chaturvedi, A. Sharma, and J. Kar, 'A research survey on applications of consensus protocols in blockchain,' Security and Communication Networks, 2021.

[11] Q. Wang, J. Huang, S. Wang, Y. Chen, P. Zhang, and L. He, 'A comparative study of blockchain consensus algorithms,' In Journal of Physics: Conference Series, IOP Publishing, 1437 (1), pp. 012007, 2020.

[12] X. Fu, H. Wang, P. Shi, 'A survey of Blockchain consensus algorithms: mechanism, design and applications,' Science China Information Sciences, 64(2), pp. 1-15, 2021.

[13] X. Zhu, 'Research on blockchain consensus mechanism and implementation,' In IOP Conference Series: Materials Science and Engineering, IOP Publishing, 569(4), pp. 042058, 2019.

[14] I. Jalal, Z. Shukur and K. A. A. Bakar, 'A Study on Public Blockchain Consensus Algorithms: A Systematic Literature Review,' 2020.

[15] J. Kaur, R. Rani, and N. Kalra, 'Blockchain-based framework for secured storage, sharing, and querying of electronic healthcare records, Concurrency, and Computation: Practice and Experience, 33(20), pp, 1-24, 2021.

[16] B. Lashkari, P. Musilek, 'A comprehensive review of blockchain consensus mechanisms,' IEEE Access, 9, pp. 43620-43652, 2021.

[17] W. Yao, J. Ye, R. Murimi, and G. Wang, 'A survey on consortium blockchain consensus mechanisms', arXiv preprint arXiv:2102.12058, 2021.

[18] M. M. Alhejazi, R. M. A. Mohammad, 'Enhancing the blockchain voting process in IoT using a novel blockchain Weighted Majority Consensus Algorithm (WMCA)', Information Security Journal: A Global Perspective, 31(2), pp. 125-143, 2022.

[19] Ö. Gürcan, 'Proof-of-Work as a Stigmergic Consensus Algorithm', In Proceedings of the 21st International Conference on Autonomous Agents and Multiagent Systems, pp. 1613-1615, 2022.

[20] P. Liu, S. Ren, J. Wang, S. Yuan, Y. Nian, and Y. Li, 'A Blockchain Consensus Optimization-Based Algorithm for Food Traceability', Mobile Information Systems, pp. 1-7, 2022.

[21] H. Xiong, M. Chen, C. Wu, Y. Zhao, W. Yi, 'Research on Progress of Blockchain Consensus Algorithm: A Review on Recent Progress of Blockchain Consensus Algorithms', Future Internet, MDPI, pp. 1-24, 14(2), 2022.

[22] V. Sharma, N. Lal and Vishal Sharma 'A novel comparison of consensus algorithms in blockchain', Advances and Applications in Mathematical Sciences, 20(1), pp. 1-13, 2020.

[23] https://www.statista.com/statistics/655492/most-valuable-virtual-currencies-globally/, accessed on May 2022.

[24] N. Wang, X. Zhou, X. Lu, Z. Guan, L. Wu, X. Du, M. Guizani, 'When energy trading meets blockchain in electrical power system: The state of the art', Applied Sciences, 9(8), pp. 1-31, 2019.

[25] https://www.statista.com/statistics/806453/price-of-ethereum/, accessed on May 2022.

[26] Y. Yuan, F. Y. Wang, 'Blockchain and cryptocurrencies: Model, techniques, and applications', IEEE Transactions on Systems, Man, and Cybernetics: Systems, 48(9), pp.1421-1428, 2018.

[27] K. Christodoulou, E. Iosif, A. Inglezakis and M. Themistocleous, 'Consensus crash testing: exploring Ripple's decentralization degree in adversarial environments', Future Internet, MDPI, 12(3), pp. 1-12, 2020.

[28] A. Polyviou, P. Velanas, J. Soldatos, 'Blockchain technology: financial sector applications beyond cryptocurrencies', Multidisciplinary Digital Publishing Institute Proceedings, 28(1), pp.1-5, 2019.

[29] M. Salimitari, M. Chatterjee, and Y. P. Fallah, 'A survey on consensus methods in blockchain for resource-constrained IoT networks', Internet of Things, Elsevier, pp. 1-19, 2020.

[30] H. Treiblmaier, A. Rejeb and A. Strebinger, 'Blockchain as a driver for smart city development: application fields and a comprehensive research agenda', Smart Cities, 3(3), pp. 853-872, 2020.

[31] C. C. Agbo, Q. H. Mahmoud, and J. M. Eklund, 'Blockchain technology in healthcare: a systematic review', In Healthcare, MDPI, 7(2), pp. 1-30, 2019.

13

Blockchain Protocols and Algorithms

H. Echchaoui, A. Miloud-Aouidate, and R. Boudour

Embedded Systems Laboratory, Annaba University, Algeria
E-mail: hanane_ech@yahoo.fr; tachouche.amal@gmail.com;
racboudour@yahoo.fr

Abstract

The fourth industrial revolution encompasses the latest technologies and communication networks in smart automation. One of the imposing innovations in Industry 4.0 is blockchain. This technology will play a significant part in the creation of factories in the 21st century. Owing to its features such as decentralization, trustworthy operations, low operational cost, and so on, blockchain can offer considerable enhancements to new applications. According to some forecasts, the blockchain market is set to grow further in the coming years. In our chapter, we will provide a guide for the scientific community looking for state-of-the-art blockchain algorithms and protocols. The goal is to provide a picture of research on blockchain protocols and algorithms to enable more consistent treatment. To this end, we investigate a wide range of blockchain algorithms and protocols' papers met in literature to understand their properties in a comprehensive classification under different criteria in the form of a tree structure.

Our chapter brings blockchain algorithms and protocols to academia and industry actors and highlights the differences between them. In addition, it is proposed that these actors who wish to suggest blockchain protocols or algorithms in the future should take notice of these latter.

Keywords: Industry 4.0, blockchain, security, algorithms, protocols.

13.1 Introduction

The fourth industry is a term concocted to ameliorate the functioning of modern factories by using recent technologies, such as those used for robotics, renewable energy, industrial cyber-physical systems (ICPS), and so on. Originally, it was initiated to upgrade the German economy in 2011. The latest High-Tech Strategy of the German Government, published in 2014, defines six fields of priority for Industry 4.0 [1]: The numeric economy and society, smart mobility, the durable economy and energy, the innovative workplace, healthy lifestyle, and civil security

It is a new step of the industrial revolution that interfaces the physical world and the numeric one using information and communication technologies making factories intelligent. Smart connected and remotely accessible machines are implemented in factories, and factories are connected to each other. It promotes the application of these technologies to enable the evolution of the plant's communication architecture [2].

Blockchain originated in 2008 with Bitcoin [3] and is one of the technologies used in Industry 4.0. According to Gartner [4], the Blockchain market will grow to $176 billion and $3.1 trillion in 2025 and 2030 respectively. In another study, Cisco foretells that blockchain will reach 10% of the global GDP by 2027 [5]. It has the power to transform several industries, including transportation, manufacturing, and agriculture. These areas form the basis of the fourth industrial revolution. Blockchain fans expect blockchain technology to disrupt many existing application areas. Securing communication, transferring value, and storing data will automate processes and remove any manual activities and third parties [6]. Therefore, it is becoming a fundamental pillar of Industry 4.0.

The consensus protocol is seen as the base and the pivot of all blockchain systems. It is the most important part of any blockchain implementation since it determines the system's behaviour and performance. A blockchain system is only as safe and reliable as the consensus process it uses. As a result, consensus algorithms have been prioritized in the development of more efficient blockchain systems. This has guided the creation of a wide range of such algorithms.

There are several well-established algorithms such as PoW, which is the best-known blockchain consensus algorithm out there used by Bitcoin. Novel consensus mechanisms have been also introduced by researchers based on their system requirements. synthesizing these consensus algorithms within the framework of a systematic study is vitally important, which is the aim of the chapter. We investigate and review well-known and alternative recently proposed blockchain consensus algorithms and discuss the possibility of their application in Industry 4.0 domains to guide practitioners to better understand and articulate blockchain consensus mechanisms.

In Table 13.1 we showed the structure of the chapter

Table 13.1 Chapter's organizational structure

Blockchain technology and Industry 4.0		
Most known blockchain consensus algorithms	**Less-known blockchain consensus algorithms**	**Blockchain protocols**
Popular algorithms	• Proof of work • Proof of stake • Federated Byzantine agreement	• Algorithms based on PoW • Algorithms based on PoS • Algorithms based on FBA • Hybrid algorithms • PoX algorithms	• Bitcoin • Ethereum • Corda • Quorum • Hyperledger • EOSIO • NEM (New Economy Movement) • Multichain
Derivative algorithms	• Algorithms based on PoW • Algorithms based on PoS • Algorithms based on FBA		
Blockchain applications		
How to develop a blockchain application	**Example of Blockchain Application in Industry 4.0**	**How to choose a blockchain consensus algorithm and protocol?**

13.2 Blockchain Technology and Industry 4.0

13.2.1 Basic concepts

Over the past decade, we have witnessed a strong evolution of blockchain technology and its applications in different fields. Blockchain can be specified as a decentralized and distributed network that assists with traceability, records management, payment applications other commercial transactions

[7]. It is a network of nodes that is not controlled by a centralized authority, keeping a record of every transaction. It can alternatively be defined as a secure agreement reached amongst users through the application of a consensus algorithm, which results in the addition of a new block to the ledger [6].

Blockchain technology has four pillars: Cryptography, smart contract, shared ledger, and finally consensus [8]. Figure 13.1 explains the blockchain's pillars.

The blockchain contains all the intended elements and characteristics to improve applications in diverse areas. It is decentralized, immutable, transparent, efficient, and offers better security as illustrated in Figure 13.2. Agreeing on a single decision across all participants is a major challenge in such distributed and decentralized systems. Therefore, a mechanism that imposes the behaviour and performs the system is required. This mechanism is called the blockchain consensus algorithm. Blockchain technology has all the required ingredients.

Industry 4.0, also sometimes called smart manufacturing refers to the fourth industrial revolution that attempts to enhance performance, reduce costs, and improve quality in several areas of industry. It is also called the smart factory in a certain manner [9]. It describes the growing detour toward automation, interconnectivity, machine learning, and data exchange within the manufacturing industry revolutionizing the way businesses grow and operate.

Cryptography
Makes sure that all data gets encrypted and only authorized user can decrypt the information.

Smart contract
Verifies and validates the participants of the network.

Shared ledger
Provides the complete details of transaction within networks.

Consensus
Dictates how a system behaves and performs

Pillars of Blockchain

Figure 13.1 Pillars of blockchain technology.

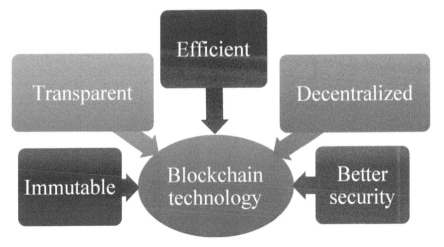

Figure 13.2   Blockchain technology characteristics.

13.2.2 Related works

Over the past 3 years, researchers conducted multiple surveys on blockchain protocols and algorithms.

Krishnamurthi and Shree provided in [10] a survey on the different consensus algorithms in which they compared some widely utilized algorithms like PoW and PoS, besides a comparative analysis based on various properties such as risks, security, and scalability.

A comparative review of the existing well-known consensus algorithms is offered in [11] where the authors suggested a four-category analytic framework: algorithms throughput, the profitability of mining, degree of decentralization, and algorithms' shortcomings.

Kaur *et al.* [12] considered broadly current consensus algorithms like PoW, PoS, DPoS, and more. They studied their performance and different features. Similarly, common and mostly used consensus mechanisms were described in [13] and [14] and analyzed in [15].

Ferdous *et al.* [16] explored and synthesized public blockchain and cryptocurrency systems consensus algorithms leading to a comprehensive categorization of structural, security, performance, and reward properties.

Salimitari and Chatterjee [17] focused on discussing consensus protocols that can be implemented in blockchain-based IoT, whereas in [18], the focal point was to summarize the most important blockchain consensus protocols, used in a private blockchain.

Bouraga [19] reviewed a large selection of new consensus algorithms and proposed a four-category classification framework.

Oyinloye *et al.* [20] on the other hand focused on highlighting lesser-known consensus protocols. They contributed by calculating their performance based on different metrics such as scalability and security; they also provided classification centered on their properties as illustrated in Figure 13.3.

13.2.3 Algorithm or protocol?

Distributed systems do not have a central authority, so the totality of the participants in the network have to work together to agree and decide. These participants do not trust each other. They need some sort of protocol to achieve a consensus that everyone has agreed upon. The complex process of reaching an agreement on a blockchain network is referred to as a consensus problem. Consensus algorithms are performed on a blockchain to achieve approval on transactions in the network between distributed nodes. Their role is to assure the validity and authenticity of the transactions. Consensus algorithms are essential for blockchain security. DoS, DDoS, and Sybil attacks are some of

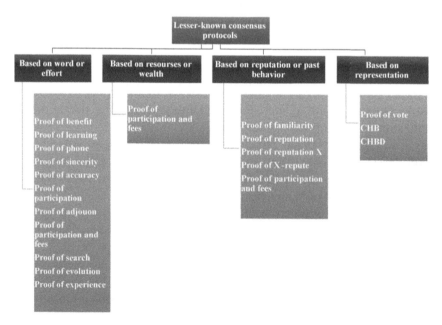

Figure 13.3   Taxonomy of alternative consensus protocols [60].

the attacks that a blockchain network can face without a consensus-reaching process [13]. Different variations of blockchain consensus algorithms exist, but they all have to hold the following properties [12]:

Termination: The process of reaching an agreement should be completed;

Agreement seeking: Each algorithm should strive to get as much network consensus as possible;

Collaborative: participants should work in unison for the greater good of the network;

Cooperative: the participants should favour teamwork over individualism.

Egalitarian: a system must be as egalitarian as possible;

Inclusive: a system should involve as many participants as possible in reaching a consensus;

Participatory: Everybody should take part in the complete consensus-reaching process;

Integrity: the majority decision is made sure to apply.

Pointing out and clarifying the difference between an algorithm and a protocol is important in the world of blockchain because they are not identical despite the fact that these terms are used interchangeably in the literature and approached by other close terms such as design, mechanism, and scheme. Researchers, especially computer scientists have to differentiate between the two.

Generally, an algorithm is a set of instructions for solving a problem or accomplishing a task [21], it tells the system what to do to produce a result output. The sequence of the instructions in an algorithm is crucial. A protocol is a set of guidelines that the system has to obey and follow for the right implementation. The chronological order of these rules is not important. In simple terms, we can define a protocol as the main rules of a chain and the algorithm as the mechanism by which these rules will be followed.

13.3 Known Blockchain Consensus Algorithms

In a blockchain environment, consensus methods are rules and procedures to achieve an agreement where the order is crucial for the network to operate correctly. Therefore, these methods are actually algorithms.

We reviewed a wide range of blockchain consensus algorithms and classified them into two big categories: The most commonly utilized algorithms and lesser-known alternatives.

13.3.1 Popular algorithms

Proof of work (PoW), proof of stake (PoS), and federated Byzantine agreement (FBA) are the most well-known algorithms out there. We classified them in Figure 13.4 and discussed them further below:

13.3.1.1 Proof of work (PoW)

PoW first appeared in 1993 as a way to combat junk emails in [22]. But in 2008, Nakamoto introduced with Bitcoin an innovative way to use PoW as a consensus algorithm [5], and it became the most known and used consensus algorithm.

The nodes in a blockchain that uses PoW consensus, are called miners. These miners will compete against each other to solve a complex puzzle using computing power. Once a miner has solved the puzzle, it will broadcast a new block to the network, where the solution will be checked by the other miners. The first miner who solves the puzzle gets rewarded with cryptocurrency. The puzzle is solved by using a Hash function. Hash is a sophisticated mathematical algorithm used to certify the transactions stored in blocks [11].

In summary, the computing power of the node will determine its chance to be chosen to add a block.

Although PoW has many advantages such as its levels of security and scalability, its main disadvantage is energy waste and high computational cost [11]. A visual illustration of the PoW consensus algorithm is shown in Figure 13.5.

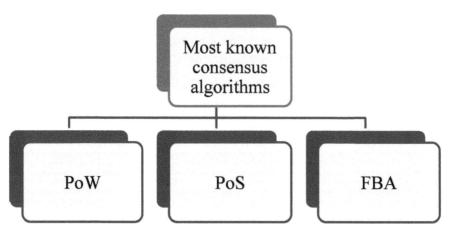

Figure 13.4   Most known consensus algorithms

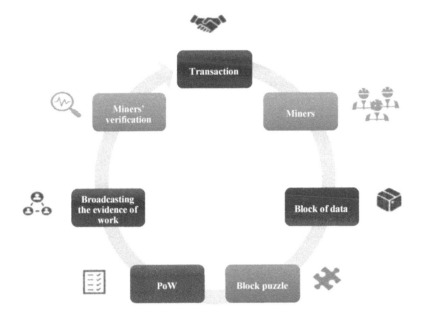

Figure 13.5   Proof of work.

13.3.1.2 Proof of stake

After PoW, the most widely known consensus algorithm is PoS. It is a technique that was first proposed by a forum user in 2011 as a way to lower transaction fees and speed up transaction confirmation, an alternative to PoW that has a number of disadvantages such as encouraging the use of mining pools that make the blockchain more centralized and requiring a large amount of energy to operate.

Rather than pitting miners against one another, PoS employs an election process to choose a random node to validate the next block. A node invests a given number of coins into the network. This stake guarantees that the node will comply with the protocol's rules. The higher the stake estimate, the more likely it is that a node will be chosen to validate the following block. The chosen node proves its ownership over the stake with a digital signature instead of solving a problem [17].

PoS consumes less energy, takes less time to process, and is less expensive than PoW [13]. However, it is prone to the nothing-at-stake attack, because there is nothing to lose and block generators can vote on multiple chains [20]. We illustrate the PoS algorithm in Figure 13.6.

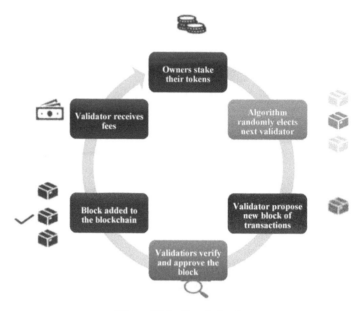

Figure 13.6   Proof of stake.

13.3.1.3 Federated byzantine agreement (FBA)

A Byzantine agreement occurs when a quorum of nodes in a blockchain network agrees on a solution to a problem, approving a block on the blockchain. FBA uses "quorum slices," to persuade specific nodes of the network to agree with them. Pairs of quorums have at least one intersecting node [20]. Nodes can select which quorum slices they trust to carry out the consensus process [15]. The Ripple blockchain pioneered FBA but the Stellar blockchain refined it and implemented the first secure FBA. The protocol ensures low latency and flexibility [16].

No such thing as a flawless consensus algorithm exists. There are always certain compromises related to different criteria from performance, and security to scalability. We summarized in Table 13.2 the pros and cons of the three most known and largely used consensus algorithms.

13.3.2 Derivative algorithms

We classified other well-known algorithms into three subcategories: algorithms based on PoW where the consensus is achieved using hardware

Table 13.2 The upsides and downsides of the most well-known consensus algorithms

Algorithms	Pros	Cons
PoW	• High level of security • Good technological maturity • More decentralized networks	• High energy consumption • Slow transaction speed • High cost
PoS	• Higher transaction speed • Small energy consumption • Superior scalability	• The rich get richer • Unproven at large scale • Less security
FBA	• Reduction of the impact of faulty nodes • High throughput • Robustness in the face of failure	• Small size consortium • Requires trust between participants • quorum slices can lead to centralization

or computational energy, based on PoS where the nodes need to deposit a stake to prove that it will behave as per rules, and based on FBA where the goal is to resolve the Byzantine generals' problem, and where a specific number of nodes must agree on a block before it can be added to the network.

13.3.2.1 Algorithms based on pow

Consensus algorithms based on PoW are algorithms that prevent data manipulation by imposing high energy and hardware control requirements to solve the puzzle and reach a consensus. In Figure 13.7, we illustrate two of the popular algorithms that are based on PoW, and we discussed them more below.

• Proof of capacity (PoC)

Proof of capacity (PoC) was presented in 2015 [23]. It is a consensus algorithm where miners need to prove that they have more storage capacity in their hard drives to mine a block. Nodes temporarily provide storage space on their hard drives as a stake. Before mining, the system generates massive datasets called plots, which are saved on the hard disk, this process is called plotting [13]. The more memory a participant allocates for plotting, the higher its chances of generating the next block.

The mechanism is considered to be energy-efficient, resource-efficient, and accessible to a broader audience because it only requires regular hardware, unlike PoW which requires special hardware [11]. However, larger

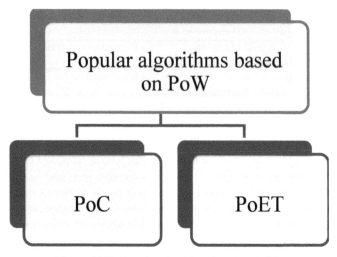

Figure 13.7   Popular algorithms based on PoW.

drives are accessible and anyone can obtain them and use them to mine the majority of the blocks.

- Proof of elapsed time (PoET)

Intel corporation proposed in 2016 the concept of PoET for blockchain construction based on their trusted Software Guard Extensions (SGX) which is a special set of CPU instructions [24].

PoET ensures a safe and random selection of "leaders" using these CPU instructions. Each node in the blockchain network must wait for a random period of time that is generated in a random way and in a range equal for all. It guarantees that nodes have an equal chance to become the next block generator [12]. The first participant to finish the waiting period is the next block generator, and the new block is broadcasted to the other nodes.

This system uses far less energy than other consensus algorithms, such as PoW, but the dependency on SGX makes the system centralized and dependent on a third party (Intel) which conflict with the blockchain decentralization philosophy.

13.3.2.2 Algorithms based on Pos

Figure 13.8 is where we classified popular consensus algorithms that are based on PoS. These algorithms are where participants use their tokens or another form of value as stake.

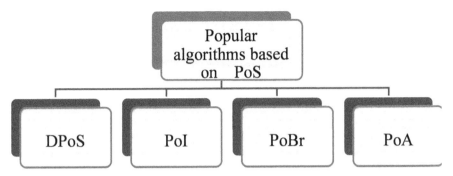

Figure 13.8 Popular algorithms based on PoS.

- Delegated proof of stake (DPoS)

The consensus algorithm DPoS was created in 2014, by a software developer called Daniel Larimer as a variant of the PoS algorithm.

The DPoS system is carried out through an election process where coin holders choose by vote their delegates who are then responsible for validating new blocks. DPoS can be looked at as a digital democracy.

Delegates, also known as witnesses, are compensated for adding blocks to the network. Each stakeholder has a number of votes that is relative to the number of coins he owns; he can assign their stake to another stakeholder who will at that point vote in their place. Witnesses take turns to generate on a block within a time frame of n seconds [10].

A witness who misbehaves or constantly fails to produce a block will be discarded. Delegates deposit a certain amount of funds that are held in reserve to prove their commitment but will be confiscated in case of the witnessed act maliciously.

In comparison to PoW, this algorithm provides higher performance and lower power usage. But it creates an environment in which participants with a big number of coins can exercise a bigger influence [16]. It is considered to be more efficient and democratic than its predecessor PoS. Both are employed as less energy-consuming alternatives to PoW algorithms.

- Proof of importance (PoI)

PoI is a blockchain consensus algorithm launched by the technology platform NEM, or New Economy Movement, building on PoS. It is used to determine which node is eligible to perform the necessary calculations to add a new block to the blockchain (harvesting) and receive the associated recompense. The possibility for an account to harvest a block is determined by its importance score. It uses a cryptocurrency called XEM [10]. A minimum of 10,000 vested XEMs in approximately 30 days is required for harvesting [25].

- Proof of burn (PoBr)

Invented by Iain Stewart, PoBr is a consensus algorithm where miners reach consensus by burning coins. It uses virtual mining rigs to validate transactions. Users burn their coins to convey that they are committed to the network [20]. The miner sends the coins to burn to a randomly generated public address that has no private key and is thus never usable [16], this means that the coins have been burned and cannot be recovered. The more coin they burn the higher the mining power they get, the greater the speed of finding a new block, and the greater the reward they receive.

Figure 13.9 illustrates the PoBr consensus algorithm. PoBr reduces rates of energy consumption in comparison to PoW, however, it can be unsustainable because burning coins is costly.

- Proof of authority (PoA)

PoA was proposed in 2015 by Gavin Wood as a variant of PoS [20] where the participants use their identity and reputation as a stake instead of tokens. PoA uses a small number of block validators known as authority master nodes [16] to verify blocks and transactions which makes the network highly scalable but also makes it tend toward centralization.

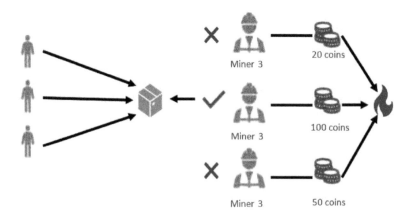

Figure 13.9 Proof of burn.

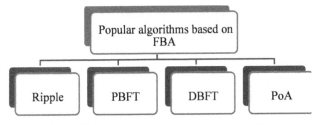

Figure 13.10   Popular algorithms based on FBA.

13.3.2.3 Algorithms based on FBA

- Ripple protocol

Ripple is a BFT consensus algorithm [20] that was first developed by Ryan Fugger and then adopted by the Ripple labs company in 2012. Ripple provides a solution for real-time and cost-effective global money transfers.

Table 13.3   Most-known consensus algorithms

Algorithm	Developed by	Year	Application	Examples
PoW	Cynthia Dwork and Moni Naor	1993	Cryptocurrency, General application	Litecoin, Ethereum
PoS	Forum user	2011	Cryptocurrency, General application	Peercoin
DpoS	Daniel Larimer	2014	Cryptocurrency, Decentralized exchange	BitShare
Ripple/XRP Protocol	Ryan Fugger	2004	Digital assets, payment, global money transfer	Ripple
FBA	David Mazières	-	General, application	Stellar blockchain
PoA	Gavin Wood	2015	Private networks	Aura
PoET	Intel Corporation	2016	Permissioned blockchains	Hyperledger Sawtooth
PoBr	Iain Stewart	-	Generalapplication	Slimcoin
PBFT	Miguel Castro and Barbara Liskov	1999	General application	Hyperledger fabric
DBFT	NEO	-	General application	NEO blockchain
PoC	Dziembowski *et al.*	2015	Cryptocurrency	Burstcoin
PoI	NEM	2015	Blockchain platform	NEM

The protocol runs on a network of servers and is based around a public shared ledger called the XRP ledger that stores information on Ripple accounts and XRP transactions.

The network has two types of nodes: server nodes responsible for the consensus and contain validators known as unique node list (UNL) and client nodes that are responsible for transferring assets [17].

Consensus is reached when 80% of the nodes within a UNL come to an agreement over a transaction. It has a low latency due to the employment of a minimal number of validators [16].

• Practical Byzantine fault tolerance (PBFT)

BFT refers to a network's capacity to reach a consensus despite the presence of hostile participants. BFT is derived from the Byzantine generals' problem, in which parties must agree on a common strategy to prevent collapse while keeping in mind that some of these parties may be corrupt.

PBFT emerged as one of the prominent optimizations of various aspects of BFT in 1999 [26]. All nodes in the system communicate with one another to reach an agreement among all honest nodes. The number of malicious nodes must be inferior to one-third of all nodes in the network during a given pass of vulnerability [12].

Clients make requests to the leader node to perform a specific service operation. The request is broadcasted to the backup nodes in the network by an elected leader node. The nodes process the requests and send the appropriate response to the client. The client would then wait for (f+1) matching responses from different nodes, with "f" representing the maximum number of potentially malicious nodes [12].

The high throughput and low computational cost of PBFT are its strengths. But it has scalability issues due to its high network overhead [17].

• Delegated Byzantine fault tolerance (DBFT)

DBFT is a NEO-developed (an open-source decentralized blockchain application platform) variation of regular BFT. Contrary to PBFT, to add a new block, this method does not necessitate the involvement of all nodes in the election process. It is an analogy of a country's governance system [20]. The speaker's proposed block must be verified by the delegates. At least 66% of the delegates must agree on the block before it is validated [10]. Unlike the PoW consensus algorithm, no expenditure of energy is needed, but the mechanism requires regulations, which include a certain level of centralization.

We summarized in Table 13.3, the most known consensus algorithms with their developers, the year of their first introduction and their field of application.

13.4 Less-known Blockchain Consensus Algorithms

Despite their widespread use, well-known algorithms nevertheless have drawbacks and limits. Efforts to address these limitations and enhance these protocols led to the nascency of a wide range of alternative, lesser-known consensus protocols that have interesting benefits and advantages and merit recognition. We classified these algorithms into the same subcategories of the most known algorithms in addition to two new categories: hybrid algorithms and PoX algorithms.

13.4.1 Algorithms based on PoW

Alternative consensus algorithms based on PoW are shown in Figure 13.11.

- Proof of experience (PoE)

PoE is a consensus algorithm that was proposed in [27] building on PoW with a substantial modification. Miners in PoE are rewarded for their computational work within a time frame by reducing the complexity of the puzzle that they have to solve. The miner will need less and less computing capability to mine a block over time and with experience [27].

- Proof of participation and fees (PoPF)

Fu *et al.* [28] proposed a PoW-based consensus algorithm named PoPF for a new distributed ledger that is based on the blockchain known as JCLedger. The algorithm improves computational cost and efficiency without scarifying the security aspect of PoW. Participants need to be candidates to have the opportunity to become a miner. Candidates are ranked by two factors: the participant's time as an accountant and the fees paid by him. The puzzle becomes easier as the rating rises.

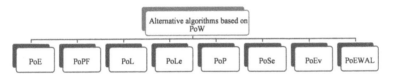

Figure 13.11 Alternative algorithms based on PoW.

- Proof of learning (PoL)

Proof of learning is a distributed consensus mechanism that validates transactions through machine learning competitions [29]. This protocol is presented to direct the energy expended in solving puzzles toward a meaningful effort. The goal is to combine two seemingly unrelated tasks: the first is validating network transactions, and the second is repurposing lost energy from the hash puzzle in the first task to accomplish useful tasks while constructing a repository of machine learning models.

- Proof of learning (PoLe)

A new consensus algorithm called proof of learning (PoLe) was proposed in [30]. Similar to PoL, PoLe indent to use the wasted computation power for a practical purpose. Its goal is to train neural network models in PoW blockchain networks. It consists of two types of nodes: The data node, which commissions machine learning tasks via the blockchain, and the consensus node, which supplies the system with computing power [30]. Consensus nodes compete to train the model and the winner receives a reward from the data node.

- Proof of phone (PoP)

By confining mining hardware to smartphones, PoP was presented as a solution to the problem of blockchains' ever-increasing operational expenses [31]. Theoretical analysis revealed that PoP reduces costs by up to 98.2% when compared to traditional blockchain [31].

- Proof of search (PoSe)

To address the issue of the large amount of energy wasted by blockchains, Shibata [32] proposed PoSe, a consensus algorithm for blockchains where the wasted electricity and computational power are oriented to search for solutions to problems submitted by any nodes. The nodes in this system are called clients. They can submit a job for solving an optimization problem. The client rewards the node that solves the problem the best.

- Proof of evolution (PoEv)

PoEv is a consensus algorithm that has the PoW features and uses the majority of the computational effort spent for mining to execute genetic algorithms (GAs) that clients submit [33]. Nodes in PoSe look for solutions to a problem that has been submitted. PoEv encourages these nodes to share their best solutions to allow miners' cooperation and add to their popularity in the network.

Table 13.4 Alternative algorithms based on PoW

Paper	Authors	Algorithm	Application
[27]	Masseport *et al.*	PoE	Blockchain networks
[28]	Bizzaro *et al.*	PoPF	JCLedger
[29]	Bravo-Marquez *et al.*	PoL	Blockchain networks
[30]	Liu *et al.*	PoLe	Blockchain networks.
[31]	Kim *et al.*	PoP	Conventional blockchain networks
[32]	Shibata *et al.*	PoSe	Blockchain networks
[33]	Bizzaro *et al.*	PoEv	Blockchain networks
[34]	Liu *et al.*	PoB	Energy transactions for electric vehicles
[35]	Raghav *et al.*	PoEWAL	Blockchain in IoT

- Proof of benefit (ONPOB)

Liu *et al.* [34] proposed a proof of benefit consensus algorithm with Online benefit generating (ONPOB) to schedule electric vehicles (EV) charging and discharging to minimize the power fluctuation level. The EVs and their drivers are the network's participants. The algorithm generates a benefit number problem and the participants solve it to mine a block [20]. The simulation results demonstrated that ONPOB can significantly lower the power fluctuation level. (PFL) [34].

- Proof of elapsed work and luck (PoEWAL)

PoEWAL is a lightweight consensus algorithm for non-cooperative blockchain in IoT environments [35]. Each participant has to partially solve a problem in a given time frame. This algorithm consumes minimal computer resources and has low latency, according to test results.

Table 13.4 summarizes the alternative algorithms that are based on PoW and their field of application.

13.4.2 Algorithms based on PoS

We illustrate seven alternative algorithms that are based on PoS in Figure 13.12 and we discussed them further right after.

- Robust Round Robin

Robust Round Robin is a consensus schema that was proposed in [36] as a solution to the leader selection problem in proof of stake systems for permissionless blockchains where the leader candidate selection is deterministic instead of a random section. The system establishes long-term identities and records them in the ledger which is the notion of stake. This process is handled in two different manners: bootstrapping from existing infrastructures and mining identities where identities function as stakes [36].

- Metaheuristic proof of criteria (MPoC)

MPoC is a novel consensus algorithm that applies meta-heuristic algorithms to increase the number of block producers and improves their decentralization compared to the DPoS protocol [37].

- Albatross

Albatross is a permissionless, open-source blockchain consensus mechanism based on PoS [38]. Taking inspiration from BFT algorithms, it was presented by Berrang *et al.* [38]. Participants who are in charge of manufacturing blocks are known as validators. Any node that stakes tokens is a potential validator. The active validator is chosen from the list of potential validators proportionally to the stake value to produce blocks over a timeframe "epoch."

- Delegated proof of stake with downgrade (DDPoS)

DDPoS was presented in [39] to decrease resource consumption and enhance efficiency and security in [39]. The system has two types of nodes: witness nodes with the most votes (stake voting) verify blocks and candidate nodes that replace witness nodes once it fails using a downgrading mechanism.

- Fantômette

Fantômette is a PoS algorithm that has components of potential independent interest: a secure leader election protocol, Caucus, and a scheme for

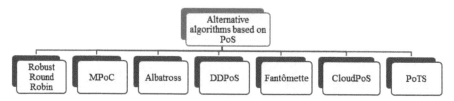

Figure 13.12   Alternative algorithms based on PoS.

incentivization [40]. To be considered as future leaders and voters, participants must cast a security deposit. The program selects the leader at random using a verifiable random function.

- CloudPoS

CloudPoS is a PoS consensus algorithm for blockchain-integrated cloud [41]. Participants stake resources in order to be candidates for the leader election where the chances to become a leader increase with the amount of stake. The resources in a cloud environment are in the form of CPU power, network data rate, or the amount of memory allocated for program execution [41].

- Proof of TEE-Stake (PoTS)

PoTS is described in [42] as a new PoS algorithm that takes advantage of the features of trusted execution environments (TEEs) to reduce the attack of fraudulent validators and posterior corruption. To run trusted code within the TEE, all validators in the network must be outfitted with security hardware.

We summarized the alternative algorithms that are based on PoS and their field of application in Table 13.5.

13.4.3 Algorithms based on FBA

In this section, we look at 10 different algorithms that are based on the FBA concept. Figure 13.13 illustrates them, and Table 13.6 summarizes them and their applications.

- BigFooT

BigFooT, is a BFT consensus algorithm that was proposed in [43]. The algorithm offers several properties such as being resilient to performance

Table 13.5   Alternative algorithms based on PoS

Paper	Authors	Algorithm	Application
[36]	Ahmed and Kostiainen.	Robust Round Robin	Permissionless blockchains
[37]	Nguyen *et al.*	MpoC	Blockchain networks
[38]	Berrang *et al.*	Albatross	Permissionless blockchains
[39]	Yang *et al.*	DDPoS	Blockchain networks
[40]	Azouvi *et al.*	Fantômette	Blockchain networks
[41]	Tosh *et al.*	CloudPoS	Blockchain integrated cloud
[42]	Andreina *et al.*	PoTS	Permissionless blockchains

degradation, and against messages being lost before synchrony is reached within the network [43].

- Delegated randomization Byzantine fault tolerance (DRBFT)

Zhan *et al.* [44] proposed a consensus algorithm named DRBFT for blockchains based on (PBFT), to improve the consensus algorithms in practice.

A random selection algorithm (RS) is presented for the fair selection of delegate nodes from all participants. DRBFT follows four steps: identify each node with a number, run a voting algorithm to generate a list of candidates from all the nodes, run the RS algorithm to select councilors from candidates, the councilors then use PBFT to reach an agreement and produce new blocks [44].

- Advanced PBFT-based consensus (ANPBFT)

Du *et al.* [45] presented an advanced PBFT-based consensus algorithm for consortium blockchain. Consensus nodes are split into active nodes that

Table 13.6 Alternative algorithms based on FBA

Paper	Authors	Algorithm	Application
[43]	Roberto.	BigFooT	Permissioned blockchains
[44]	Zhan *et al.*	DRBFT	Asynchronous network environment
[45]	Du *et al.*	ANPBFT	Bidding consortium blockchains
[46]	Chen *et al.*	EGES	Permissioned blockchains
[47]	Crain *et al.*	DcBFT	Blockchain networks
[48]	Miller *et al.*	HoneyBadgerBFT	Cryptocurrency
[49]	Yin *et al.*	HotStuff	Blockchain networks
[50]	Jalalzai *et al.*	VBFT	Blockchain networks
[51]	Jalalzai *et al.*	Proteus	Blockchain networks
[52]	Muratov *et al.*	YAC	Blockchain networks
[53]	Chen *et al.*	MSig-BFT	Private blockchains

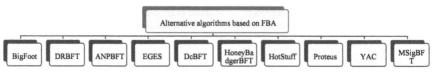

Figure 13.13 Alternative algorithms based on FBA.

verify messages and passive nodes that are chosen at random. The consensus algorithm is based on PBFT algorithm combined with tree topology communication network and message aggregation technology.

- EGES

EGES is a consensus algorithm that was proposed in [46]. EGES is resistant to DoS. The selection of a different committee for confirming each block is achieved by adopting the dynamic committee merit. While achieving consensus, the real committee is concealed by a fake one to prevent DoS attackers from distinguishing between them.

- Democratic Byzantine fault tolerance (DcBFT)

DcBFT is a leaderless BFT consensus algorithm that still operates even if the leader of the blockchain is faulty [47]. This algorithm does not use a traditional leader, but rather a weak coordinator, allowing rounds to be executed optimistically rather than waiting for a specific message [47].

- HoneyBadgerBFT

HoneyBadgerBFT is an asynchronous BFT protocol that guarantees liveness without requiring any timing assumptions [48]. It is a wrapper for the common subset agreement that builds on existing primitives. Every node proposes a random pack of encrypted transactions and when an agreement is reached, the finality is added to the network as a block after nodes collaborate to decrypt the transactions.

- HotStuff

It was proposed in [49] as a leader-based BFT algorithm. The algorithm works in a succession of views and each view has a unique leader. In two rounds of communication exchanges, a leader can reach a consensus decision. The first phase guarantees the uniqueness of the block and the second ensures that the next leader will be able to persuade replicas to vote in favour of a safe measure [49].

- Fast B4B

Fast B4B consensus protocol, achieves consensus during normal protocol operation in just two communication steps [50]. It can reach a consensus with two communication rounds where the maximum number of faults tolerated is small.

- Proteus

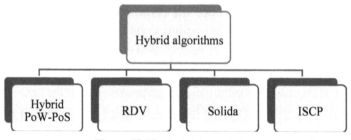

Figure 13.14   Hybrid algorithms.

Proteus is a blockchain consensus algorithm based on BFT that improves latency and throughput that was presented in [51]. The number of network failures does not effect the algorithm's performance. It selects a subset of nodes to form a root committee, which then constructs a block using a particular BFT-based technique. The result of the consensus is checked by nodes that are not a part of the root committee.

• Yet another consensus (YAC)

YAC is a consensus algorithm that addresses the issues of inefficient message passing and strong leaders that plague traditional BFT consensus algorithms [52]. It is based on block hash voting. It serves two purposes. The first is ordering, which involves grouping transactions into proposals and sending them to the network. And the second is consensus where a leader collects the nodes' votes and sends them the results. The nodes validate the response before generating the block to the network.

• Multisignature-BFT (MSigBFT)

MSig-BFT is a witness-based consensus algorithm introduced in [53]. This algorithm introduces the witness nodes that have the role of supervising the leader, the node that proposes the blocks. At least one of the witnesses or the leader must be a non-faulty node [53].

The alternative algorithms that are based on FBA are briefed in Table 13.6.

13.4.4 Hybrid algorithms

The four hybrid consensus algorithms discussed below are illustrated in Figure 13.14.

• Hybrid PoW-PoS

Gosh and Dutta suggested a hybrid PoW-PoS consensus algorithm framework in response to the constraints of both the PoW and PoS algorithms. The locational marginal pricing (LMP) of electricity is used to choose mining nodes [54].

- Register, deposit, vote (RDV)

RDV was introduced as a new consensus method for blockchain as an alternative to PoW. It is based on the "RDV: Register, Deposit, Vote" [55] distributed voting process. Both internal and external costs are used to raise the cost of attacks.

- Solida

Solida is a decentralized blockchain algorithm suggested by Abraham *et al.*, [56] based on recognized Byzantine consensus and proof of work [56]. It is based on PoW and Byzantine consensus to address the problem of limited throughput and long transaction confirmation time. Solida can endure up to 33% of malicious nodes.

- Improved SCP protocol (ISCP)

The work [57] presents an enhanced blockchain consensus mechanism, ISCP, that is said to be more secure and efficient. A hybrid PoW/BFT consensus protocol was designed to improve SCP algorithms.

We summarized the discussed hybrid consensus algorithms in Table 13.7.

13.4.5 PoX algorithms

Proof of X algorithms represent alternative algorithms of proof of something. To claim their position as legitimate miners, most of them require that participants expend some labour or asset other than coin [58]. We chose eight PoX algorithms to discuss in this chapter. Figure 13.15 illustrates them, and Table 13.8 summarizes them.

- Proof of contribution (PoCO)

Table 13.7   Hybrid algorithms

Paper	Authors	Algorithm	Application
[54]	Ghosh and Dutta.	Hybrid PoW-PoS	Blockchain networks
[55]	Solat.	RDV	Blockchain networks
[56]	Abraham *et al.*	Solida	Blockchain networks
[57]	Li *et al.*	ISCP	Blockchain networks

Song *et al.* [58] presented PoCO, a novel blockchain consensus algorithm based on participants' contributions. The node that has the highest contribution value calculated based on the users' behaviour and action information is responsible for generating the next block.

- Proof of service power (PoSP)

It is a blockchain consensus algorithm for cloud manufacturing [59]. The service power of the providers in the cloud is calculated and used for consensus instead of the workload in PoW.

- Proof of sincerity (PoSn)

Mobile-user-friendly and lightweight consensus mechanism, called PoSn was presented by Zaman *et al.* [60]. Selecting the mining node is dependent on how honest and trustworthy a node has been in the blockchain network.

- Proof of reputation (PoR)

Zhuang *et al.* [61] presented a three-step consensus reputation-based protocol called PoR. The steps include leader selection and building blocks, a consensus algorithm based on reputation, and reputation updating.

- Proof of X-repute

It is a blockchain consensus protocol presented in [62] that enables the blockchain system to reach a consensus quickly and safely.

- Proof of reputation X (PoRX)

Wang *et al.* [63] focused on improving PoX protocols for IIoT, and they presented PoRX, which allows each participant to share a broad perspective on reputation during the consensus process.

- Proof of familiarity (PoF)

PoF was proposed in [64] as a consensus-gathering algorithm for the assimilation of healthcare stakeholders' medical decisions. PoF is said to

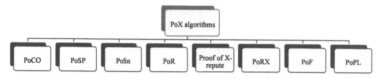

Figure 13.15 PoX algorithms

Table 13.8   PoX algorithms

Paper	Authors	Algorithm	Application
[58]	Song et al.	PoCO	Intellectual property
[59]	Zhang et al.	PoSP	cloud manufacturing
[60]	Zaman et al.	PoSn	Mobile blockchains
[61]	Zhuang et al.	PoR	Blockchain networks
[62]	Wang et al.	Proof of X-repute	Blockchain-based internet of things (IoT) applications
[63]	Wang et al.	PoRX	IIoT
[64]	Yang et al.	PoF	Medical decision-making
[65]	Yuen et al.	PoPL	Blockchain-based Peer-to-peer gaming system

lessen the risk of personal information leakage during medical decision-making collaboration.

- Proof of play (PoPL)

PoPL is a new consensus algorithm for peer-to-peer gaming systems that was described in [65]. The aim was to design a blockchain system that uses the blockchain to form consensus while not jeopardizing the network's general features.

13.5 Blockchain Protocols

Blockchain protocols are a fundamental component of blockchain networks that ensure that the data is shared securely and accurately within the nodes. Blockchains are built on protocols that determine how the system works, how nodes interact in the network and how data is allowed to be transferred not to be confused with consensus algorithms that ensure that the rules required by the protocol are respected.

There are numerous blockchain protocols available on the market. Figure 13.16 illustrates the most commonly used ones:

13.5.1 Bitcoin

Bitcoin is the first form of decentralized currency that eliminated the need for an intermediary to trade money between individuals and validated transactions by the entire network. Satoshi Nakamoto [5] was the first to introduce it in 2008. He proposed Bitcoin as a way of developing a digital currency system that would solve the double spending problem without the requirement for a central authority.

All transactions are recorded on the blockchain and are public to everyone, preventing them from being altered. Bitcoin is the origin and source of many cryptocurrencies, including Ethereum and Dash. It operates as a pure transaction ledger unlike some of the descendants that support smart contract programming [66]. It uses the PoW algorithm to reach a consensus and add a block to the ledger. Figure 13.17 shows that the total number of transactions on the Bitcoin blockchain is expanding every day.

13.5.2 Ethereum

It is a Do-It-Yourself open-source platform for implementing decentralized applications (Dapps) that builds on Bitcoin innovation and goes beyond just enabling a digital currency. It allows participants to conduct transactions without the intervention of a mediator.

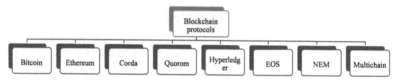

Figure 13.16   Popular blockchain protocols.

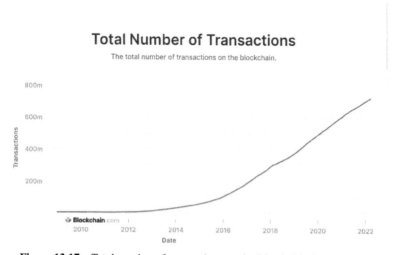

Figure 13.17   Total number of transactions on the Bitcoin blockchain [67].

Launched in 2015 but the idea was first introduced by the founder of Ethereum, Vitalik Buterin, in 2014 to develop a new protocol for developing decentralized applications (Dapps) [68].

It uses a built-in programming language called Solidity to write smart contracts the logic that runs Dapps, and the (ETH) currency, the second largest cryptocurrency out there. Ethereum, like Bitcoin, presently utilizes a PoW consensus mechanism, although it intends to switch to a PoS algorithm in the near future [69]. The main four components of Ethereum are [70]:

- **Smart contracts:** computer programs that transform the terms of agreements into lines of code [71].
- **EVM:** developers use it to create Dapps on Ethereum.
- **Dapps:** decentralized applications that use Ethereum blockchain for data storage.
- **Performance:** increase scalability using Merkle trees [70].

13.5.3 Corda

Corda is an enterprise-level private blockchain platform that was built by R3 banking consortium to primarily be the blockchain for business but also applicable to all industries [72]. To verify and sign contracts, notary nodes might use a variety of consensus algorithms. Corda decentralized applications or CorDapps are binary jars that are stored in the nodes. Each node consists of three types of objects:

- **The states** represent the blocks.
- **The contracts** are business rules that are used to validate transactions operating within java virtual machine (JVM) written in java or kotlin.
- **Flows** are codes that define business logic.

13.5.4 Quorum

Developed by JP Morgan, Quorum is an open-source, enterprise-focused, private blockchain. The Quorum party consists of two components: the node and constellation which consists of transaction managers responsible for the transactions' privacy and enclaves responsible for strengthening the privacy, encrypting and decrypting the transaction and storing the private key [73]. The Quorum party and its sub-components are represented in Figure 13.18.

13.5.5 Hyperledger

Hyperledger is a Linux Foundation development project that began in 2015. Developers, companies, and industry leaders in banking, finance, supply chain, and more work to build blockchain frameworks in this environment [74]. It is a hub or an umbrella for frameworks, standards, and tools for solving corporate tasks. Currently, Hyperledger hosts six graduated Hyperledger projects, nine incubating, and two dormant ones [75].

Table13.9 summarizes some of the Hyperledger frameworks [70, 74, 76].

13.5.6 EOSIO

An open-source protocol that was first launched in 2018. It was created to support scalable, easy-to-use, and secure blockchain applications. Transactions are made within seconds. It is powered by webAssembly that allows developers to build fast applications using popular programming languages such as C++ and DPoS consensus algorithm [77].

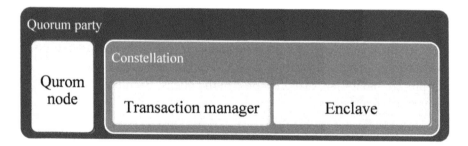

Figure 13.18   Quorum party.

Table 13.9   Hyperledger frameworks

Framework	Contributor	Consensus algorithm	Usage
Hyperledger Sawtooth	Intel	PoeT	Tokenize logistics, sales chains
Hyperledger Fabric	IBM	BFT	Smart contracts
Hyperledger Indy	Sovrin Foundation	RBFT	Decentralized identities
Hyperledger Burrow	Monax	PoET	Smart contracts

13.5.7 NEM (New Economy Movement)

NEM is a permissioned private blockchain that was designed from scratch specifically for business enterprises, incorporating the proof of importance consensus algorithm, which is intended to make it easier to reward community participants. It enables the easy transfer of any digital asset including contracts, files, and tokens. XEM is the network's native coin used as a digital asset and spending currency.

13.5.8 Multichain

Multichain is also an open-source platform developed by Coin Sciences [78], created to build private blockchain applications that function among different organizations. Companies use MultiChain technology to conduct financial transactions. It allows the use of different programming languages.

Table 13.10 lists some of the most prominent advantages and disadvantages of these various protocols.

13.5.9 Consensus algorithms and protocols classification

At the end and to fix the ideas, we synthesize the algorithms and the protocols in the form of a tree structure. Figure 13.19 illustrates this synthesis.

13.6 Blockchain Applications

13.6.1 How to develop a blockchain application?

Blockchain technology is becoming more and more common and the number of its applications is growing rapidly in different sectors and domains not only in the financial sector, but also in government, healthcare, and other sectors. Blockchain applications are currently ruling every industry because their solutions bring more security, connectivity, traceability, tracking, immutability, efficiency, and transparency. As a result, businesses and entrepreneurs are interested in learning more about the blockchain development process.

Developing a blockchain application is a process that we illustrate in Figure 13.20. We show the steps of this process and the key points of each step, and we discuss them more below.

Table 13.10 Pros and cons of blockchain protocols

Protocol	Pros	Cons	Applications
Bitcoin	High level of security due to the large number of users Most popular and highly capitalized	Slow speed (7 transactions/sec). High energy consumption	
Ethereum	Demonstrated security Compatibility	Sluggish transaction speed PoS update pending	Legal contracts Financial services Hosting and cloud storage
Corda	Interoperability Well-known programming languages	Customized to the financial sector Involvement of a third party	Permissioned enterprise blockchains
Quorum	Scalability Privacy	Centralization Less general adaptability	Permissioned enterprise blockchains
Hyperledger	Supports plug-in components Scalability and high performance	Lack of use cases Not a network fault-tolerant	Permissioned enterprise blockchains
EOS	High throughput High commercial scalability	Not enough decentralization	Public and private blockchains
NEM	Lower computational requirements Transaction speed	Require users to have 10,000 vested XEMS to harvest Only one wallet supported Low number of users	Public and private blockchains
Multichain	Customizable High speed	Lack of support of smart contracts	Private blockchains

Although the process of developing a blockchain application may vary based on the requirements, the key steps of this process are as follows:

- Find the idea

Before starting to develop a blockchain application from a technical point of view or any application for that matter, you first think about the problem you want to solve and whether blockchain technology is the best answer for solving it. The developer needs to denote all the needs and uses cases that

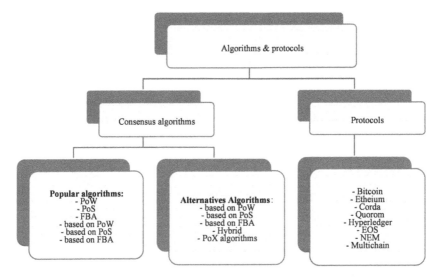

Figure 13.19 Blockchain consensus algorithms and protocols.

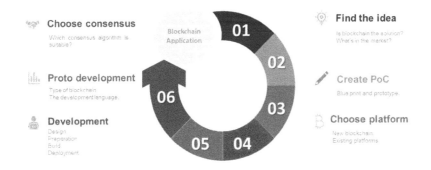

Figure 13.20 Steps to develop a blockchain application.

this application is required to achieve. Then, brainstorm the idea and refine it considering various possibilities and use cases. Moreover, the developer needs to take a look at the market and make an analysis of the existing projects before building the solution;

- Create proof of concept (PoCt)

PoC is a method to assess the practical applicability of a blockchain application in the real world. it could be a theoretical blueprint where the developer explains different parameters of the project, or a prototype to showcase designs, mock-ups, and architecture. The prototype needs to be tested and analyzed to incorporate the top features into the final solution;

- Choose a suitable blockchain platform

After developing the idea and deciding that blockchain is the right answer for it, the next step is to choose a suitable platform for developing it taking into consideration the development's budget. This choice will dictate the skills you will need to implement the application;

- Create a new blockchain

This option means creating a new blockchain framework from scratch where everything is decided by the developer, algorithms, transaction validation, and their fees. Although it is intriguing, this way of developing the application has a high cost and is time-consuming;

- Integrating popular blockchain platforms

It is considered more efficient to clone popular open-source platforms and deploy their code on one's servers. Many frameworks for implementing blockchain solutions exist, and each one has different aims and pros and cons. Quasim *et al.* [70] presented a thorough analysis of available blockchain frameworks. from which we can cite: Ethereum, Hyperledger, and Quorum;

- Choose the appropriate consensus algorithm

This chapter examined a wide range of consensus algorithms. It enables trust among participants or devices and should be chosen to suit the needs of the application;

- Proto development

Blockchain development is expensive. It is better to develop a proto application before deploying it starting with the simplest version of the application by following the minimum viable product (MVP) approach. During this stage, final decisions about the type of the blockchain, the language to use, and the consensus algorithm are made;

- Development

The development step covers different stages, designing the app, preparing the development environment and architecture, building the app and its code, and the deployment.

13.6.2 Example of blockchain application in industry 4.0

To overcome the bottlenecks of the traditional manufacturing industry supply chain, such as the tedious and time-consuming task of tracing back the history and information of purchased products, Xu et. al thought about integrating blockchain in manufacturing industry supply chain management due to its features that can lead to more security, traceability, and collaboration.

They presented an Ethereum blockchain-based supply chain management solution employing smart contracts and Node.js for the manufacturing industry [79]. The aim was to fulfil the supply chain participants' requirements from ordering and trading to information tracking. The system is intended to improve supply chain information sharing, increase overall efficiency, and verify the authenticity of data [80].

The network has two types of nodes: administrator nodes that manage the database server and supply chain organizations nodes such as producers and distributors.

The choice of Ethereum protocol and blockchain was justified by: its support of smart contracts that can be employed in numerous logical services and Solidity which is a Turing Complete programming language, which provides a convenient business support channel for all parties in the supply chain [79].

A consortium blockchain is chosen for the implementation of this application because participants need to go through a registration process that is regulated by administrative nodes to access the system. We summarized in Table 13.11 the different technical choices for the blockchain in this system.

The system mainly works following these steps

- Nodes generate the transaction and share it with the network;
- All the nodes compete to calculate the signature and hash (Proof of work);
- The first node that completes the task and broadcasts the findings to the network is the one that records the transaction and broadcasts the new block;
- The remaining nodes validate the block they received and add it to the chain.

13.6.3 How to choose a blockchain consensus algorithm and protocol?

Choosing the right consensus algorithm and protocol for one's blockchain application is crucial for guaranteeing the safety, security, and fault tolerance of the system. Making a bad choice can result in low performance, consensus failure, or blockchain forks. Consensus mechanisms provide their blockchains with unique qualities, and each has its own set of qualities and limitations.

There are a lot of frameworks and protocols, none of them is flawless and all of them are slightly different in terms of consensus algorithms or currency.

The developer should take into account these factors when making a decision about a consensus algorithm for blockchain:

- The number of miners or validators;
- The speed of writing a block;
- Level of trust in the nodes;
- Resiliency against node failures;
- Node behaviour.

Considering the type of blockchain is essential while making the choice to meet the requirements of the application scenario of each type. Is it a permissioned, consortium, or permissionless blockchain?

If the application deals with private information to be shared among the users, the right answer is a private network. However, if the app promotes business trust and transparency, the data has to be stored in a public blockchain.

Table 13.11   The application's technical choices

Idea	Consensus algorithm	Protocol	Type of blockchain	Programming language
Ethereum-blockchain based management system's designing scheme for the manufacturing industry supply chain	Proof of work	Ethereum platform	Consortium permissioned blockchain	Solidity

Bitcoin and Ethereum platforms are better suited for public permission-less blockchains. In this type of blockchain, participants are anonymous, numerous, and initially considered untrusted so the chosen consensus algorithm should guarantee security from potential malicious nodes. PoW algorithms that require a significant amount of energy from the nodes are suited for this purpose.

While PoW is reliable in this situation, PoW algorithms are chosen if the network relies on computing power to achieve consensus. They are not well suited for applications that require less energy consumption and high transaction speed. In that case, PoS algorithms can be the right choice for the system.

Multichain and Hyperledger Fabric are more suitable for private blockchains where the participants need to be authorized to be a part of the consensus process. These nodes are verified beforehand and considered trustworthy. They are also limited in number therefore, BFT or RAFT algorithms are a better match for the requirements of this type of blockchain than PoW algorithms which are not the best choice here. Permissioned blockchains usually use PBFT and voting algorithms such as Robin Round instead of PoW algorithms.

There are other factors that take part in choosing the right protocols and frameworks for the distributed network:

- Community activity;
- Technology adoption;
- Privacy/confidentiality;
- Scalability, throughput, and latency;
- What programming language is mastered
- Is the smart contracts feature needed or not? Some platforms do not support smart contracts.

13.7 Conclusion

In this chapter, we presented the most advanced blockchain consensus algorithms and protocols. We enlightened readers on their importance. This chapter is organized into four main parts. The first part introduced elaborated consensus algorithms of the blockchain. The second part described fewer known consensus algorithms of blockchain technology. The third part of our chapter concentrated on the protocols developed in the blockchain for several applications. The fourth part emphasized the blockchain applications development in Industry 4.0. We used an application example to illustrate the algorithm and protocol of blockchain. Nevertheless, we underline the difficulty that there are very few technical details in the literature for the development of blockchain applications.

Currently, our team conducted two projects, the first relates to electronic voting in Algeria and the second to civil status documents.

Finally, blockchain technology continues to grow. At each time, there are more platforms with additional features, which allow the creation of highly scalable applications. Although the application of this technology is still in its infancy, many challenges remain, but it is already a source of much envy around the world.

References

[1] Fanis, The Fourth Industrial Revolution, Nov. 2017, https://farahhhanis702.wordpress.com/2017/11/05/the-fourth-industrial-revolution/, [Retrieved Feb, 2022].

[2] T. M. Fernández-Caramés and P. Fraga-Lamas, 'A Review on the Application of Blockchain to the Next Generation of Cybersecure Industry 4.0 Smart Factories,' IEEE Access, vol. 7, pp. 45201-45218, 2019.

[3] Nakamoto, S. 2008. 'Bitcoin: A peer-to-peer electronic cash system', whitepaper.

[4] D. Furlonger, and R. Valdes, Practical Blockchain: A Gartner Trend Insight Report, mar. 2017.

[5] J. Heal, Cisco report estimates 10% of global GDP will be stored on blockchains by 2027, mar. 2019, https://coinrivet.com/cisco-report-estimates-10-of-global-gdp-will-be-stored-on-blockchains-by-2027/, [Retrieved Feb, 2022].

[6] S. B. ElMamy, H. Mrabet., H. Gharbi, A. Jemai, and D. Trentesaux, 'A survey on the usage of blockchain technology for cyber-threats in the context of industry 4.0,' Sustainability, vol. 12, p. 9179, 2020.

[7] M. Javaid, A. Haleem, R. P. Singh, S. Khan, and R. Suman, 'Blockchain technology applications for Industry 4.0: A literature-based review,' Blockchain: Research and Applications, vol 2, p. 100027, 2021.

[8] M. Singh, A. Singh and S. Kim, 'Blockchain: A game-changer for securing IoT data,' 2018 IEEE 4th World Forum on Internet of Things (WF-IoT), Singapore, 2018, pp. 51-55.

[9] N. Mohamed and J. Al-Jaroodi, 'Applying Blockchain in Industry 4.0 Applications,' 2019 IEEE 9th Annual Computing and Communication Workshop and Conference (CCWC), Las Vegas, NV, USA, 2019.

[10] R. Krishnamurthi, T. Shree, 'A Brief Analysis of Blockchain Algorithms and Its Challenges', Architectures and frameworks for developing and applying blockchain technology, pp. 69-85, 2019.

[11] S. Mojtaba H. Bamakan, A. Motavali, and A. B. Bondarti, 'A survey of blockchain consensus algorithms performance evaluation criteria,' Expert Systems with Applications, vol 154, p. 113385, 2020.

[12] S. Kaur, S. Chaturvedi, A. Sharma, and J. Kar, 'A research survey on applications of consensus protocols in blockchain,' Security and Communication Networks, vol. 2021, 2021.

[13] S. Aggarwal and N. Kumar, 'Cryptographic consensus mechanisms,' Introduction to blockchain., Advances in Computers, vol 121, pp. 211-226, 2021.

[14] J. Jayabalan and J. N, 'A Study on Distributed Consensus Protocols and Algorithms: The Backbone of Blockchain Networks,' 2021 International Conference on Computer Communication and Informatics (ICCCI), Coimbatore, India, 2021.

[15] S. Zhanga and J. H. Leeb, 'Analysis of the main consensus protocols of blockchain,' ICT Express, Online, vol. 1, 2019.

[16] M. S. Ferdous, M. J. M. Chowdhury, and M. A. Hoque, 'A survey of consensus algorithms in public blockchain systems for crypto-currencies,' Journal of Network and Computer Applications, vol. 182, p. 103035, 2021.

[17] M. Salimitari and M. Chatterjee, 'A Survey on Consensus Protocols in Blockchain for IoT Networks,' arXiv:1809.05613v4, 2019.

[18] S. Pahlajani, A. Kshirsagar and V. Pachghare, 'Survey on Private Blockchain Consensus Algorithms,' 2019 1st International Conference on Innovations in Information and Communication Technology (ICIICT), Chennai, India, 2019.

[19] S. Bouraga, 'A taxonomy of blockchain consensus protocols: A survey and classification framework,' Expert Systems with Applications, vol 168, p. 114384, 2021.

[20] D. P. Oyinloye, J. Sen Teh, N. Jamil, and M. Alawida, 'An Overview of Alternative Protocols,' Symmetry, vol. 13, p. 1363, 2021.

[21] L. Downey, Algorithm, 2021,'https://www.investopedia.com/terms/a/algorithm.asp', [Retrieved Mar, 2022.]

[22] C. Dwork, M. Naor, 'Pricing via Processing or Combatting Junk Mail,' In: Brickell E. F. (eds) Advances in Cryptology CRYPTO' 92. CRYPTO 1992, Lecture Notes in Computer Science, vol 740, Berlin, Heidelberg, 1993.

[23] S. Dziembowski1, S. Faust, V. Kolmogorov, and K. Pietrzak, 'Proofs of space,' Annual Cryptology Conference, Berlin, Heidelberg, 2015.

[24] L. Chen, L. Xu, N. Shah, Z. Gao, Y. Lu, and W. Shi, 'On security analysis of proof-of-elapsed-time (poet)', International Symposium on Stabilization, Safety, and Security of Distributed Systems, p. 282-297, 2017.

[25] NEM Technical Reference, https://nemplatform.com/wp-content/uploads/2020/05/NEM_techRef.pdf, 2018.

[26] M. Castro and B. Liskov, 'Practical byzantine fault tolerance,' OsDI, 1999, p. 173-186.

[27] S. Masseport, B. Darties, R. Giroudeau, and J. Lartigau, 'Proof of Experience: Empowering Proof of Work protocol with miner previous work', In Proceedings of the 2020 2nd Conference on Blockchain Research & Applications for Innovative Networks and Services (BRAINS), Paris, France, 2020.

[28] X. Fu, H. Wang, P. Shi, and H. Mi, 'PoPF: A Consensus Algorithm for JCLedger,' 2018 IEEE Symposium on Service-Oriented System Engineering (SOSE), Bamberg, Germany, 2018.

[29] F. Bravo-Marquez, S. Reeves, and M. Ugarte, 'Proof-of-Learning: A Blockchain Consensus Mechanism Based on Machine Learning Competitions,' 2019 IEEE International Conference on Decentralized Applications and Infrastructures (DAPPCON), Newark, USA, 2019.

[30] Y. Liu, Y. LAN, B. LI, C. Miao, and Z. Tian, 'Proof of Learning (PoLe): Empowering neural network training with consensus building on blockchains,' Computer Networks, vol. 201, p. 108594, 2021.

[31] J. M. Kim, J. Won Lee, K. Lee, and J. Huh, 'Proof of Phone: A Low-cost Blockchain Platform,'2019 IEEE International Conference on Consumer Electronics (ICCE), Las Vegas, NV, USA, 2019.

[32] N. Shibata, 'Proof-of-search: combining blockchain consensus formation with solving optimization problems,' IEEE Access, vol. 7, pp. 172994-173006, 2019.

[33] F. Bizzaro, M. Conti, and M. S. Pini, 'Proof of Evolution: leveraging blockchain mining for a cooperative execution of Genetic Algorithms,' 2020 IEEE International Conference on Blockchain (Blockchain), Rhodes, Greece, 2020.

[34] C. Liu, K. K. Chai, X. Zhang, and Y. Chen, 'Proof-of-Benefit: A Blockchain-Enabled EV Charging Scheme,' 2019 IEEE 89th Vehicular Technology Conference (VTC2019-Spring), Kuala Lumpur, Malaysia, 2019.

[35] Raghav, N. Andola, S. Venkatesan, and S. Verma, 'PoEWAL: A lightweight consensus mechanism for blockchain in IoT', Pervasive and Mobile Computing, vol 69, p. 101291, 2020.

[36] Ahmed, M., & Kostiainen, K. (2018). Don't Mine, Wait in Line: Fair and Efficient Blockchain Consensus with Robust Round Robin. arXiv: Cryptography and Security.

[37] B. M. Nguyen, T. Nguyen, T. Nguyen, and B. -L. Do, 'MPoC - A Meta-heuristic Proof of Criteria Consensus Protocol for Blockchain Network,' 2021 IEEE International Conference on Blockchain and Cryptocurrency (ICBC), Sydney, Australia, 2021.

[38] P. Berrang, P. von Styp-Rekowsky, M. Wissfeld, B. França and R. Trinkler, 'Albatross – An optimistic consensus algorithm,' 2019 Crypto Valley Conference on Blockchain Technology (CVCBT), Rotkreuz, Switzerland, 2019.

[39] F. Yang, W. Zhou, Q. Wu, R. Long, N. N. Xiong, and M. Zhou, 'Delegated Proof of Stake With Downgrade: A Secure and Efficient Blockchain Consensus Algorithm With Downgrade Mechanism,' IEEE Access, vol. 7, pp. 118541-118555, 2019.

[40] S. Azouvi, P. McCorry, and S. Meiklejohn, 'Betting on Blockchain Consensus with Fantômette,' CoRR, arXiv:1805.06786v2, 2018.

[41] D. Tosh, S. Shetty, P. Foytik, C. Kamhoua, and L. Njilla, 'Cloud-PoS: A Proof-of-Stake Consensus Design for Blockchain Integrated Cloud,' 2018 IEEE 11th International Conference on Cloud Computing (CLOUD), San Francisco, CA, USA, 2018.

[42] S. Andreina, J. -M. Bohli, G. Karame, W. Li, and G. A. Marson, 'PoTS: A Secure Proof of TEE-Stake for Permissionless Blockchains,' in IEEE Transactions on Services Computing, 2020.

[43] R. Saltini, 'BigFooT: A robust optimal-latency BFT blockchain consensus protocol with dynamic validator membership,' Computer Networks, vol 204, p. 108632, 2022.

[44] Y. Zhan, B. Wang, R. Lu, and Y. Yu, 'DRBFT: Delegated randomization Byzantine fault tolerance consensus protocol for blockchains,' Information Sciences, vol. 559, pp 8-21, 2021.

[45] L. Du, Y. Tao, T. Chen, Q. Wang, and H. Lv, 'An Advanced PBFT-based Consensus Algorithm for a Bidding Consortium Blockchain', the

3rd International Conference on Blockchain Technology (ICBCT '21), Association for Computing Machinery, New York, USA, 2021.

[46] X. Chen, S. Zhao, J. Qi, J. Jiang, H. Song, C. Wang, T. O. Li, H. Chan, F. Zhang, X. Luo, S. Wang, G. Zhang, and H. Cui, 'Efficient and DoS-resistant Consensus for Permissioned Blockchains,' Performance Evaluation, vol 153, p. 102244, 2022.

[47] T. Crain, V. Gramoli, M. Larrea, and M. Raynal, 'DBFT: Efficient Leaderless Byzantine Consensus and its Application to Blockchains,' 2018 IEEE 17th International Symposium on Network Computing and Applications (NCA), Cambridge, MA, USA, 2018.

[48] A. Miller, Y. Xia, K. Croman, E. Shi, and D. Song, 'The Honey Badger of BFT Protocols,' In Proceedings of the 2016 ACM SIGSAC Conference on Computer and Communications Security (CCS '16), Association for Computing Machinery, New York, NY, USA, 2016.

[49] M. Yin, D. Malkhi, M. K. Reiter, G. G. Gueta, and I. Abraham, 'HotStuff: BFT Consensus in the Lens of Blockchain,' CoRR, arXiv:1803.05069v6, 2019.

[50] M. M. Jalalzai, C. Feng, and V Lemieux, 'Fast B4B: Fast BFT for Blockchains ('patent pending'),' CoRR, arXiv:2109.14604v2, 2021.

[51] M. M. Jalalzai, C. Busch, and G. G. Richard, 'Proteus: A Scalable BFT Consensus Protocol for Blockchains,' 2019 IEEE International Conference on Blockchain (Blockchain), Atlanta, GA, USA, 2019.

[52] F. Muratov, A. Lebedev, N. Iushkevich, B. Nasrulin, and M. Takemiya, 'YAC: BFT Consensus Algorithm for Blockchain,' CoRR, arXiv:1809.00554v1, 2018.

[53] C. W. Chen, J. W. Su, T. W. Kuo, and K. Chen, 'MSig-BFT: A Witness-Based Consensus Algorithm for Private Blockchains,' 2018 IEEE 24th International Conference on Parallel and Distributed Systems (ICPADS), Singapore, 2018.

[54] S. Ghosh and R. S. Dutta, 'A Hybrid Blockchain Consensus Algorithm Using Locational Marginal Pricing for Energy Applications,' 2021 IEEE International IOT, Electronics and Mechatronics Conference (IEMTRONICS), Toronto, Canada, 2021.

[55] S. Solat, 'RDV: An Alternative To Proof-of-Work And A Real Decentralized Consensus For Blockchain', CoRR, arXiv:1707.05091v5http://arxiv.org/abs/1707.05091, 2017.

[56] I. Abraham, D. Malkhi, K. Nayak, L. Ren, and A. Spiegelman, 'Solida: A Blockchain Protocol Based on Reconfigureurable Byzantine Consensus,' CoRR, arXiv:1612.02916v2, 2017.

[57] Z. C. Li, J. H. Huang, D. Q. Gao, Y. H. Jiang, and L. Fan, 'ISCP: An Improved Blockchain Consensus Protocol,' International Journal of Network Security, vol.21, pp. 359-367, 2019.

[58] H. Song, N. Zhu, R. Xue, J. He, K. Zhang, and J. Wang, 'Proof-of-Contribution consensus mechanism for blockchain and its application in intellectual property protection,' Information Processing & Management, vol 58, p. 102507, 2021.

[59] Y. Zhang, L. Zhang, Y. Liu, and X. Luo, 'Proof of service power: A blockchain consensus for cloud manufacturing,' Journal of Manufacturing Systems, vol 59, pp. 1-11, 2021.

[60] M. U. Zaman, T. Shen, and M. Min, 'Proof of Sincerity: A New Lightweight Consensus Approach for Mobile Blockchains,' 2019 16th IEEE Annual Consumer Communications & Networking Conference (CCNC), Las Vegas, NV, USA, 2019.

[61] Q. Zhuang, Y. Liu, L. Chen, and Z. Ai, 'Proof of Reputation: A Reputation-based Consensus Protocol for Blockchain Based Systems,' In Proceedings of the 2019 International Electronics Communication Conference (IECC '19), Association for Computing Machinery, New York, NY, USA, 2019.

[62] E. K. Wang, R. Sun, C. Chen, Z. Liang, S. Kumari, and M. K. Khan, 'Proof of X-repute blockchain consensus protocol for IoT systems', Computers & Security, vol 95, p. 101871, 2020.

[63] E. K. Wang, Z. Liang, C. Chen, S. Kumari, and M. K. Khan, 'PoRX: A reputation incentive scheme for blockchain consensus of IIoT', Future Generation Computer Systems, vol 102, pp. 140-151, 2020.

[64] J. Yang, M. M. H. Onik, N. Lee, M. Ahmed, and C. Kim, 'Proof-of-familiarity: a privacy-preserved blockchain scheme for collaborative medical decision-making', Applied Sciences, vol. 9, no 7, p. 1370, 2019.

[65] H. Y. Yuen, F. Wu, W. Cai, H. Chan, Q. Yan, and V. Leung, 'Proof-of-Play: A Novel Consensus Model for Blockchain-based Peer-to-Peer Gaming System,' In Proceedings of the 2019 ACM International Symposium on Blockchain and Secure Critical Infrastructure (BSCI '19), Association for Computing Machinery, New York, NY, USA, 2019.

[66] Y. J. Lin, P. W. Wu, C. H. Hsu, I. P. Tu, and S. W. Liao, 'An Evaluation of Bitcoin Address Classification based on Transaction History Summarization,' 2019 IEEE International Conference on Blockchain and Cryptocurrency (ICBC), Seoul, Korea (South), 2019.

[67] 'https://www.blockchain.com/charts/n-transactions-total,'   [Retrieved Mar, 2022].

[68] 'https://ethereum.org/en/what-is-ethereum/,' [Retrieved Mar, 2022].

[69] 'https://ethereum.org/en/developers/docs/consensus-mechanisms/#top,' [Retrieved Mar, 2022].

[70] M. T. Quasim, M. A. Khan, F. Algarni, A. Alharthy, and G. M. M Alshmrani, 'Blockchain Frameworks,' In Decentralised internet of things, Springer, Cham, pp. 75-89, 2020.

[71] J. Chen, X. Xia, D. Lo, J. Grundy, X. Luo, and T. Chen, 'Defining Smart Contract Defects on Ethereum,' IEEE Transactions on Software Engineering, vol. 48, pp. 327-345, 2022.

[72] R. C. Brown, 'The corda platform: An introduction,' Retrieved, vol. 27, 2018.

[73] R. M. Nadir, 'Comparative study of permissioned blockchain solutions for enterprises,' 2019 International Conference on Innovative Computing (ICIC), Lahore, Pakistan, 2019.

[74] V. Dhillon, D. Metcalf, and M. Hooper, 'The Hyperledger Project,' In: Blockchain Enabled Applications. Apress, Berkeley, CA, 2017.

[75] 'https://www.hyperledger.org/,' [Retrieved Mar, 2022].

[76] E. Elrom, 'Hyperledger,' In: The Blockchain Developer. Apress, Berkeley, CA, 2019.

[77] Y. Zhao, J. Liu, Q. Han, W. Zheng, J. Wu, 'Exploring EOSIO via Graph Characterization,' In: Blockchain and Trustworthy Systems, Communications in Computer and Information Science, vol 1267, 2020.

[78] J. Polge, J. Robert, and Y. L. Traon, 'Permissioned blockchain frameworks in the industry: A comparison,' ICT Express, vol 7, pp. 229-233, 2021.

[79] Z. Xu, Y. Liu, J. Zhang, Z. Song, J. Li, and J. Zhou, 'Manufacturing Industry Supply Chain Management Based on the Ethereum Blockchain,' 2019 IEEE International Conferences on Ubiquitous Computing & Communications (IUCC) and Data Science and Computational Intelligence (DSCI) and Smart Computing, Networking and Services (SmartCNS), Shenyang, China, 2019.

[80] M. Asante, G. Epiphaniou, C. Maple, H. Al-Khateeb, M. Bottarelli, and K. Z. Ghafoor, 'Distributed Ledger Technologies in Supply Chain Security Management: A Comprehensive Survey,' in IEEE Transactions on Engineering Management, 2021.

14

Cybersecurity in Autonomous Driving Vehicles

Manoj Kumar Sharma, Ruchika Mehta, and Rajveer Singh Shekhawat

Manipal University Jaipur, India
E-mail: m0918.sharma@gmail.com; ruchika.mehta@jaipur.manipal.edu;
rajveer.shekhawat.in@gmail.com

Abstract

At present around 750 million people are living on the earth and the growth of urbanization and industrialization is very high. As we are developing ourselves as individuals or as society, more and more facilities and precision of work are important. Nowadays, a number of new technologies are being invented to help us to do more and more work and create a comfortable environment. Due to rapid urbanization and increasing incomes, more and more vehicles are roaming on the highways and city roads, which is a cause of increasing driving security threat, because increasing traffic on the roads has increased the chances of human driving errors (i.e., around 95% road incidents are due to human error). So, autonomous driving has been the hot cake research for the industries and academics. We have the power of deep learning algorithms and computer vision to develop a robust autonomous driving vehicle solution. In autonomous driving vehicle solutions various IoT sensors (air pressure sensor, parking sensor, LIDAR, ultrasonic, Radar, Camera, GPS, etc.) and other electronic devices are used to update the different stages of the vehicle operations. These devices, sensors acquire a lot of run time information and send that data to the processing unit which machine learning model analyzes and gives an appropriate instruction to the vehicle control system. However, the data collected by sensors and other devices and processed by machine learning tools has a great chance to hack the signal and malfunction the vehicle. So the cybersecurity threat to the autonomous vehicles has been a great concern to the researchers. In this chapter we are exploring different cybersecurity threats, case studies, and their possible prevention in autonomous vehicles.

Keywords: IoT, cybersecurity, sensor, autonomous vehicle, deep learning.

14.1 Introduction

Technology development for autonomous vehicles has continued since 1920 and the first autonomous car was introduced in 1980 with "Carnegie Mellon University's Navlab." Presently in the western world the automotive industry has been very significant and it has almost 14% capital investment in production and its manufacturing. It is very clear in our mind that autonomous vehicles are the future of transportation and will be having drastic influence on the businesses. The growing technology for the autonomous vehicle will soon replace the transportation employees and corporate fleets for deliveries. At the same time, it will reduce the financial burden of the insurance industry by increasing more and more transportation safety and reducing accidents. With the help of sophisticated autonomous driving vehicles in coming years the road accidents can drop to 80%. Important components of the autonomous vehicles are optical cameras, sensors, ultrasonic devices, radars, and robust computer vision algorithms [1]. On the basis of driver attentiveness and required intervention while driving instead of vehicle capabilities, we can classify the autonomous vehicles in certain categories "(National Highway Traffic and Safety Administration, 2014)" and "Standard SAE J3016, 2016."

(i) **Class 0**: In this class there is no automation and consider the vast majority of transportation vehicles. Brake, throttle, and steering are handled by the driver. All the environmental signals and accordingly navigation is also handled by the driver.

(ii) **Class 1**: In this class of vehicles, brake, throttle, and steering are somewhat handled by the vehicle itself but in certain critical circumstances the driver has to be ready to take control.

(iii) **Class 2**: In this class of vehicles, brake, throttle, and steering are handled by the vehicle itself and in certain circumstances over the road which cannot be handled by the vehicle it informs the driver to take control.

(iv) Class 3: In this class of vehicles, they themselves monitor the surrounding environments and handle brake, throttle, and steering. In certain environments drivers must be attentive to intervene as and when required or requested by the vehicle.

(v) Class 4: In a wide range of environments, this class of vehicles can themselves monitor the surrounding environments and handle brake, throttle, and steering. However, in certain severe conditions the driver must intervene.

(vi) Class 5: It is about a fully automated driving vehicle, we have to set the destination and forget; rest of the things is taken care of by the vehicle itself.

Advancement in autonomous vehicle technology and with their potential, industry and society both can enjoy certain benefits.

- Enhanced safety features and disciplined traffic on the road are the reasons for substantial decrease in road accidents, resulting in saving humans, even animal lives and reducing compensation cost and most importantly fuel consumption which is directly related to the environmental pollution and economic expanses of the person and country.
- A disciplined traffic on the road enhances the roadway capacity, less traffic congestion, reduced need of safety gaps, and smooth traffic flow.
- With the increasing use of autonomous vehicles, the rash driving and over speeding issues will also be automatically overcome. It can decrease the safety gap on the road, parking space requirement, which will provide more space for driving and parking.
- Disciplined traffic will reduce the need of police monitoring and will reduce vehicle insurance premium and importantly increase theft security.
- Removal of steering wheel increases the leg space and with auto-navigation and control at list in class 5 autonomous vehicles, passengers can finish other important tasks during traveling time.

14.1.1 Industry 4.0

A categorization of the Industrial Revolution starts with the exploration of steam engines, which boost industrial production and is known as Industry 1.0. In the second revolution electric power was used to enhance industrial growth and is known as Industry 2.0. In the Industry 3.0, extensive use of electronic circuits enhanced the industrial growth and production. Now, we are in the age of Industry 4.0 revolution, where instead of technical revolution, automation of the businesses is through new communication and digitization technologies like IoT, cloud computing, edge computing, fog computing, social media (to enhance social connect), etc. A simple representation of the Industrial Revolutions is given in Figure 14.1 [2]. The fundamental principles of Industry 4.0 are enhancing the virtualization of the processes and resources, increasing flexibility in the production and enhancing use of modern communication platforms, etc. Thus, a lot of automation of the industries and their processes are done these days, mobile applications are controlling and handling the machines.

However, Industry 4.0 is more than application-based automation; it is not the technological version update. An extensive use of internet and related technologies not only creating a connected environment (i.e., IoT-based network) for the devices and machines involved in the production but it is helping to create new production capabilities, optimum utilization of the resources and most importantly helps to enhance global relationships and partnerships through social connect. This is not only enhancing the production capabilities with quality and at the same time reducing the running cost of the business [2].

14.1.2 Autonomous vehicles and industry 4.0

Mobility-as-a-service and self-driving vehicles are the future of transportation. IoT-enabled sensors along with cognitive computing capabilities are helping to create cyber-physical systems (CPS). In CPS, sensors and other

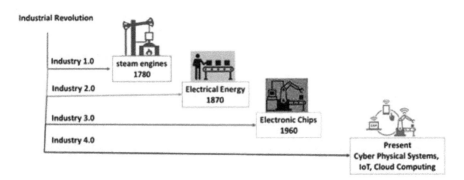

Figure 14.1   Industrial Revolution [2].

electronic devices are used to collect run-time environmental parameters and which are efficiently synchronized between cyber computational devices and physical components and devices of the vehicle. Advanced data analytics helps autonomous vehicles to enhance realistic coordination among different operations and decisions, which are usually required to drive safely [3]. A simple autonomous mobile operation using artificial intelligence algorithms is given in Figure 14.2. Globally, different initiatives are taken for the integration of automated shuttles and existing industrial environments. However, it is very early to discuss their positive–negative implications. Some of the use cases we are discussing here. Gao [4] has discussed a 10-billion dollar plant in south Milwaukee. In this plant, some road lanes are reserved for autonomous vehicles only, which connects the Milwaukee's International Airport to Foxconn factory and this mechanism helps them to automate transportation of people and goods with great safety.

Russo G. *et al.* [5] are exploring the self-driving "Baidu Apolong mini buses" which were given to Japan in 2019. These buses are operational in the Fukushima nuclear power plant for employee transportation. Traub *et al.* [6] have explained about Magride, which is a short-range intelligent campus transportation provider (i.e., at airports, industrial campus, etc.) and their "L4 shuttle bus G100" is in operation in "Vanke Architecture Research Center," China. Rassolkin *et al.* [7] have discussed Tsintel technology and their pilot project which is running in "Tianjin Huaming High-Tech Industrial Park" of China. They are operating L4 autonomous vehicles in a semi-open environment. This bus service provides transportation on fixed routes.

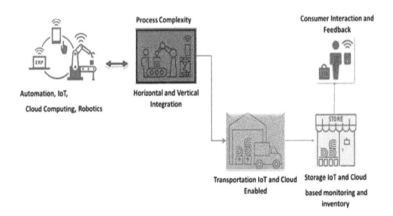

Figure 14.2 Manufacturing technology platform vision for 2030 [3].

14.1.3 Criteria to use autonomous vehicles in industrial operation

However, it is true that increasing use of automated vehicles in the industrial operations is enhancing the safety and security aspects of the operational machinery as well as employees and on other hand it is a cheapest way of transportation. Thus, there are certain criteria to use autonomous vehicles in industrial operations [3].

  (i) **Capacity**: So far the use of autonomous vehicles in industrial operations is possible with certain restrictions and in a controlled environment, like vehicle load, path, sale of open space, parking lot, etc., must be predefined.

 (ii) **Inspection**: It has already been clarified that so far we can use autonomous vehicles in controlled environments, so there must be a periodical inspection to have error-free operations.

(iii) **Maintenance**: Installation cost and running cost has to be monitored during installation of such systems. So, the operating status can be confirmed by taking decisions about power (electricity), route, load etc.

 (iv) **Optimization**: It is very clear that autonomous vehicles depend on computer vision and IoT sensors. So, operational optimization can be achieved by using advanced 3D mapping algorithms, which can reduce energy as well as time requirement.

  (v) **Assessment**: Continuous monitoring of the vehicle behavior is very important for fault detection and maintenance, which can help in avoiding unwanted collisions or operational failures.

14.2 IoT and Cybersecurity

In today's world, digital technology has been the driving force for the society, industry, and economy. Traditional business models are transforming into new business models (i.e., transportation through autonomous vehicles, healthcare, critical operations of patients from remote location, power grid management, financial managements, etc.) which require consistent internet facility for uninterrupted data and instruction transmission. On one side, this transformation is very fruitful not only for the businesses but also for end users. On other hand it has increased the potential threat of cyberattacks on such technology-based businesses. A single cyberattacker can collapse the complete business model, which may cause not only the financial loss, supply chain of the product but also can be a threat for human lives. Now, researchers and industries are very keen in the development of IoT and AI-enabled autonomous vehicles which can make their own decisions by assessing the surrounding environment and a single cyberattack can corrupt the environmental variables, change the route plane, can be a cause of accidents on road and loss of human lives. So, it has been a challenging issue to the cyber experts to create a massive IoT-enabled network which is free from denial of service attack. However, it is not possible to stop cyberattacks but at least we need to have a robust mechanism which can detect those which have already encountered the present system and we need to develop an efficient risk assessment mechanism which can estimate the possible cyberattack risks and possible interruption in the services. In terms of cybersecurity, sometimes dynamic environmental shifts are responsible for cyberattacks. So, the validation and other threat detection mechanisms also must be dynamic in nature to assess the framework risk. In continuation of this in autonomous driving vehicles risk assessment can be classified as; justification of the risk profile, model selection, training for risk assessment, real-time threat detection, dynamic validation and authentication, etc. It has observed that IoT-enabled businesses are adopting holistic cybersecurity approaches. This is a perfect collaboration of technology, consumers, preventive strategies to guard against cyberattacks; such dynamic preventive strategies do real-time security threats monitoring, system and data privacy, etc.

14.2.1 Cybersecurity—risk assessment

For the risk assessment of a cybersecurity, both business and technical policies are important. However, it is very important to have complete knowledge of business use cases and regulatory requirements to design robust IoT-based business frameworks. This helps us to incorporate business processes and their identification while making a robust IoT network. Risk assessment and security are integral aspects of technical framework. However, in the IoT-based cybersecurity of autonomous vehicles, continuous assessment of the environment variables and reacting over their feedback help greatly to assess the risk involved in the system. In the risk assessment, we consider analysis of the possible threats and business strategies. On the other hand, in security we try to uncover system or security flows, vulnerabilities which can finally turn into security breaches. In the risk assessment, initially we include IoT network devices like sensors, edge or frog computing nodes, cloud-based distributed or centralized services, etc. A simple cybersecurity risk assessment framework is given in Figure 14.3 which is presenting a risk assessment of a cybersecurity framework by assessing IoT devices and network, mobile edge network, cloud-based centralized services, etc. [8]. In this framework data is collected from different sources which are stored in the form of archival and event. This system needs processing capabilities and must be flexible enough to assess security directives. This is a framework to analyze the behavior of the application and assess the possible risk. IoT has a huge connected device (i.e., sensors, phones, TVs, vehicles, etc.) network, these devices are communicated to each other by sending and receiving data. IoT has significant applications in autonomous vehicles [9]. However, security is a major concern in wireless IoT devices. Cyberattacks like spoofing, falsification, and eavesdropping can easily occur, which may be the cause of accidents, loss of property, and even loss of human life. The resilient cyberattack framework consists of preparation of the assessment strategies, detection of the possible threats for the security, counter response against the attacks, and finally a holistic review of the processes adopted and achieved goals [8].

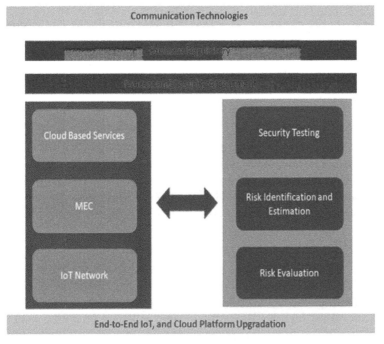

Figure 14.3   Cybersecurity risk assessment framework [8].

T

14.3 Autonomous Driving Applications

In autonomous driving, real-time object detection and further classification of detected objects in scenes are very basic and very important operations. In autonomous driving, traffic signs and other objects detection is very important. Some of them are described below [10].

- **Traffic sign**:In traffic signs, we follow the international standard and can be classified into warnings, prohibition, and mandatory signs. Each traffic sign is represented with a unique color.
- **Traffic light**:Traffic light is the most important way to represent traffic sign; it has six classes (i.e., green, yellow, red, person, bike, etc.). Pedestrian and non-motor vehicle classes are represented even without any color or whatever color.
- **Vehicle**: Simple motor vehicles are classified in vehicle class.

In autonomous vehicle driving, the most important thing is computer vision and image processing, which are successfully handled by convolutional neural networks [10]. In autonomous driving vehicles, the driving assistant system is very important to control the vehicle behavior. However, it is not easy to develop a driving assistant because it has to pass through thousands of random testing scenarios and the risk factor for the test personnel is extremely high during the test and cost of equipment and expenses is also very high. However, an autonomous vehicle technology does not mean that it picks you from your home and drop at your office and vice versa, it is a multi-disciplinary technology which is Inter-industry commercial applications in navigation, industrial processes, agriculture, in aviation industry and many more (e.g., delivery vehicle, shuttles and robotics etc.). Autonomous transportation vehicles have been popular in recent years, which comes under level-4 autonomy of the vehicles. Autonomous transport vehicles will help to increase passenger load and will help to reduce around 90–95% road accidents which mostly happen due to behavioral issue or negligence of the driver which can be substantially decreased with enhanced safety feature and disciplined traffic on the road, which will be resultant to save not only human lives but also animal lives. This will help to reduce compensation cost and most importantly fuel consumption which is directly related to the environmental pollution and economic expenses of the person and country. A disciplined traffic on the road enhances the roadway capacity, less traffic congestion, reduced need of safety gaps, and smooth traffic flow. With the increasing use of autonomous vehicles, rash driving and over speeding issues also will automatically be overcome. It can decrease the safety gap on the road, parking space requirement, which will provide more space for driving and parking. Disciplined traffic will reduce the need of police monitoring and will reduce vehicle insurance premium and importantly increase theft security. Removal of the steering wheel increases the leg space and with auto-navigation and control at least in class 5 autonomous vehicles, passengers can finish other important tasks during traveling time. Trucking is the life line of industrial transportation, which has a great challenge in terms of long driving and long hours of driving, which increases risk to both goods and drivers. However, autonomous driving vehicles can easily minimize such risks. In autonomous driving vehicles, we have advanced technologies like LIDAR, data analytic algorithms, and prediction through deep learning algorithms which help to make transportation safer not only for that specific vehicle but also for other vehicles which directly increase the productivity of the business. Autonomous technology is not only enhancing the productivity of transport vehicles but also proving itself in automation of the industrial process, operation and maintenance of heavy industrial machinery, farming and agriculture, construction and mining, etc. Autonomous vehicle technology has a wide range of applications in the aviation industry also. Unmanned aerial vehicles have proved themselves not only in domestic applications but also in industrial and security areas. It is involved in the border patrol, biological or chemical radio-activity detection, anti-drug warfare, surveillance, rescue missions, infrastructure mapping, fisheries, resource monitoring, disasters monitoring, forestry, emergency services, firefighting, airborne communication, weather activity monitoring, data collection, etc. [11]–[13].

14.3.1 Autonomous vehicle road behavior

For an autonomous vehicle it has great possibility to hit the front vehicle due to sudden state change of the front vehicle (i.e., lane change, braking, slow down etc.). It not only increases the chances of hitting the front vehicle but also to collide with the back vehicle. A real-time reactive mechanism can easily handle such road activities and minimize the road accidents. Here, lateral and longitudinal dynamics play an important role in emergency obstacle avoidance with the great coordination of brake and navigation. Computer vision has the real-time capability to estimate the relative position of the vehicle to the required trajectory. There are certain stages which help to detect the trajectory [14].

  (i) **Image filtering:** In computer vision, real-time information is captured through cameras which contain noise factors which can affect the quality of decisions. So, noise reduction techniques are very important in the image capturing process of autonomous vehicles.

 (ii) **Scene edge detection:** Edge detection from the scenes is the next operation to segment the specific portion of the real scene to focus. For edge detection, different filters, gradient magnitude calculation, direction detection of the edge, and threshold techniques are important to determine potential edges.

(iii) **Contour extraction:** Edges of the scenes not only contain path information but also contain other environmental data. Such unwanted environmental data value creates problems in detection of continuous edge contour of the scene.

(iv) **Morphology:** Over the extracted edge contours certain expansion and correction operations are required to correct intermittent edges.

 (v) **Model fitting:** Finally, in computer vision the characteristics of the path can be established through a geometrical model.

(vi) **Object detection:** Object detection is very important work of the computer vision tools so the vehicles can be easily identified on the road from its background.

(vii) **Object tracking:** Similarly, tracking of the moving objects on the road by subtracting them from other objects and from the background is the next level of computer vision work. With object tracing we can easily establish the behavior activities of the vehicle on the road.

14.4 Autonomous Driving Vehicle: Enabling Technologies

Technological advancement is a key force behind autonomous driving vehicles. Autonomous vehicle enabling technologies can be classified into certain categories which address the vehicle positioning, scene localization and mapping, sensor fusion, vehicle guidance, control, deep learning models, communication, perception and sensing, and privacy and ownership. However, computational power and reliability are two critical challenges in the development of emerging technologies. Autonomous vehicles communicate to each other through electronic signals and data exchange, which help them learn the surrounding environment. Thus, cybersecurity is another challenge for the emerging autonomous vehicle technologies [15]. On the basis of driver attentiveness and required intervention while driving instead of vehicle capabilities, we can classify the autonomous vehicles into certain categories "(National Highway Traffic and Safety Administration, 2014)" and "Standard SAE J3016, 2016."

(i) **Class 0**: In this class, there is no automation and consider the vast majority of transportation vehicles. Brake, throttle, and steering are handled by the driver. All the environmental signals and accordingly navigation is also handled by the driver.

(ii) **Class 1**: In this class of vehicles, brake, throttle, and steering are somewhat handled by the vehicle itself but in certain critical circumstances the driver has to be ready to take control.

(iii) **Class 2**: In this class of vehicles, brake, throttle, and steering are handled by the vehicle itself and in certain circumstances over the road which cannot be handled by the vehicle it informs the driver to take control.

(iv) **Class 3**: In this class of vehicles, they themselves monitor the surrounding environments and handle brake, throttle, and steering and in certain environments drivers must be attentive to intervene as and when required or requested by the vehicle.

(v) **Class 4**: In a wide range of environments this class of vehicles can themselves monitor the surrounding environments and handle brake, throttle, and steering. However, in certain severe conditions the driver has to intervene.

(vi) **Class 5**: It is about a fully automated driving vehicle, we have to set the destination and forget; rest of the things is taken care of by the vehicle itself.

14.4.1 Functional framework of autonomous vehicle

In Figure 14.4, a functional framework of the autonomous vehicle is given which is broken down to perception, path, and behavioral planning, state estimation, safety functionalities, etc. [15]. A simple functional framework of the autonomous vehicle is given in Figure 14.4.

14.4.1.1 Position, mapping and localization

These three aspects are related to each other to provide robust and efficient navigation to the vehicle. An autonomous vehicle must be able to navigate like a human. Global navigation satellite system (GNSS) along with GPS L1/L2, GLONASS, etc., help to compute the actual position of the vehicle. However, these systems can give accuracy in a range of one or two meter. However, fixed based station broadcast correction signals are very accurate to compute the accurate position of the vehicle. However, a possibility of noise in the GNSS signals is very obvious because it connects to the satellite

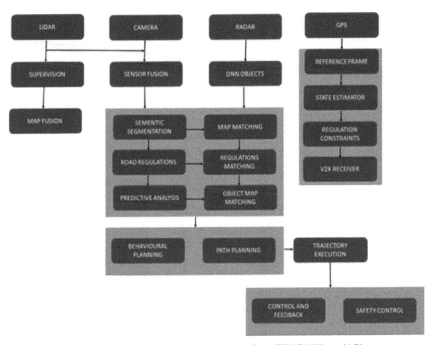

Figure 14.4 Functional architecture of the EDI DbW car [15].

and ground level reference station also. Also, high computational capabilities are needed because data captures from a number of sensors and have to make a broad range map. Thus the mapping must be in a coordinated way for both individual vehicles and cloud-based processing [15], [16].

14.4.1.2 Deep learning algorithms, sensor fusion, guidance and control

Computer vision approaches are very prominent for autonomous vehicles. Navigation is handled by the position, mapping, and localization framework and real-time scene identification; assessment of cybersecurity threat and vehicle behavior analysis is done by sensor fusion, guidance and control system, and deep learning algorithms. Image depth estimation is very hard in 2D images so LIDAR and RADAR sensors are used to fuse the data. On the other hand, deep learning algorithms can represent a large amount of data and patterns to predict the efficient navigation and decision-making by analyzing a complex traffic environment. DL algorithms adopt hierarchical learning to handle complex high-dimensional data representation. Deep learning (DL) algorithms can easily handle both clustering and classification data patterns in large amounts of data. In unsupervised modeling, a high computational power is needed to train the model. However, for an efficient guidance and control system, high computation power is required along with other cloud-based computing and storage, secure communication channels to prevent cyberattacks on the vehicle, etc. [15], [16].

14.4.1.3 Communications

Communication is a strong bond among autonomous vehicles to create a large traffic network which can easily be controlled automatically or by humans. Vehicles can exchange their positional and navigational information with other vehicles to regularize the traffic flow. With the help of IoT sensors, the critical health monitoring of the vehicle hardware as well as behavioral aspects can be done which can be easily shared with others so the necessary warning and preventing broadcasting can be done by the vehicles on the road.

A hybrid communication can help to handle heterogeneous and dynamic traffic with great efficiency and accuracy. However, "Short-Range-Communication" works well with 10 Hz frequency with approximate 300 m distance, but such channels can exhaust quickly with increasing traffic which results in information loss and miscommunication. For long-range data communication, road-side units and repeaters are useful to communicate filtered information. However, 3G, 4G, 5G, and LET technologies are also useful to enhance the communication range. Effective information communication channels help the vehicles to moderate their road behavior like lane change, overtaking, following traffic rules, assessing congestion on the road, etc., [15], [16]. A simple communication framework of autonomous vehicle is given in Figure 14.5.

14.4.1.4 Sensing and perception

Development of hardware and software algorithms are directly proportional; both must be compatible to each otherwise decision-making can be affected which can be the cause of disasters. The combination of camera, RADAR and LIDAR helps to recognize vehicles in close vicinity. On the other hand, communication devices provide a broad range of information about non–line-of-sight objects. However, sensing and perception are still challenging issues in autonomous vehicle technology and real-time operational information is handled with the fusion of vision, LIDAR and RADAR data [15], [16]. A prospective view of perception and sensing is given in Figure 14.6.

14.4.1.5 Data ownership and privacy

In autonomous vehicle development data privacy and ownership are very important. In the autonomous vehicle network, we share the data among pears for route estimation, to pass information about road conditions and traffic load, data collection for model training. Privacy of the data and authenticity of the information in the vehicular network is very important. Cybersecurity mechanisms have the responsibility to minimize possibilities of cyberattacks

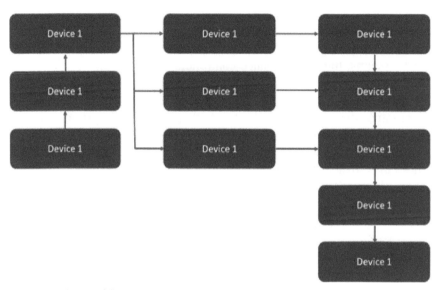

Figure 14.5   Intercommunication and software framework [15], [6].

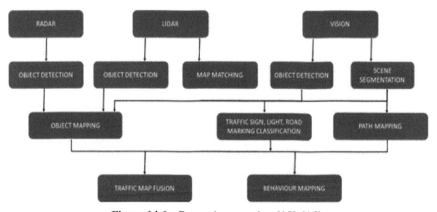

Figure 14.6   Perception overview [15], [16].

and their influence over the decision and quality of the information. Otherwise, the whole network can collapse and high level damage of property and human lives can happen on the roads. However, decentralization of the processing and decision-making nodes can help a lot to prevent damage from cyberattacks. The validation of the shared information and their source is very important to make safe and secure decisions at run time [15], [16].

14.4.1.6 Embedded systems

Technological advancement in the autonomous vehicle hardware and software is always anticipated and it distinguishes the safety controllers and system performance.

In such systems, high performance controllers are based on extendable embedded systems which can run algorithms and fail-operational and fail-aware activities. In such systems, the safety controller uses the vehicle state data to prioritize the control signals if a high performance system fails. A simple safety controller software-framework of the embedded system is presented in Figure 14.7. A specialized safety controller mechanism is highly desirable to achieve fail-operational and fail-safe systems, which utilize software to utilize sensor data to achieve desired goals (i.e., health monitoring of the system, emergency braking system, etc.). Inspection of

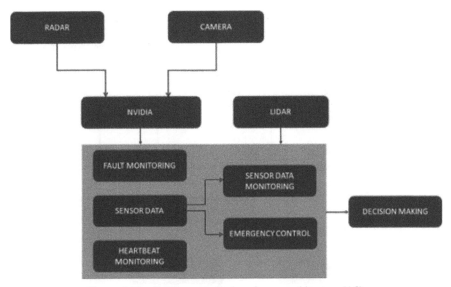

Figure 14.7   Safety controller's software architecture [16].

sensors, timestamps data, and their physical connection are very important to monitor the system health by analyzing the heartbeat monitoring of the system. Data from the LIDAR and RADAR systems is very important to perform emergency operations (i.e., emergency braking, etc.).

14.4.1.7 Variable data acquisition (VDA)

It is very important to have a "variable data acquisition unit" (VDA) to optimally utilize system resources (i.e., computation dominance, network bandwidth, etc.). Driving scenario is responsible for the behavioral planning and data flow redirection from IoT sensors to the edge base or centralized cloud-based processing unit.

14.5 Autonomous Vehicle Security

We are living in the technological edge, where IoT networks and cloud computing technologies are drastically influencing the business frameworks and social security infrastructure. Autonomous vehicles are a future of transportation where IoT networks along with wired or wireless cloud computation facilities and storage can play an important role. For this increasing IoT and cloud networks, security has been a biggest challenge, because a number of electronic devices and sensors are part of such networks which used to acquire data from different environmental conditions and activities, which further transmit the storage and processing units and that time they can be easily exposed to the security attackers. Cybersecurity attacks not only can steal the transmitting data but also influence the electronic sensors and other hardware systems to give wrong or fabricated data for the analysis and on the other hand they can control the data analytic tools and can give wrong prediction and instruction. This kind of intervention in the data collection, processing, and transmission can be a cause of disaster. A strong cybersecurity mechanism is necessary to protect sensitive information and operations of the system. In the autonomous vehicles it is very important to take some preventive measures against security threats like unlegislated data and system access, damaged and fabricated data, influence an appropriate action. So, driving assistance technologies (i.e., different warning messages, auto emergency braking, safety communication, etc.) have been an integral part of autonomous vehicle security. A strong driver assistance can help the drivers to avoid road crashes, loss of property and lives [15], [16].

14.5.1 Protection methods

A multi-layer cybersecurity framework reduces the possibilities of cyberattacks on autonomous vehicle networks and components. In this framework, a priority-based risk detection and protection mechanism is needed for the safety critical control mechanism. A real-time response system is needed to address the cyberattacks on the autonomous vehicle network and IoT components. A reliability and recovery mechanism is required to recover the system from the potential damage to the system and data due to cyberattack and there must be a cybersecurity awareness mechanism which can help the automotive industry to understand and adopt new preventive technologies and measures. Thus, it has proved that autonomous vehicles are more secure and less prone to road accidents as compared to manual driving vehicles. However, continuous security patch updation is mandatory to prevent them from the cyberattack which can control the vehicle and can be a cause of road accidents. Attackers can use such vehicles as a weapon and for this there is no need for attackers to be present physically, they can remotely control the vehicle and can control safety critical systems and functionalities of the vehicle. Through cyberattacks it is very simple to control a single vehicle or the entire fleet. So, there must be an intensive secreting vetting for such autonomous vehicles. To better understand, the preventive measure vehicular cybersecurity can be divided into three aspects: (i) Software and hardware components of the safety critical systems, which are responsible for data exchange. (ii) Another aspect is operation of the vehicle which must be secure enough and must have capability to repel cyberattacks. (iii) As we know autonomous vehicles can be operated from remote centers which must be secure enough from the cyberattacks [17], [18].

14.5.2 Potential security risks of AV crowdsourcing

This is an established truth that crowd-based technology development for the autonomous vehicle can expedite the innovations but they can provide a soft access to the cyberattackers to influence the system. So, we have to establish the identity validation mechanism of the contributors. There must be a trustworthy mechanism to establish a contributor objective. There must be a mechanism to ensure the objective of the contributor is to improve the technology so it does not have any malicious objective [17], [18].

14.5.2.1 Attacks on automotive control systems

Cybersecurity attacks over the autonomous vehicle can be termed as control system attacks, system components attack, and communication attacks. On the other hand, the preventive measures against such attacks can be termed as security framework, intrusion detection and prevention, and operational and behavioral anomaly detection and prevention. However, the role of data analytics and deep learning tools has been strongly established in the prevention of abnormalities of the system [19]. A list of cybersecurity attacks on the autonomous vehicle is given in Figure 14.8.

14.5.2.2 Electronic control unit attack

In the autonomous driving vehicle, electronic control unit is responsible for the car state monitoring, driver and passengers, and their coordinated behavior. Cyberattackers can control the electronic devices of the system and can generate false data/signals from them.

14.5.2.2.1 In-vehicle network attack

Initially, attackers study the data communication network of the autonomous vehicles and they can jam it by performing replay attacks. Vehicle network attacks can be both passive (i.e., monitoring the information and stealing it) and active (i.e., modification of the data, injecting malicious data, etc.). Injecting the unauthenticated messages in the data flow in the network helps the attackers to control the vehicle operations, stealing the secure information, and fully controlling the vehicle for their specific purpose.

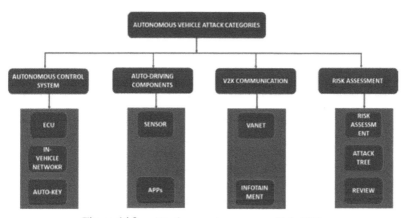

Figure 14.8   Attacks on autonomous vehicle [19].

14.5.2.2.2 Automotive key attack

In this type of attack, an attacker acquires the keyless entry in the system or starts an autonomous vehicle with the pass key, this comes under passive attack. A replay attack can easily perform automotive key attack by replaying the duplicate signal in the system network and it can take place in both wired and wireless networks. Such attacks can easily steal the actual key by making few authentication attempts. An encrypted authentication mechanism can prevent such automotive key attacks.

14.5.2.2.3 Attacks on autonomous driving system components

Cyberattacks on autonomous vehicles are quite different from the traditional cyberattacks. Cyberattackers can control the hardware electronic devices and can easily transmit false or malicious data from them.

(i) **Sensor attack**: Sensors are electronic devices which are the most important component of the IoT-enabled network whether it is a vehicular network or other. Sensors sense the environmental data from different environments of the vehicle (i.e., distance measure, health of the engine, heat of the engine, parking sensors, safety locks, door and windows lock sensors, seat belt sensors, etc.). Cyberattackers can easily inject malicious data in the network which can control the sensors or the network and can transmit false or fabricated data to the data analytics and decision-making systems which can be the cause of simple system and control failure to the critical safety issues to the vehicle as well as passengers.

(ii) **Software attack**: Vehicle network is connected with control devices and centers in wireless mode and different smartphone applications can easily control the operations of the vehicle. A malicious mobile application can easily capture the long-range wireless network. Such attacks can easily overrule the authenticated instruction to the vehicle systems. A strong preventive tool is needed to counter such attacks and save the system from the damage due to such application-based attacks.

14.5.3 Attacks on V2X communications technologies

Different cyberattacks on the communication technologies can influence the system functionality and integrity of the information.

14.5.3.1 VANET attack

In VANET, transmission of data takes place through wireless networks. Data can be transmitted and accepted from outside of the network which exposes it to the attackers. VANET attack can be any of "bogus information attack," "man-in-the-middle attack," "DoS," "malicious code injection," "replay attack," and "location tracking." VANET attacks can happen on electronic control units which can be code modification, fuzzing, phishing, etc.

14.5.3.2 Bluetooth attack

Remote access of the autonomous vehicle network and operational system through Bluetooth connectivity can give a backdoor entry to the attacker to control the operations of the autonomous vehicle. In the modern vehicles, we have various devices which have Bluetooth connectivity (i.e., speaker, radio, etc.). Attackers initially can connect with such devices and from there it can hack the security protocols of the network and can control the devices. There must be a secure authentication system to connect a third-party device with Bluetooth devices of the vehicle.

14.5.4 Autonomous vehicle defense

However, attackers have different possibilities to attack on autonomous vehicle networks or systems.

Certainly we need to have some preventive measures and systems to provide safeguard to the systems. The attack defense can be divided into architecture level security, intrusion detection and prevention, and AI and big data tools for security. A brief detail of defense of autonomous vehicles is given in Figure 14.9.

14.5.4.1 Security architectures for autonomous defense

There must be security architecture for the autonomous driving vehicles which can prevent various cybersecurity attacks and can provide a robust defense against such attacks.

14.5.4.1.1 Control area network and electronic control unit security

Control area network is a standard which allows communication between autonomous vehicles and mechanisms. CAN protocol can protect the network from the cyberattacks (i.e., replay attack, masquerade, etc.). In CAN, confidentiality and reliability are two important aspects which measure the

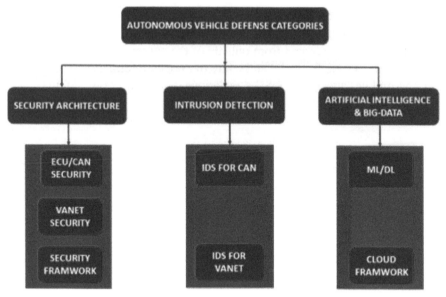

Figure 14.9   Autonomous vehicle defense [19].

security risk of the autonomous vehicle. Similarly, a remote authentication can prevent the electronic control unit tampering. In this, MAC computes authentication messages to ensure reliability and integrity. Authentication protocols are used to broadcast communication in the CAN bus and with this existing node can be modified with maintaining the compatibility. In the embedded autonomous vehicles, it is possible to have hundreds of electronic control units to monitor and control different operations and conditions of the vehicle (i.e., braking system, real-time object detection, engine temperature control and monitoring, fuel status, etc.). Initially ECUs were to accept the command without validating it. Now, we are at the edge of new technology and ECU devices also require protection from cyberattacks. There must be a robust mechanism to validate the command passed to the ECU unit otherwise the attacker can manipulate the operation of ECU like break fail, steering jam, etc. Gateway implementation between ECU devices is the best way to prevent outside inference in the ECU network, only the devices which are part of the network. Other ECU devices are usually accessed by open ports and by protecting test and debugging interfaces can avoid cyberattacks on such circuits.

14.5.4.1.2 Vehicular ad hoc networks security (VANET)

VANETs are mobile ad hoc networks with mobile routing protocols. It is frequently used in autonomous transport vehicles for inter-vehicle communication [20]. It ensures the safety of the vehicle and passengers, improved driving experience, finds roadside services, monitoring and controlling highway traffic, etc. In the VANETs network certain criteria can ensure the vehicle safety from the cyberattacks [21].

(i) **Authentication**: Before responding to the communication it has to verify the identity and authenticity of the communicating vehicle.
(ii) **Data verification**: On the receivable of communicated data the sender has to verify the originality of the received data.
(iii) **Availability**: There must be alternative arrangements when the network is under cyberattack, so the vehicle can communicate to others with any performance degradation.
(iv) **Data integrity**: Over the transmission, we have to transmit encrypted data only so the middle man attack cannot happen and cannot alter the data.
(v) **Non-repudiation**: After data transmission, both the sender and receiver cannot deny that they are the original sender and receiver so the identity of the vehicles can be established.
(vi) **Privacy**: The private and secret data must be under the authorized access only.
(vii) **Real-time constraints**: Vehicle has to maintain the real-time constraints when it is connected with VANET for a short duration.

14.5.4.1.3 Security design and process

As new technologies are developed for the autonomous vehicle and IoT-enabled frameworks are developed to operationalize the vehicle. This is creating various security issues in the vehicle as reliance on the operations, and interaction among smart vehicles through intelligence devices has exposed the vehicular network to attackers. The security by design can be enhanced for autonomous vehicles [22].

14.5.5 Security by design

In security by design, security objectives are the most important factors to take care of by the designer. However, security requirements usually start from people. However, the cost of the cybersecurity attacks and security control mechanisms are equally important. The prime security concern of autonomous vehicles is resilience of the functions and their autonomy which demands the highest level of security features because human lives are at stake. A layered security management for autonomous vehicles is given in Figure 14.10.

Traditionally, security policies focus on integrity of the data, confidentiality of security operations and information, availability of the system, etc. In autonomous vehicles, certain threats (i.e., identity spoofing, data tempering while storage or communication, repudiation, breach of confidentiality, DoS, unauthorized access etc.) for the security are considered [22].

14.5.5.1 Adversarial model

It defines the attacker's capabilities (i.e., hearing the traffic, intercepting network, replaying and synthesizing packets, etc.). However, this model is used for network security analysis and data privacy. The analysis of the adversarial model for autonomous transport systems can be done on the basis of following aspects.

(i) **Attack objective**: Cyberattackers always attempt to breach more and more security objectives (i.e., control of the vehicle, etc.).

(ii) **Communication capability**: In adversarial model, attacker communication capability is also defined (i.e., intercept network packets, breach the confidentiality of the packets, inject malicious data, etc.).

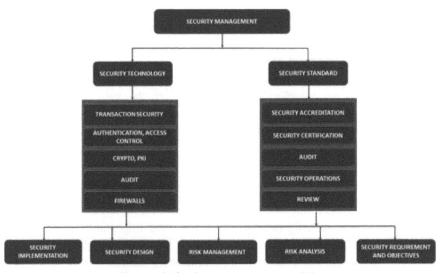

Figure 14.10   Security management [22].

(iii) **Computing power**: After getting unauthorized access in the autonomous vehicle, it is the prime objective of the attacker to control the vehicle, for which it requires capabilities of cryptanalysis, decoding, and reverse the security critical parameters. Attackers require high computation capabilities to perform such tasks.

14.5.5.2 Trust model and security

It has already been mentioned that maintaining data confidentiality, authenticity, and integrity are the main security objectives of cryptography-based security mechanisms. In cryptographic models, key management is very important (i.e., third-party authentication, certification, etc.). The design of all security critical models start with a realistic trust model. When third-party validation or verification is needed for security of the cyber-physical systems, then the trustworthiness of the third party is highly required. Suppose vehicles are communicating to each other through VANET and data packets are encrypted well but the key management of the cryptographic algorithm is not good then it will be prone to the attacks.

14.5.5.3 Security mechanisms

Communication channels are very important for autonomous vehicle security framework. However, trustworthiness of the IoT devices and cryptographic keys is also very important. In ad hoc network, every time we have to establish secure communication for new communicators. Some of the trustworthy components are explained here [22].

(i) **Trustworthy infrastructure**: For the autonomous vehicle security policies, detection of the newly admitted malicious nodes is very important.

(ii) **Secure routing**: For a secure routing trustworthy station it helps to perform encryption and authentication operation, and at the same time for multipath and secure geographic protocols secure routing is very important.

(iii) **Integration of heterogeneous network**: In IoT-enabled framework heterogeneity of the communication devices is required, which needs data format scaling for communication and such scaling can happen only when we have the knowledge of message payload and this knowledge is prone to cyberattacks. So there must be a trustworthy mechanism to handle such heterogeneity.

(iv) **Secure resources**: In the network resources, data storage and computation nodes must be accessed through the authentication process only.

Data remanence attacks can happen over unauthenticated data storage. However, an encrypted authentication system is most secure from cyber threats.

(v) **Trustworthy identification**: In the autonomous vehicles before the communication a valid identity establishment is necessary. In particular, for the low-end devices it has been very compulsory to establish their identity because they are certainly attack prone as compared to other devices.

(vi) **Cryptography-based secure transmission**: Most of the time, cyber-attacks happen while data is transmitted over the network, and in autonomous vehicular networks vehicles exchange information for smooth driving experience. Cryptographic secure transmission is the only way to prevent cyberattacks on transmitting data.

(vii) **Secure authentication**: In the VANET, transmission of data takes place through wireless networks. Data can be transmitted and accepted from outside of the network which exposes it to the attackers. VANET attack can be any of "bogus information attack," "man-in-the-middle attack," "DoS," "malicious code injection," "replay attack" and "location tracking." VANET attacks can happen on electronic control units which can be code modification, fuzzing, phishing, etc. A strong authentication mechanism can easily identify the malicious nodes in the network, either they are already members of the network or newly joined. Similarly, remote access of the autonomous vehicle network and operational system through Bluetooth connectivity can give a backdoor entry to the attacker to control the operations of the autonomous vehicle. In the modern vehicles, we have various devices which have Bluetooth connectivity (i.e., speaker, radio, etc.). Attackers initially can connect with such devices and from there it can hack the security protocols of the network and can control the devices. There must be a secure authentication system to connect a third-party device with Bluetooth devices of the vehicle.

14.5.5.4 Autonomous vehicle operational model and security objectives [22]

In autonomous driving vehicle, communication is the important component which uses wireless or wired networks to support firmware and data transmission. It periodically transmits the vehicle logs to the control centers for vehicle maintenance requirements and life cycle management. Traffic management and remote control of the vehicle, especially emergency cases, is also handled by the communication channels. However, IoT devices are used to sense the environmental variable to operationalize the vehicle. Such sensed data help to take navigational decisions of the vehicle and ensure level 5 autonomy. For the level 5 autonomy to the driving and control operations a real time updation of the required data like route updation, steering adjustment according to the road conditions, speed regulation for the safety, braking system in normal condition and emergency condition is very important. There are certain security objectives of the autonomous vehicle, some of them are explained here.

(i) Integrity of the data and remote operation is very important in the autonomous vehicle so the cyberattacker cannot take advantage of the vehicle operation or the full control over the vehicle. Navigation is directly related to the vehicle and passenger safety which depends on the integrity of sensor systems. The safety critical operations like braking, steering, and speed control are performed according to the sensor data analysis which must be original.

(ii) As we are aware that autonomous vehicles exchange information among their own network for better safety, traffic management, etc. So, confidentiality of inter vehicles is most important and should not be tampered. At the same time the log transmission to the manufacturer or to the control center also needs high level confidentiality to maintain vehicle life cycle.

14.5.5.5 Safety standards for autonomous vehicle

Safety of the vehicle and passengers is the utmost requirement of autonomous vehicles. Cybersecurity mechanism is directly responsible to ensure the safety parameters. However, standard safety standards are available for the autonomous vehicles, some of them are explained here [22]:

(i) **ISO 26262**: It was derived from IEC-61508 and was invented for electrical-electronic systems safety and integrity. It specially targets autonomous vehicles. The safety lifecycle of this standard is "management, development, production, operation, service, and decommissioning."

(ii) **ISO PAS 21448**: This targets the safety of intended functionalities whose misuse can create hazards.

(iii) **SAE J3061**: This standard directly targets the cybersecurity of the system. It enhances the general awareness of the terminology used in autonomous vehicles. This standard emphasizes the distinction between cybersecurity and system security.

14.5.5.6 Security perimeter

Security perimeter distributes autonomous vehicles in different threat segments, which help to validate the trust model and allow analysis of vehicle security which can classify in physical security and cybersecurity. In the layered security mechanism of the autonomous vehicle, there must be sufficient flexibility to reinforce the layer boundaries to analyze the design and its security mechanism.

14.5.6 Autonomous vehicle security management

Security risk in the autonomous vehicle is systematically reduced by the security by design principle and it is mandatory to address the security challenges of the vehicle throughout its life cycle. Certain security incidents of the autonomous vehicle are discussed here. A rigorous testing of the cybersecurity systems is a prime condition of a security vehicle because in the standard procedure quantification of the known threats is possible. Standard SAE-J3061 is responsible for risk assessment and threat analysis [22].

   (i) **Attack tree analysis**: Attack tree analyzes the threat and a worst case threat scenario is also formulated to calculate the risk factor.
  (ii) **TVRA**: It is the standard for cyber-physical system risk assessment, vulnerability assessment, and threat assessment, where system assets keep the threat.
 (iii) **Software vulnerability analysis**: As we are aware that the implementation and development environments of software can be totally different. Autonomous vehicles are developed by considering the influence of malicious attackers.

14.6 Cyberattacks on Autonomous Vehicle

Autonomous driving vehicles are now the reality of safe and intelligent transportation. However, still we have to address some critical issues which could be the hurdle in deployment of such vehicles. In autonomous vehicles data communication is very important which happens in V2V and V2I or I2V form. Security of the sensitive-information and the interference in autonomous operations are the prime concerns of the developers. It is already clear to everyone that for autonomous vehicles, we have to create an IoT-enabled network which can assess the environment and can take driving decisions on the basis of such assessed data. Cyberattacks are the prime threat for such systems. Some attack models are explained here [23].

(i) **External v/s internal attacks:** Intruders can identify the weak nodes or flaw of security protocols to inject malicious data in the vehicular network and can remotely steal or modify the data or can completely control the vehicle operations. Internal attackers are the authenticated members of the network, who can easily reach the secure data and can easily communicate with other peers.

(ii) **Rational v/s malicious attacks:** Rational attackers try to be benefited through stealing information, modifying operations, etc. However, the objective of the malicious attacks is not being profited personally but to harm the system and capture the system controls.

(iii) **Passive v/s active attacks:** Passive attacks neither corrupt the data nor try to control the vehicle operations. These are silent attacks which mainly steal the secret data or listen to the data from the transmission. However, active attacks are easily identifiable because they directly affect the functionality of the system, corrupt the data, and try to control the system operations.

14.6.1 Security and privacy threats: case of autonomous automated vehicles

An IoT network is created to automate the operation of driving vehicles which consists of sensors, RADAR, LIDAR, GPS, and camera devices. Some of the attack surfaces are discussed here [23].

(i) **Traffic sign**: Nowadays, traffic signs are also dynamic in nature and controlled by the sensing devices which can be modified by the cyberattackers and navigation can crash.

(ii) **Computer vision**: Autonomous vehicle uses cameras for scene capturing and object detection on the road and roadside. Cyberattackers can easily capture the camera which either can stop working or can send still images to the computer vision models to wrongly detect and predict the objects, path, traffic signs, obstacles, etc., on the road.

(iii) **Global positioning system**: Positioning and localization of the vehicle is done by GPS which helps in navigation. Cyberattackers can easily interrupt the GPS signals and false signals can be transmitted which can be the cause of vehicle crashes.

(iv) **Electronic control device**: Electronic control circuits like IC have inbuilt instructions to operationalize the vehicle driving actions. On the other hand, various devices are connected with third-party devices (i.e., Bluetooth speakers, Radio, USB ports, etc.). A cyberattacker can easily inject the malicious codes in the electronic control circuits through these devices because most of the time these devices don't ask for any authentication establishment before connecting.

(v) **Acoustic sensor**: Acoustic sensors are very useful to monitor the health of a vehicle by identifying different known audio signals (i.e., trigger airbags by listening to crash sound etc.). Cyberattackers can modify the input and output of such sensors, which not only be the cause of miscommunication about vehicle health but also be the cause of vehicle crash.

(vi) **RADAR and LIDAR**: RADAR transmits radio waves to detect objects and LIUDAR returns infrared to detect the object. Cyberattackers can easily manipulate the signal of these devices (i.e., there are some cases of airplane crash due to wrong estimation of vertical distance between plane and ground, etc.).

(vii) **Security system**: Usually, authentication-based security provides for different devices and even for vehicles for data exchange. However, basic authentication can be bypassed by the cyberattackers and they can easily inject malicious code in the network which can either corrupt the normal function of the devices or can easily capture the whole control of the vehicle remotely.

14.7 Conclusion

Autonomous driving has been the hot cake research for the industries and academicians and obility-as-a-service and self-driving vehicles are the future of transportation. IoT-enabled sensors along with cognitive computing capabilities are helping to create cyber-physical systems (CPS). In CPS, sensors and other electronic devices are used to collect run-time environmental parameters and which are efficiently synchronized between cyber computational devices and physical components and devices of the vehicle. A powerful deep learning and computer vision algorithms have been the helping hand to develop a robust autonomous driving vehicle solution. IoT has been an integral part of our daily life and industrial development. Dream of autonomous vehicles cannot happen without IoT and cloud computing framework. IoT network has different types of sensors to assess environmental variables in real time and processing nodes process that sensed information to perform driving operations. Autonomous vehicles can be classified into five classes as per their autonomy of operations. For an autonomous vehicle, it has a great possibility to hit the front vehicle due to sudden state change of the front vehicle (i.e., lane change, braking, slow down, etc.). It not only increases the chances of hitting the front vehicle but also to collide with the back vehicle. A real-time reactive mechanism can easily handle such road activities and minimize the road accidents. Cybersecurity is a biggest challenge in the deployment of secure and safe autonomous vehicles. Attackers not only can steal the transmitting data but also influence the electronic sensors and other hardware systems to give wrong or fabricated data for the analysis and on the other hand they can control the data analytic tools and can give wrong prediction and instruction. This kind of intervention in the data collection, processing and transmission can be a cause of disaster. A strong cybersecurity mechanism is necessary to protect the sensitive data and operations of the system and in the autonomous vehicles such mechanisms are very important which can help to take some preventive measures against such security threats, unlegislated data and system access, damaged and fabricated data, influence an appropriate action. So, driving assistance technologies (i.e., different warning messages, auto emergency braking, safety communication, etc.) have been an integral part of autonomous vehicle security. So far it is not possible to design a fully robust autonomous vehicle against cyberattacks but by ensuring basic secure authentication and cryptography-based secure data transmission we can somehow prevent most of the security threats. In future, we can have such security protocols which can easily identify all malicious data and activities in the cyber-physical system and can secure the system from their bad intentions.

References

[1] Ján Ondruš, Eduard Kolla, Peter Vertal, Željko Šarić, 'How Do Autonomous Cars Work?', in Transportation Research Procedia, Vol.44, pp. 226–233, 2020.

[2] L. Bassi, 'Industry 4.0: Hope, hype or revolution?', in 3rd Int. Forum on Research and Technologies for Society and Industry (RTSI), pp. 1-6, 2017.

[3] S. Raivo, R. Anton, W. Ruxin, O. Tauno, 'Integration of autonomous vehicles and Industry 4.0', in Proceedings of the Estonian Academy of Sciences, Vo.68, No.4, pp. 389–394, 2019.

[4] P. Gao, 'Disruptive trends that will transform the auto industry', McKinsey & Company. https://www.mckinsey.com/industries/automotive-and-assembly/our-insights/disruptive-trendsthat-will-transform-the-auto-industry.

[5] G. Russo, E. Baccaglini, L. Boulard, D. Brevi, R. Scopigno, 'Video processing for V2V communications: A case study with traffic lights and plate recognition', in Proc. of 2015 IEEE 1st Int. Forum on Research and Technologies for Society and Industry Leveraging a better tomorrow (RTSI), pp. 144–148, 2015.

[6] M. Traub, V. Hans-Jörg, X. Eric, S. Thilo, H. Jérôme., 'Digitalization in automotive and industrial systems', in Proc. of the 2018 Design, Automation and Test in Europe Conference and Exhibition (DATE), pp. 1203–1204, 2018.

[7] A. Rassõlkin, S. Raivo, L. Mairo, 'Development Case Study of the First Estonian Self-Driving Car', in Electr. Control Commun. Eng., Vol.14, No.1, pp. 81–88, 2018.

[8] S. K. Datta, 'DRAFT-A Cybersecurity Framework for IoT Platforms', in Zooming Innovation in Consumer Tech. Conf. (ZINC), pp. 77-81, 2020.

[9] D. K. Alferidah, N. Z. Jhanjhi, 'Cybersecurity Impact over Bigdata and IoT Growth', in Int. Conf. on Computational Intelligence (ICCI), pp. 103-108, 2020.

[10] Y. Li, W. Jue, X. Tengfei, L. Tianlu, C. Li, K. Su, 'TAD16K: An enhanced benchmark for autonomous driving', in IEEE Int. Conf. on Image Processing (ICIP), pp. 2344-2348, 2017.

[11] Feasibility Study for an Unmanned Aerial System Mission Supported by Integrated Space Systems, ESA Portal, July 2009.

[12] H. Song, 'The Application of Computer Vision in Responding to the Emergencies of Autonomous Driving', in Int. Conf. on Computer Vision, Image and Deep Learning (CVIDL), pp. 1-5, 2020

[13] C. H. Yu, Y. Z. Chen, I. C. Kuo, 'The benefit of Simulation Test Application on the Development of Autonomous Driving System', in Int. Automatic Control Conference (CACS), pp. 1-5, 2020.

[14] J. Guo, P. Hu, R. Wang, 'Nonlinear Coordinated Steering and Braking Control of Vision-Based Autonomous Vehicles in Emergency Obstacle Avoidance', IEEE Trans. on Intelligent Transportation Systems, Vol. 17, No. 11, pp. 3230-3240, 2016.

[15] R. Novickis, A. Levinskis, R. Kadikis, V. Fescenko and K. Ozols, 'Functional Architecture for Autonomous Driving and its Implementation', in 17th Biennial Baltic Elect. Conf. (BEC), pp. 1-6, 2020.

[16] C. Englund, E. John, J. Juhani, M. John, 'Enabling Technologies for Road Vehicle Automation', in Meyer G., Beiker S. (eds) Road Vehicle Automation 4. Lecture Notes in Mobility. Springer, Cham., pp. 177 – 185, 2017.

[17] R. Razdan, 'Tesla Deceptions ? Is Automotive Cybersecurity A National Defense Issue?', in Forbes, 2020.

[18] Zora Mirko, 'Are we doing enough to protect connected cars? Help Net Security', 2020.

[19] K. Kim, J. S. Kim, S. Jeong, J. H. Park, H. K. Kim, 'Cybersecurity for autonomous vehicles: Review of attacks and defense', in Computers & security, Vol.103, pp.1-27, 2021.

[20] V. H. La, A. Cavalli, 'Security Attacks and Solutions in Vehicular Ad hoc Networks: A Survey, International Journal on AdHoc Networking Systems (IJANS) Vol. 4, No. 2, April 2014

[21] M. Raya, J. P. Hubaux, 'The security of vehicular ad hoc networks', in Proc. of 3rd ACM workshop on Security of ad hoc and sensor networks (SASN '05), 2005.

[22] A. Chattopadhyay, K. Y. lam, Y. Tavva, 'Autonomous Vehicle: Security by Design', IEEE Trans. On Intelligent Transpo. Systems, Vol.22, No.11, pp.7015-7029, 2021.

[23] J. Petit, S. E. Shladover, 'Potential Cyber-attacks on Automated Vehicles', in IEEE Trans. on Intelligent Transportation Systems, Vol.16, No.2, pp.546-556, 2015.

Index

About the Editors

Abhinav Sharma is presently working as an Assistant Professor (Selection Grade) in the department of Electrical & Electronics Engineering in University of Petroleum & Energy Studies (UPES). He received his B.Tech. degree from H. N. B. Garhwal University, Srinagar, India in 2009 and the M.Tech. and Ph.D. degrees from Govind Ballabh Pant University of Agriculture and Technology, Pantnagar, India in 2011 and 2016. Dr. Sharma has a rich teaching and diversified research experience. The areas of his research interests include signal processing and communication, smart antennas, artificial intelligence and machine learning. He has published research articles in SCI/Scopus indexed Journals and in National and International conferences.

Arpit Jain is currently working as an Technical Project Manager at QpiAI India Pvt. Ltd., India. A multidisciplinary engineer having experience in machine learning, data science, control systems and fuzzy logic systems with an excellent vision toward industry-focused education and state–of-the-art consulting solutions. A seasoned academician having 12+ years of diverse experience in academics, edtech, and IT consulting domain. Dr. Jain has worked as an Assistant Professor for 10 years at UPES India, a part of Global University System (GUS), Netherlands. Research profile includes Indian patents, research articles in SCI/ Scopus indexed Journals, and edited books with IEEE, Emerald, RIVER, CRC, and many other reputed publishing house. He received his B.Eng. degree from SVITS, Indore, in 2007 and M.Eng. from Thapar University, Patiala in 2009, and Ph.D. from UPES, India in 2018.

 **Paawan Sharma** received his Ph.D. (Engineering) from Homi Bhabha National Institute, Mumbai, M.Tech. (Communication Systems) from SVNIT, Surat, and B.E. (ECE) from University of Rajasthan, Jaipur. His research area is multi-disciplinary in nature spanning applications/solution development in various domains such as signal processing, embedded systems, pattern recognition, machine vision, and artificial intelligence. He has five Indian patents (published) and more than 40 publications. He has guided/co-guided three Ph.D. scholars focusing on disaster management technology and smart grid analysis. Mr. Sharma has also worked in Wipro Technologies in VLSI domain. He is a Senior Member, IEEE and Member to ACM, ISSIA, and a life member of IAPR.

 **Mohendra Roy** received his Ph.D. in Electronics and Information Engineering from Korea University, South Korea, M. Tech in Bioelectronics from Tezpur Central University, M. Sc. in Physics from Tezpur Central University. He was a postdoctoral fellow at Delta-NTU corporate Lab of Cyber Physical System of Nanyang Technological University, Singapore. His research area is artificial intelligence and its applications in various domains. He has awarded with 13 National and International awards/recognitions, two patents, and published more than 26 publications.

9 788770 228053